Armen H. Zemanian

Transfiniteness

For Graphs, Electrical Networks, and Random Walks

1996
Birkhäuser
Boston • Basel • Berlin

Armen H. Zemanian
Department of Electrical Engineering
State University of New York at Stony Brook
Stony Brook, NY 11794-2350
USA

Library of Congress Cataloging In-Publication Data

Zemanian, A. H. (Armen H.)
Transfiniteness for Graphs, Electrical Networks, and Random Walks / Armen H. Zemanian.
p. cm.
Includes bibliographical references and index.
ISBN 0-8176-3818-0 (alk. paper). -- ISBN 3-7643-3818-0 (alk. paper)
1. Graph theory. 2. Transfinite Numbers. I. Title.
QA166.Z46 1995 95-24601
511'.5--dc20 CIP

Printed on acid-free paper
© 1996 Birkhäuser Boston

Birkhäuser 

Copyright is not claimed for works of U.S. Government employees.
All rights reserved. No part of this publication may be reproduced, stored in a retrieval system, or transmitted, in any form or by any means, electronic, mechanical, photocopying, recording, or otherwise, without prior permission of the copyright owner.

Permission to photocopy for internal or personal use of specific clients is granted by Birkhäuser Boston for libraries and other users registered with the Copyright Clearance Center (CCC), provided that the base fee of $6.00 per copy, plus $0.20 per page is paid directly to CCC, 222 Rosewood Drive, Danvers, MA 01923, U.S.A. Special requests should be addressed directly to Birkhäuser Boston, 675 Massachusetts Avenue, Cambridge, MA 02139, U.S.A.

ISBN 0-8176-3818-0
ISBN 3-7643-3818-0
Reformatted from author's disk by Texniques, Boston, MA
Printed and bound by Quinn-Woodbine, Woodbine, NJ
Printed in the U.S.A.

9 8 7 6 5 4 3 2 1

CONTENTS

PREFACE

"What good is a newborn baby?"

Michael Faraday's reputed response when asked,
"What good is magnetic induction?"

But, it must be admitted that a newborn baby may die in infancy. What about this one — the idea of transfiniteness for graphs, electrical networks, and random walks? At least its bloodline is robust. Those subjects, along with Cantor's transfinite numbers, comprise its ancestry.

There seems to be general agreement that the theory of graphs was born when Leonhard Euler published his solution to the "Konigsberg bridge problem" in 1736 [8]. Similarly, the year of birth for electrical network theory might well be taken to be 1847, when Gustav Kirchhoff published his voltage and current laws [14]. Ever since those dates until just a few years ago, all infinite undirected graphs and networks had an inviolate property: Two branches either were connected through a finite path or were not connected at all. The idea of two branches being connected only through transfinite paths, that is, only through paths having infinitely many branches was never invoked, or so it appears from a perusal of various surveys of infinite graphs [17], [20], [29], [32]. Our objective herein is to explore this idea and some of its ramifications. It should be noted however that directed graphs having transfinite paths have appeared in set theory [6, Section 4.4 and Chapter 7], but, by virtue of that directedness and the sets the graphs represent, the question of connections at infinite extremities of graphs is either avoided or trivially resolved. This question is however a substantial issue in our theory of undirected arbitrary transfinite graphs and is resolved by the invention of a new kind of node.

Research on graphs, electrical networks, and random walks continues to thrive, and new ideas for them are proliferating. It therefore is incumbent upon us to argue why still another may be of interest. One reason is offered in Section 1.4, where it is shown that a transfinite network is the natural outcome of an effort to close a heretofore unaddressed lacuna in the theory of infinite electrical ladders. A more compelling consideration is that transfiniteness introduces radically new constructs and expands the three aforementioned topics far beyond their conventional boundaries. For instance, a transfinite graph is not a graph in any prior sense but is instead an extension roughly analogous to Cantor's extension of the natural numbers to the transfinite ordinals. It is undoubtedly way too much to expect that the introduction of transfinite graphs will eventually be as important as Cantor's invention of the transfinite numbers. However, even a small fraction of that success would be ample justification for our endeavor.

Transfiniteness for undirected graphs and electrical networks was first introduced through some embryonic ideas in [33] and [34] and then in all its generality in [35, Chapters 3 and 5] and [37]. This book expands upon those prior works with a variety of new results. It is in fact a sequel to "Infinite Electrical Networks" [35] but can be read independently; there is no need to read that book before embarking on this one. All definitions and results within the mainstream of this present book are thoroughly developed herein. Some peripheral ideas that broaden our discussions but are not critical to them are borrowed from several sources, including "Infinite Electrical Networks." Moreover, most of that book was restricted to conventional infinite networks, and the lesser part of it on transfinite networks focussed upon the first rank of transfiniteness, that is, upon "1-networks." Now, however, we exclusively examine transfinite graphs and networks with due attention to higher ranks of transfiniteness. Moreover, we expand our theory considerably with the introduction of several new topics related to connectedness problems, discrete potential theory, uncountable networks, and random walks.

A critical issue in the development of transfinite graphs and networks was the choice of appropriate definitions. Much effort was expended in pursuing likely possibilities, many of which failed because of inherent contradictions or unmanageable entanglements. An example of the latter is described in Section 1.2. The definitions we have chosen appear to be free of those obstacles and yet sufficiently general for our purposes. To be sure, we could simplify a few definitions by weakening them, but this again leads to complications because of the variety of situations that then arise. We believe that we have arrived at a satisfactory compromise between the complexity of definitions and the complexity of subsequent discussions.

On the other hand, we introduce many new concepts. This is due to the unavoidable circumstance that transfinite undirected graphs are multifaceted structures and entirely novel. We cannot borrow from established theories but instead must construct everything we need. This is exacerbated by the many ranks of transfiniteness, one for each countable ordinal. Were we to restrict ourselves to the first rank, that is, to "1-graphs," the number of definitions would be comparable to that for conventional graphs. But the extensions to higher ranks through recursion expand that number severalfold, especially since some of the definitions needed for a limit-ordinal rank are significantly different from those for a successor-ordinal rank.

Almost all the ideas about transfinite graphs introduced in "Infinite Electrical Networks" are employed herein without alteration, but there are a few differences. One of them concerns the specification of a path, which previously involved some ambiguity — with the result that a path could be represented in different ways. The tighter definition we now use prevents this. A related matter is that we now make a distinction between a path "reaching" a node and "meeting" a node, the former concept being weaker than the latter. Still another alteration is that we abandon the idea of a "reduced graph" and use instead a "subgraph." A reduced graph is obtained from a given transfinite graph by choosing just some of the branches and interconnecting them in accordance with the given graph. It is so defined that a reduced graph is a transfinite graph too, but this requires the introduction of other nodes. A subgraph uses the same interconnections of the chosen branches but does not proliferate nodes; however, it is in general something other than a transfinite graph. This replacement of reduced graphs by subgraphs simplifies our theory.

Our book is arranged as follows. Chapter 1 is introductory. After explicating notations and terminology in Section 1.1, we argue why transfinite graphs and networks should be of interest and how they are needed to close a gap in the theory of conventional infinite networks. We also indicate why we adopt an unusual approach to conventional graphs in order to enable a simpler transfinite generalization.

Transfinite graphs are treated in Chapters 2 through 4, electrical networks in Chapters 5 and 6, and random walks in Chapter 7.

Chapter 2 is devoted almost entirely to the many definitions needed for the transfiniteness of graphs. Here — and indeed throughout this book — we illustrate new concepts with many examples. Connectedness problems are taken up in Chapter 3. We show that transfinite connectedness need not be transitive as a binary relation between branches. Conditions under which transitivity does hold are then obtained. Chapter 3 closes with a discussion of the cardinality of the branch set of a transfinite graph. In Chapter 4 we develop the idea of "finite structuring" for transfinite graphs, which mimics local-finiteness for conventional infinite graphs. This will be needed when we discuss discrete potential theory and random walks. Here we also present a transfinite generalization of König's classical lemma on the existence of one-way infinite paths.

Chapter 5 presents some fundamental theorems asserting the existence and uniqueness of voltage-current regimes in transfinite electrical networks under very broad conditions. No restrictions on the transfinite graph of the network are imposed. These theorems are based on a generalization of Tellegen's equation, since Kirchhoff's laws are no longer inviolate principles in the transfinite case. Furthermore, node voltages need not be unique nor even exist. The chapter ends by establishing additional conditions under which node voltages are uniquely existent. Maximum principles for node voltages and branch currents are established in Chapter 6. This last result makes full use of the finite structuring developed in Chapter 4 and therefore holds for a class of transfinite networks that is smaller than that for the fundamental theorems.

Finally, a theory for random walks on a finitely structured transfinite network is established in Chapter 7. The theory extends nearest-neighbor random walks by allowing them to pass through transfinite nodes. This is accomplished by exploiting the relationship between electrical network theory and random-walk phenomena.

The appendices present brief surveys of three topics of importance to our theory, namely ordinal and cardinal numbers, summable series, and irreducible and reversible Markov chains. With this information our book should be accessible to anyone familiar with the basic ideas about graphs, Hilbert spaces, and resistive electrical networks.

ACKNOWLEDGMENTS. Much of the research on the subject of this book was funded by the U.S. National Science Foundation; that support is gratefully acknowledged. I am also obliged to the National Research Council of Italy for sponsoring a series of lectures at the University of Milan and the University of Rome "Tor Vergata." These presentations, along with a special-topics course given at the University at Stony Brook, allowed me to test early versions of the manuscript. In particular, my discussions with Professors L. DeMichele, M. Picardello, P.M. Soardi, and W. Woess, and also with the students of that course, namely A. Alexeyev, Y.R. Chan, V.A. Chang, J.K. Fwu, H.T. Li, and H. Sundaram, led to many revisions and improvements in this book.

Finally, to my wife Edna, for her unfailing patience, encouragement, and support, once again and always — thanks.

Chapter 1

Introduction

This preliminary chapter introduces some suggestive ideas that lead to the versions of transfiniteness that we shall be examining in this book. The second section herein speculates about the possibility of connecting infinite graphs together at their infinite extremities to get a structure that is indescribable as a conventional graph. It points out that the conventional definitions of a graph lead to unwieldy constructions of the desired transfinite structure, one that is unnecessarily complicated. The third section then presents an alternative — and possibly novel — definition of a graph that is more amenable to transfinite generalization. The last section discusses a lacuna in the theory of infinite electrical networks that can be closed only by introducing connections to the network at infinity. This is the first step toward transfiniteness for electrical networks. But, first of all, we start by defining the symbols and phrases we will be using in this book.

1.1 Notations and Terminology

The symbols and phrases defined in this section are for the most part standard. They are reviewed here to remove possible ambiguities in our subsequent discussions. However, some of our definitions are unusual. One need merely skim through this section, noting the unfamiliar definitions, and then refer back to it as the need arises. There is an index of symbols at the end of this book, which may also be of help.

Let X be a set or, synonymously, a collection. X is called *denumerable* if it has infinitely many members that can be a arranged into a one-to-one correspondence with the natural numbers. X is called *countable* if it is either finite or denumerable, and called *uncountable* otherwise. The set of elements in X for which a statement $P(x)$ about x is true is denoted by $\{x \in X : P(x)\}$, or simply by $\{x : P(x)\}$ when X is understood. Quite frequently, in this book we will be constructing infinite sets, such as transfinite paths, by selecting elements from infinite collections of sets; when doing so, we will be tacitly invoking the axiom of choice [6]. The *cardinality* of a set is denoted by $\overline{\overline{X}}$. $\emptyset$ is the void set. A set with exactly one element x is called a *singleton* and is denoted by $\{x\}$. Given two sets X and Y, $X \subset Y$ means that X is a subset of Y and may possibly be equal to Y. (We will not use the symbol $X \subseteq Y$.) $Y \backslash X$ is the *difference of* Y *over* X, that is, the set of elements of Y that are not in X. If $X \subset Y$ and $Y \backslash X$ is not void, then X is called a *proper* subset of Y. The notation $x, y, z, \ldots \in X$ means that all the elements $x, y, z, \ldots$ are members of X, and $x, y, z, \ldots \notin X$ means that none of those elements is a member of X. A *partition* of X is a set of pairwise-disjoint, nonvoid subsets of X whose union is X.

In this book, we will be often occupied with sequences, so let us now specify their symbols and terminology in some detail. $\{x_i\}_{i \in I}$ denotes a set of indexed elements where the index i traverses the set I. A *sequence* is an indexed set where I is a set of integers; the sequence is assigned the total ordering corresponding to the order of the integers. A sequence is called *nontrivial* if it has two or more members. Such a sequence is also called a *vector*, and its members are called *components*. When the x_i are numbers, the support $\operatorname{supp} \mathbf{x}$ of a sequence $\mathbf{x} = \{x_i\}_{i \in I}$ is the set of indices i for which $x_i \neq 0$. It may happen that the i are also indices for branches in a network, in which case we may view the support as a set of branches. In Section 3.4 we will introduce *transfinite sequences*, but these are by definition sequences whose members are themselves sequences; in this book it is important that a sequence is always indexed by some or all of the integers — and not by a transfinite set of ordinals. A sequence is called *finite* or *two-ended* if I is finite; it is called *one-way infinite* or *one-ended* if I is bounded above (resp. below) and is unbounded below (resp. above); and it is called *two-way infinite* or *endless* if I is unbounded both below and above. An *orientation* is assigned to a sequence; it is the direction in which the indices increase. The *reverse orientation* is the opposite direction. One proceeds *rightward* (resp. *leftward*) along a sequence in the direction of the orientation (resp. reverse orientation). If a sequence terminates on either side, its member of least index (resp. highest index) is called its *first* or *leftmost* (resp. *last* or *rightmost*) member. Similarly, a member x_i of a sequence is *before* (resp. *after*) another member x_k if $i < k$ (resp. $i > k$). If $I = \{0, 1, 2, \ldots\}$ is the set of natural numbers, various alternative notations for sequences are used, such as $\{x_i\}_{i=0}^{\infty}$, $\{x_0, x_1, x_2, \ldots\}$, or $(x_0, x_1, x_2, \ldots)$. If the sequence is finite with m members, we let $I = \{1, 2, \ldots, m\}$ and write $(x_1, \ldots, x_m)$; in this case, it is more usual to call it a *vector* instead of a sequence and to think of it as a vertical array $[x_1, \ldots, x_m]^T$, where the superscript T denotes matrix transpose. A vector with two, three, or n components is called a *doublet*, *triplet*, or *m-tuple*, respectively.

Let X and Y be two sets again. The *Cartesian product* $X \times Y$ is the set of all doublets (x, y) with $x \in X$ and $y \in Y$. As a special case, we have the Cartesian product $X \times X$ of X with itself. A *binary relation* R on X is a subset of $X \times X$. R is called *reflexive* if $(x, x) \in R$ for all $x \in X$. R is called *symmetric* if, for all $x, y \in X$, $(x, y) \in R$ implies that $(y, x) \in R$. R is called *transitive* if, for all $x, y, z \in X$, $(x, y) \in R$ and $(y, z) \in R$ together imply that $(x, z) \in R$. R is called an *equivalence relation* if it is reflexive, symmetric, and transitive. When this is so, R induces *equivalence classes* in X; the *equivalence class of any* $x \in X$ is the set $\{y \in X : (x, y) \in R\}$. Any member of an equivalence class is called a *representative* of that class. Furthermore, the equivalence classes partition X.

A *function* f from X to Y is a rule that assigns exactly one element $y \in Y$ to each element $x \in X$. Synonymous with function is *mapping* or *operator*. The alternative notations $fx = y$ or $f(x) = y$ or $f : x \mapsto y$ display typical elements x and y for this assignment by f. Also, x is called the *independent variable* and y the *dependent variable* for f. At times, f is replaced by a specific formula in which the independent variable is suppressed; in such a case, the dot notation $f(\cdot)$ is used to show where the independent variable should be. The set X of all x for which $f(x)$ is defined is called the *domain* of f. Also, $f(X)$ denotes the *range* of f, that is, the set of all $y = f(x)$ with $x \in X$; $f(X)$ may be a proper subset of Y. If the range of f is contained in the real line R^1, f is called a *functional*. If W is a proper subset of X and if g is a function defined only on W and is such that $g(x) = f(x)$ for all $x \in W$, then g is called the *restriction of f to W*. The notation $f : X \rightsquigarrow Y$ is used to display the domain X of f and a set Y containing the range of f. $f : X \rightsquigarrow Y$ is called *one-to-one* or *injective* or an *injection* if the equation $f(x) = f(u)$ implies that $x = u$. $f : X \rightsquigarrow Y$ is called *onto* or *surjective* or a *surjection* if $f(X) = Y$. $f : X \rightsquigarrow Y$ is called *bijective* or a *bijection* if it is both injective and surjective. If f is bijective and if $y = f(x)$ for $x \in X$, $f^{-1}\colon y \mapsto x$ denotes the *inverse function* of f.

R^m denotes real m-dimensional Euclidean space. Thus, any $x \in R^m$ is a vector $x = [x_1, \ldots, x_m]^T$ of m real numbers x_i, and the *norm* $\|x\|$ of x is $(\sum_{i=1}^m x_i^2)^{1/2}$. The inner product of R^m is denoted by $\langle \cdot, \cdot \rangle_m$. On the other hand, the inner product for certain infinite-dimensional Hilbert spaces is denoted by $(\cdot, \cdot)$. Actually, the last notation is also used for a doublet, but the context in which we use this symbol should dispel any possible confusion. A mapping of a Euclidean space into a Euclidean space has a representation as a matrix, given the natural basis for such spaces. We use the same symbol for both the mapping and the matrix.

By an *element* of a graph we mean either a branch or a node. On the other hand, an *electrical element* of a network is either a resistor or a source. Electrical units and their symbols are as follows: volt V, ampere A, watt W, ohm Ω, and mho $\mho$. These are defined in Section 5.1.

A prominent role in this book is taken by the ordinal and cardinal numbers; these are discussed in Appendix A. A not uncommon usage is to write "ordinal" instead of "ordinal number," and "cardinal" instead of "cardinal number." This we usually do, but we never write "natural" as a replacement for "natural number." (Some authors make a distinction between ordinal number and ordinal [25, page 176], but we do not.)

A structure often arising in transfinite graphs is a finite regression of sets, that is, a set, whose elements are sets, each of which has sets as elements, and so on finitely many times. Any set $\mathcal{A}$ in this regression will be said to *embrace* itself, all its elements, all elements of its elements, and so on. In order not to overwork the word "embrace," we will at times simply say that an embraced set $\mathcal{E}$ of $\mathcal{A}$ is *in* or *of* $\mathcal{A}$ without referring to $\mathcal{E}$ as being embraced even when $\mathcal{E}$ is not an element of $\mathcal{A}$; it will be clear from the context that $\mathcal{E}$ is embraced by $\mathcal{A}$.

Throughout this book we will be discussing a variety of objects such as "tips," "nodes," "paths," "sections," "subsections," "graphs," and "networks." They will appear in a hierarchical structure. The position of any such object in that structure will be designated by an assigned *rank* ρ, which will be displayed by referring to the object as a *ρ-object.* ρ will be either a countable ordinal or an entity that precedes a countable limit ordinal. In fact, ranks form the totally ordered set

$$\mathcal{R} = \{\vec{0}, 0, 1, 2, \ldots, \vec{\omega}, \omega, \omega + 1, \ldots, \overrightarrow{\omega \cdot 2}, \omega \cdot 2, \omega \cdot 2 + 1, \ldots\}$$

where ω denotes the first infinite ordinal as always. $\mathcal{R}$ consists of all the countable ordinals with in addition a rank $\vec{\lambda}$ inserted before each countable limit ordinal λ. Such ranks $\vec{\lambda}$ are called *arrow ranks* and are not ordinals. $\vec{0}$ is the smallest of all ranks, and $\vec{\omega}$ is the next arrow rank. $\vec{\omega}$ is larger than every natural-number rank and less than ω. $\mathcal{R}$ is a *well-ordered set*; that is, it is totally ordered, and every subset has a least member. Ranks will be symbolized with lower-case Greek letters. Also, there will be occasions where we use $\rho - 1$ to symbolize the predecessor of some arbitrary ordinal rank ρ; in the event that ρ is a limit ordinal λ, $\lambda - 1$ will be taken to designate $\vec{\lambda}$. Thus, $\vec{\lambda}+1$ will mean λ. In particular, in terms of ranks, we have $0-1 = \vec{0}$ and $\vec{0}+1 = 0$, and also $\omega - 1 = \vec{\omega}$ and $\vec{\omega}+1 = \omega$.

Furthermore, with regard to any rank β no less than 1, the symbols $\beta-$ and $\beta+$ have special meanings. $\beta-$ denotes any rank less than β and larger than $\vec{0}$, and $\beta+$ denotes any rank no less than β. Thus, when we are dealing with a graph or network of rank ν, all ranked objects have ranks no larger than ν, and we have $0 \leq (\beta-) < \beta \leq (\beta+) \leq \nu$. For example, $\vec{\omega}-$ is any unspecified natural number. We use the symbol $\beta-$ or $\beta+$ when we wish to work in general with some object having an unspecified rank.

Let us point out a change from previous terminology. In [35] and other prior works, a path or a loop whose branch resistances had a finite sum was called *perceptible*. We now say instead *permissive*. The reason for this is that "permissive" is more descriptive of the physical idea we are trying to convey and in addition serves as a better antonym for the dual concept of "restraining" (see Section 5.3). Henceforth, we take permissive and perceptible to be synonyms in our theory of transfinite electrical networks but will use permissive throughout this book.

Finally, the symbol ♣ will be used to mark the end of a proof or example.

1.2 Conventional Graphs — a Poor Start

There are several ways of defining graphs, each being equivalent to the others. None of them appears to be amenable for generalization to transfinite graphs. As will be explained below, difficulties arise when one tries to extend transfinitely the idea of a "branch." However, an unconventional definition of a graph can be devised which leads naturally to transfinite graphs merely through extensions of the idea of a "node." This will be presented in the next section. In this section we state a conventional definition of a graph in order to discuss why it is a poor starting point for a theory of transfinite graphs.

Let $\mathcal{N}$ and $\mathcal{B}$ be two (finite or infinite) sets, and let $\mathcal{N}'$ be the set of all one-element and two-element subsets of $\mathcal{N}$. Also, let G be a mapping from all of $\mathcal{B}$ into $\mathcal{N}'$. Then, the triplet $\mathcal{G} = \{\mathcal{B}, \mathcal{N}, G\}$ is called a *graph*. It is called an *infinite graph* if either or both of $\mathcal{N}$ and $\mathcal{B}$ are infinite sets. The elements of $\mathcal{B}$ are called *branches*, and the elements of $\mathcal{N}$ are called *nodes*. A branch b and a node n are said to be *incident* if n is a member of $G(b)$. If $G(b)$ is a singleton member of $\mathcal{N}'$, then b is called a *self-loop*. Moreover, if for two branches b_1 and b_2 we have $G(b_1) = G(b_2)$, then b_1 and b_2 are said to be *in parallel* and are called *parallel branches*. Furthermore, if the node n is not a member of $G(b)$ for every b, then n is called an *isolated* node. Another standard definition of a graph can be obtained simply by dropping the mapping G and identifying each branch as being a one-element or two-element subset of $\mathcal{N}$. This leads, however, to a difficulty when parallel branches are to be distinguished, unless a branch-indexing scheme is used, but the latter leads back to the mapping G. In any case, both definitions — and in fact all conventional definitions — treat branches as being identified by their incident nodes if parallel branches are disallowed.

A *subgraph* $\mathcal{G}_s$ of $\mathcal{G}$ is obtained by choosing a subset $\mathcal{B}_s$ of $\mathcal{B}$ and a subset $\mathcal{N}_s$ of $\mathcal{N}$ such that $\mathcal{N}_s$ contains the set $G(\mathcal{B}_s)$ and then setting $\mathcal{G}_s = \{\mathcal{B}_s, \mathcal{N}_s, G_s\}$, where G_s is the restriction of G to $\mathcal{B}_s$.

Let us now explore heuristically how a graph might be extended transfinitely. In order to do so, it will be helpful to introduce the concept of "connectedness," which in turn requires a "path" to be defined. A *path* in $\mathcal{G}$ is an alternating sequence of distinct nodes and branches:

$$P = \{\ldots, n_m, b_m, n_{m+1}, b_{m+1}, \ldots\}, \qquad (1.1)$$

where each branch b_m is incident to the two nodes n_m and n_{m+1} adjacent to b_m in the sequence. That the elements of P are distinct means that no element appears more than once in the sequence. P may be a finite sequence, a one-way infinite sequence, or a two-way infinite sequence, and P is then called *finite* and also *two-ended*, *one-ended*, or *endless*, respectively. If P terminates on either side, it is required to terminate at a node. A *subpath* of P is a subsequence of (1.1) which by itself is a path in $\mathcal{G}$.

If n_a and n_b are two terminal nodes of a finite path P, n_a and n_b are said to be *connected by* P, and P is called an $n_a n_b$-path. Given any two nodes, either there exists a path connecting them or there is no such path. If every two nodes of $\mathcal{G}$ are connected by some path, $\mathcal{G}$ itself is called *connected.*

Two graphs may be combined into a larger graph by joining them at a node. More precisely, if $\mathcal{G}_1 = \{\mathcal{B}_1, \mathcal{N}_1, G_1\}$ and $\mathcal{G}_2 = \{\mathcal{B}_2, \mathcal{N}_2, G_2\}$ are two graphs with $\mathcal{B}_1 \cap \mathcal{B}_2 = \emptyset$ and $\mathcal{N}_1 \cap \mathcal{N}_2 = \emptyset$, we may take a node $n_1 \in \mathcal{N}_1$ and a node $n_2 \in \mathcal{N}_2$ and identify them as being the same node: $n = n_1 = n_2$. This yields the graph $\mathcal{G} = \{\mathcal{B}_1 \cup \mathcal{B}_2, \mathcal{N}_{12}, G\}$, where $\mathcal{N}_{12}$ and G are defined as follows. $\mathcal{N}_{12}$ is the union of $\mathcal{N}_1$ and $\mathcal{N}_2$ except for the replacement of n_1 and n_2 by the single node n. Let $\mathcal{N}'_{12}$ be the set of one-element and two-element subsets of $\mathcal{N}_{12}$. G is the mapping from $\mathcal{B}_1 \cup \mathcal{B}_2$ into $\mathcal{N}'_{12}$ whose restriction to $\mathcal{B}_1$ (resp. $\mathcal{B}_2$) is G_1 when $n = n_1$ (resp. G_2 when $n = n_2$). If $\mathcal{G}_1$ and $\mathcal{G}_2$ are each connected graphs, then $\mathcal{G}$ is connected too. Indeed, given any $n_a \in \mathcal{N}_1$ and any $n_b \in \mathcal{N}_2$, there is an $n_a n_b$-path consisting of an $n_a n$-path followed by an $n n_b$-path. The result of all this is simply another graph $\mathcal{G}$ having a larger branch set $\mathcal{B}_1 \cup \mathcal{B}_2$.

However, if $\mathcal{G}_1$ and $\mathcal{G}_2$ are infinite connected graphs, something new might be achieved if we somehow join $\mathcal{G}_1$ and $\mathcal{G}_2$ at their infinite extremities. For instance, if $\mathcal{G}_1$ is simply a one-ended path

$$P_1 = \{n_1, b_1, n_2, b_2, \ldots\}$$

and $\mathcal{G}_2$ is another one-ended path

$$P_2 = \{n_a, b_a, n_b, b_b, \ldots\},$$

we might reverse the order of P_2 and append it to P_1 to get the structure

$$\{n_1, b_1, n_2, b_2, \ldots, b_b, n_b, b_a, n_a\}. \tag{1.2}$$

This is not a sequence, for it is impossible to consecutively index the elements using the integers while maintaining the natural order of the integers. We seem to have a "transfinite path." It appears to be a graph, for there is implicitly a mapping from each branch to the pair of nodes adjacent to that branch in (1.2), but there is something more in that structure, namely, the joining of the two extremities of P_1 and P_2. We might view that joining as a new kind of node — let's call it a "1-node."

More generally, if $\mathcal{G}_1$ and $\mathcal{G}_2$ are infinite connected graphs containing one-ended paths, we might choose a one-ended path P_1 in $\mathcal{G}_1$ and another one-ended path P_2 in $\mathcal{G}_2$ and then perhaps connect $\mathcal{G}_1$ and $\mathcal{G}_2$ together by joining the infinite extremities of P_1 and P_2 through a 1-node. The result is a new kind of graph that is transfinitely connected, but not connected in the conventional sense; indeed, a node of $\mathcal{G}_1$ and a node of $\mathcal{G}_2$ are now connected only through a transfinite path of the form of (1.2) where a 1-node is understood to appear somewhere among the "dots."

How can we extend the conventional definition of a graph to encompass our transfinite structure? The difficulty that arises in this regard is that there is no branch incident to the 1-node and therefore no way of introducing the 1-node through a mapping from the branch set $\mathcal{B}_1 \cup \mathcal{B}_2$ into the collection of one-element and two-element subsets of an extended node set that contains the 1-node.

Nonetheless, this gambit inspires still another kind of transfinite structure. Perhaps a 1-node and an ordinary node can be united, that is, the infinite extremity of one graph might be joined to an ordinary node n_a of a second graph to obtain a different kind of 1-node n^1. Then, any branch $\{n_a, n_b\}$ of the second graph that is incident to n_a would become a new kind of branch — an "extended branch" that is identified with the two-element set $\{n^1, n_b\}$. In fact, still another kind of extended branch might be identified with a set of two different 1-nodes, where now either or both of the 1-nodes may simply represent the extremities of one-ended paths. Furthermore, a node of still higher rank, say, a "2-node," might be envisaged at an infinite extremity of a structure obtained by connecting together infinitely many infinite graphs. This would lead to other kinds of extended branches that are two-element sets, where one or both of the elements are 2-nodes. The process might even be repeated indefinitely to obtain a whole hierarchy of transfinite structures. This was the approach to transfinite graphs that was initiated in [34] but was eventually abandoned because it led to a cumbersome proliferation of different kinds of extended branches.

In summary, the conventional definition of a graph, whereby branches are identified by their incident nodes, is not conveniently generalized to the transfinite case. In the next section, we introduce a different definition, which reverses the roles of the nodes and branches. Instead of the branches being defined by the nodes, the nodes are defined by the branches. Transfinite generalizations again lead to nodes of higher ranks as infinite extremities of various ranks are joined, but extended branches are avoided without any loss of generality.

1.3 0-Graphs

From the perspective of electrical circuits, it is the branches — not the nodes — that are the fundamental elements. This is because energy conversion takes place in the branches. The nodes simply serve to connect branches together, thereby allowing current to flow from one branch to another. The structure produced by those branches and connections is a graph. With regard to the graph alone, the only thing of importance about a branch is that it has two terminals, at which the said connections are made. In fact, we need merely specify how the terminals are grouped together into nodes in order to define the graph. This then is our alternative approach to a graph; it will lead naturally to transfinite graphs having a hierarchy of generalized nodes but only one kind of branch.

We start with a nonvoid set $\mathcal{B}$ of branches, where each branch is simply a pair of terminals. We call each terminal an *elementary tip* or, synonymously, a $\vec{0}$-*tip*. $\vec{0}$ denotes the *rank* of the tip and may be referred to verbally as "zero arrow"; it is the first of the "arrow ranks," something we shall see more of. Each $\vec{0}$-tip belongs to exactly one branch. In order to conform with subsequent terminology, we shall speak of branches in certain ways that presently might seem quite arbitrary. In particular, we shall view a branch as a primitive sort of path, which we call an *endless* $\vec{0}$-*path*. With b being the pairing of the two elementary tips $t_1^{\vec{0}}$ and $t_2^{\vec{0}}$, we refer to $t_1^{\vec{0}}$ and $t_2^{\vec{0}}$ as the *terminal tips* of b. We also say that b *traverses* (but does not embrace) each of its terminal tips and that b is a *representative* of each of its terminal tips.

Next, let $\mathcal{T}^{\vec{0}}$ denote the set of all $\vec{0}$-tips of all the branches in $\mathcal{B}$. With $\overline{\overline{\mathcal{X}}}$ denoting the cardinality of a set $\mathcal{X}$, we have $\overline{\overline{\mathcal{T}^{\vec{0}}}} = 2\overline{\overline{\mathcal{B}}}$ if $\mathcal{B}$ is a finite set, and $\overline{\overline{\mathcal{T}^{\vec{0}}}} = \overline{\overline{\mathcal{B}}}$ if $\mathcal{B}$ is an infinite set. In general, we allow $\mathcal{B}$ to have any cardinality. As the next step, we partition $\mathcal{T}^{\vec{0}}$ into subsets and call each subset a 0-*node*. Motivated by electrical terminology, we also say that there is a *short* between a 0-node and all its elements or that these entities are *shorted together* (or simply *shorted*) when the 0-node is a nonsingleton. On the other hand, if a 0-node is a singleton, it and its sole $\vec{0}$-tip are said to be *open*. The *degree* of a 0-node n^0 is simply its cardinality. In accordance with the definition of "to embrace" given in Section 1.1, n^0 *embraces* itself and each of its elementary tips. A *self-loop* is a branch having both of its $\vec{0}$-tips in a single 0-node, that is, shorted. If a 0-node n^0 contains one or both $\vec{0}$-tips of a branch b, we say that n^0 and b are *incident* or that b *reaches* n^0 *with* each such tip. However, b does not embrace n^0. If two branches b_1 and b_2 are incident to the same 0-node, we say that b_1 and b_2 are *adjacent*. Moreover, two branches, which are not self-loops, are said to be *in parallel* or are called *parallel branches* if they are incident to the same two 0-nodes.

With a given partitioning of $\mathcal{T}^{\vec{0}}$, we have a set $\mathcal{N}^0$ of 0-nodes. We define the 0-*graph* $\mathcal{G}^0$ as the pair

$$\mathcal{G}^0 \;=\; \{\mathcal{B}, \mathcal{N}^0\}. \tag{1.3}$$

$\mathcal{G}^0$ is called *finite*, *infinite*, *countable*, and *uncountable* if its branch set $\mathcal{B}$ or equivalently the set $\mathcal{T}^{\vec{0}}$ has these respective properties. $\mathcal{G}^0$ is also called *locally finite* if every one of its 0-nodes has finite degree.

A 0-graph is a conventional graph in the sense of the preceding section because there is implicitly a well-defined mapping from $\mathcal{B}$ into the set of one-element and two-element subsets of $\mathcal{N}^0$ provided by the incidences between branches and 0-nodes. Thus, many of the ideas pertaining to graphs can be carried over to 0-graphs.

For instance, a path may be defined exactly as it was in the preceding section with regard to (1.1). Now, however, we will call it a 0-*path* and assign to it the *rank* 0 in order to distinguish it from the transfinite paths we shall introduce later on. Moreover, we shall denote 0-paths by

$$P^0 \;=\; \{\ldots, n_m^0, b_m, n_{m+1}^0, b_{m+1}, \ldots\} \tag{1.4}$$

to emphasize that it and its nodes are of rank 0. A 0-path is called *trivial* if it consists of only a single 0-node; thus, it is *nontrivial* if it has at least one branch. A 0-*loop* is defined exactly like a finite 0-path except that the two terminal nodes are required to be the same node. (Actually, the 0-loop can be written as a sequence in twice as many ways as it has 0-nodes, for each of the latter can be taken as the terminal nodes and each sequence can be reversed; to be more precise, the 0-loop should be viewed as the class of these equivalent sequences. We will not be so meticulous.) As was defined in Section 1.1, a 0-path or a 0-loop *embraces* itself, all its branches and 0-nodes, and all the $\vec{0}$-tips embraced by its 0-nodes. We also say that P^0 *traverses* the elementary tips of its branches. As in the preceding section, if (1.4) terminates on either side, it terminates at a 0-node n_t^0. Let b_t be the branch in (1.4) adjacent to n_t^0. The $\vec{0}$-tip with which b_t reaches n_t^0 is called a *terminal* $\vec{0}$-*tip of* P^0. (The only other possibilities are that P^0 is one-sided or endless. In these cases, we shall say that P^0 has one or two terminal "0-tips," an idea that will be introduced in the next chapter.)

We turn now to our version of a "subgraph." Given the 0-graph (1.3), let $\mathcal{B}_s$ be any nonvoid subset of $\mathcal{B}$. Remove from $\mathcal{N}^0$ every 0-node that does not have at least one $\vec{0}$-tip belonging to a branch of $\mathcal{B}_s$. The remaining 0-nodes comprise a nonvoid subset $\mathcal{N}_s^0$ of $\mathcal{N}^0$ consisting of those 0-nodes of $\mathcal{N}^0$ that are incident to branches of $\mathcal{B}_s$. Then, $\mathcal{G}_s^0 = \{\mathcal{B}_s, \mathcal{N}_s^0\}$ is the *subgraph of* $\mathcal{G}^0$ *induced by* $\mathcal{B}_s$.

This definition of a subgraph conforms with that of the preceding section because the mapping G_s is now implicitly defined by the incidences between the branches of $\mathcal{B}_s$ and the nodes of $\mathcal{N}_s^0$. Here, too, a subgraph *embraces* itself, all the branches in $\mathcal{B}_s$, all the 0-nodes in $\mathcal{N}_s^0$, and all the $\vec{0}$-tips of all those 0-nodes. Also, note that a 0-path P^0 is a subgraph of $\mathcal{G}^0$ induced by the branches in P^0.

In general, $\mathcal{G}_s^0$ is not a 0-graph because a node of $\mathcal{N}_s^0$ may contain a $\vec{0}$-tip that does not belong to any branch of $\mathcal{B}_s$, that is, the union of all the branches of $\mathcal{B}_s$ may be a proper subset of the union of all the 0-nodes of $\mathcal{N}_s^0$ — in violation of our definition of a 0-graph. Actually, a related idea, namely a "reduced graph," is defined in [35, page 8]; it is something like a subgraph but is moreover a 0-graph. However, in this book we shall make do with subgraphs because reduced graphs complicate subsequent discussions due to the fact that they introduce many new nodes, the "reduced nodes," which must then be taken into account.

The usual definitions relating to graphs are easily modified for 0-graphs. Let us state two of them. The *union* (resp. *intersection*) of two subgraphs of a 0-graph is the subgraph induced by the union (resp. intersection) of their branch sets. Intersections may be void, of course.

A subgraph is said to *traverse* a $\vec{0}$-tip if it embraces the branch having that $\vec{0}$-tip. Two subgraphs or a subgraph and a 0-node are said to *meet* or to be *incident* if they embrace a common 0-node. Otherwise, they are called *totally disjoint*. At this point "meeting" and "incidence" are identical ideas, but later on when we introduce transfinite graphs they will be different; "meeting" will be a stronger concept than "incidence."

Next, let $\mathcal{B} = \bigcup_{k\in K} \mathcal{B}_k$ be a partitioning of the branch set $\mathcal{B}$ of $\mathcal{G}^0$ into the subsets $\mathcal{B}_k$; thus, $\mathcal{B}_k \cap \mathcal{B}_l = \emptyset$ if $k \neq l$. Let $\mathcal{G}_k^0$ be the subgraph of $\mathcal{G}^0$ induced by $\mathcal{B}_k$. Then, the set $\{\mathcal{G}_k^0\}_{k\in K}$ is called a *partitioning* of $\mathcal{G}^0$.

Two branches (or two 0-nodes, or a branch and a node) of a 0-graph $\mathcal{G}^0$ are said to be 0-*connected* if there is a finite 0-path that embraces them. If this is true for all pairs of branches of $\mathcal{G}^0$, we say that $\mathcal{G}^0$ is 0-*connected*. A *component* of $\mathcal{G}^0$ is a subgraph of $\mathcal{G}^0$ induced by a maximal set of branches that are pairwise 0-connected; in this case, that subgraph is a 0-graph. A *forest* is a 0-graph having no 0-loops, and a *tree* is a forest with only one component. A *spanning tree* in $\mathcal{G}^0$ is a subgraph of $\mathcal{G}^0$ that is a tree and is incident to every 0-node of $\mathcal{G}^0$. A *series circuit* is a 0-path P^0 every node of which is of degree 2 except possibly for its terminal nodes; here, the degree of a node is taken with respect to the graph in which P^0 lies. A *parallel circuit* consists of two nodes and finitely or infinitely many branches, each of which is incident to both nodes.

1.4 Why Transfiniteness?

As was pointed out in the Preface, until just a few years ago all graph theory during the 250 years since its inception and all electrical network theory during its 150 years have shared one immutable principle: Any two branches either are connected by a finite path or are not connected at all. However, when the graph or network has an infinity of branches, another idea suggests itself. Perhaps two branches might somehow be connected through an infinitely long path while not being connected by any finite one.

This idea has a physical basis in electrical networks. Consider the ladder shown in Figure 1.1. It extends infinitely to the right and has two input nodes a and b. The "shunting" resistances r_k (k odd) and the "series" resistances $r_k/2$ (k even) all have positive values. Let us hope that a nonzero voltage v impressed upon the nodes a and b produces a unique current i as shown. When this is truly the case, $R_i = v/i$ will be called the *input resistance* or synonymously the *driving-point resistance* of the ladder. Moreover, if the standard rules for combining finitely many parallel and series resistances continue to hold for our infinite ladder, then R_i will be given by the infinite continued fraction

$$R_i = \cfrac{1}{g_1 + \cfrac{1}{r_2 + \cfrac{1}{g_3 + \cfrac{1}{r_4 + \ddots}}}} \tag{1.5}$$

where $g_k = 1/r_k$ (k odd) represents the kth "conductance."

Now this continued fraction may or may not converge depending upon the values of r_k and g_k. That convergence is determined by the behavior of its even truncations

$$R_n = \cfrac{1}{g_1 + \cfrac{1}{r_2 + \ddots + \cfrac{1}{r_n}}} \quad (n \text{ even}) \tag{1.6}$$

and of its odd truncations

$$R_n = \cfrac{1}{g_1 + \cfrac{1}{r_2 + \ddots + \cfrac{1}{g_n}}} \quad (n \text{ odd}). \tag{1.7}$$

As $n \to \infty$, the even truncations increase monotonically and are all bounded above by any odd truncation; thus, the even truncations converge to a limit R_e. The odd truncations are all positive and decrease monotonically; hence, they too converge to a limit R_o. If $R_e = R_o$, (1.5) is called *convergent* and has the value $R_i = R_e = R_o$. This happens when the series $g_1 + r_2 + g_3 + r_4 + \cdots$ diverges [30, page 120]. With regard to our speculation in Section 1.2 about possible connections at infinite extremities, the convergence to R_i is quite independent of anything that might be connected to the ladder at infinity. What is happening here electrically is that the infinite size of either the total shunting conductance $g_1 + g_3 + \cdots$ or of the total series resistance $r_2 + r_4 + \cdots$ prevents the transmission to infinity of a nonzero portion of the power generated by the finite source at the input, and therefore whatever is connected to the ladder at infinity will have no influence upon the voltage-current regime. Figuratively speaking, we might say that the "source cannot see infinity" or that "infinity is imperceptible."

On the other hand, the continued fraction (1.5) may not converge. This will occur when the series $g_1+r_2+g_3+r_4+\cdots$ converges [30, page 120]. In this case, R_e will be less than R_o, and the nonconvergence of (1.5) will exhibit itself through an oscillating sequence of its truncations. To interpret this in still another fashion, consider a truncation of the ladder terminated by a loading resistor R_l, as shown in Figure 1.2. Its input resistance is

$$R_n = \cfrac{1}{g_1 + \cfrac{1}{r_2 + \ddots + \cfrac{1}{g_{n-1} + \cfrac{1}{r_n + R_l}}}} \quad (n \text{ even}). \tag{1.8}$$

The even truncation (1.6) is the value of (1.8) when R_l is a short, that is, when $R_l = 0$. Thus, the limit R_e of the even truncations can be viewed as the result of expanding a shorted truncation into an infinite ladder — or more intuitively as the result of connecting a short at infinity to the ladder of Figure 1.1. Similarly, the odd truncation (1.7) is the value of (1.8) when R_l is an open, that is, when $R_l = \infty$. After expanding an open truncation into an infinite ladder, we may view the limit R_o as the result of having an open at the infinite extremity of the ladder of Figure 1.1.

All this suggests another possibility: R_l can be chosen to have any positive value in (1.8). In the limit as $n \to \infty$, we may view R_l as being connected to the ladder at infinity as indicated in Figure 1.3 and may take the limiting value $R_i = \lim_{n\to\infty} R_n$ of (1.8) as the input resistance of that loaded infinite ladder. (The last limit truly exists [35, Theorem 6.11-1].) Moreover, we will have $R_e < R_i < R_o$ when $0 < R_l < \infty$.

Example 1.4-1. Here is a numerical example. In Figure 1.1 let $r_k = 1$ for all k. Then, with v fixed, there is a unique voltage-current regime in that infinite ladder for which the total power dissipated in all the resistances is finite [35, Theorem 3.7-1]. Moreover, we now have a uniform ladder, and its input resistance R_i is its characteristic resistance $(-1+\sqrt{5})/2$ [35, pages 164-165], a value that is independent of any loading resistance at infinity. We might say that "infinity is imperceptible."

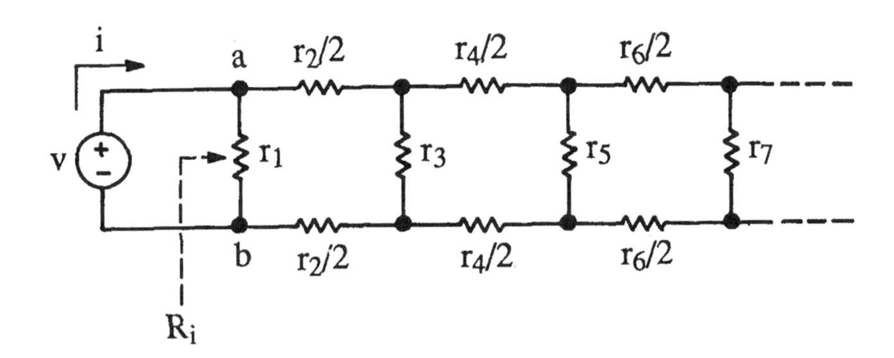

Figure 1.1. A one-way infinite ladder network. All the resistances $r_k (k = 1, 2, \ldots)$ are positive numbers, and $R_i = v/i$ is the input resistance.

Next, by way of contrast, let $r_k = 10^{(k+1)/2}$ for k odd and $r_k = 10^{-k/2}$ for k even in Figure 1.1. Thus, $r_1 = 10,\ r_3 = 100,\ r_5 = 1000,\ \ldots$ and also $r_2 = .1,\ r_4 = .01,\ r_6 = .001,\ \ldots$. In this case, the input resistance R_i does depend upon what is connected to the ladder at infinity, or, to be more precise, the input resistance of the loaded truncations (Figure 1.2) approaches different limits for different values of R_l. For example, the first four digits of the limit are $R_e = .1098$ when $R_l = 0$, $R_i = .9909$ when $R_l = 1$, and $R_o = 9.001$ when $R_l = \infty$. Now, "infinity is perceptible." ♣

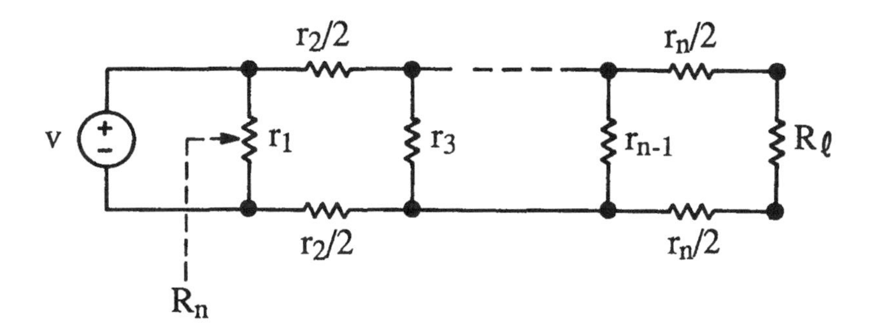

Figure 1.2. A truncation of the ladder of Figure 1.1 with a loading resistor R_l connected at the point of truncation.

All this suggests a bipartite classification of all 0-connected, resistive, infinite electrical networks: A network is either affected or unaffected by passive connections at infinity; in other words, its voltage-current regime either does or does not depend upon what loading resistances are connected to various infinite extremities of the network. When it does so depend, one is obliged to specify what those loading resistances are (shorts and opens being possibilities) and to what "parts of of infinity" they are connected if a unique regime is to be obtained. Introducing transfiniteness in this way is not generalization just for the sake of generalization. We have here a lacuna in the theory of conventional infinite electrical networks, and to eliminate it we are forced to go transfinite with specifications of connections "at infinity."

A body of research has revolved around this dichotomy in the behavior of conventional infinite networks, much of it in disguised form — as, for example, in the examination of recurrence or transience for random walks on infinite networks [7], [27], [32]. That dichotomy can be avoided (but not eliminated) by always specifying the connections at infinity — whether or not such a specification is needed — because then every regime is unique, but we are not discounting the said body of research. The dichotomy remains an important consideration even though it can be dodged. Indeed, the divergence (resp. convergence) of $g_1 + r_2 + g_3 + r_4 + \cdots$ as a condition for the uniqueness (resp. nonuniqueness) of the response of the ladder of Figure 1.1 is not devalued by the elimination of nonuniqueness through the specification of R_l in Figure 1.3.

Another matter: So far we have overlooked a difficulty concerning the connection of R_l at the infinite extremity of the ladder of Figure 1.3. To what can R_l be connected? The ladder extends indefinitely, and no branch or node of it is "at infinity." What is needed now is the invention of a new kind of node. Let us borrow the idea of a "1-node" suggested in Section 1.2. It might combine the infinite extremity of the one-ended path along the upper horizontal branches of Figure 1.3 with an elementary tip of the branch for R_l and thereby allow current to flow along that path and into R_l. Another "1-node" combining the infinite extremity for the lower horizontal branches with the other elementary tip of the branch for R_l might allow current to flow out of R_l and along those lower branches. Once this is accomplished, we will have again a "transfinite network," something that is fundamentally different from conventional infinite networks in that R_l is connected to the branches of the ladder only through infinite paths.

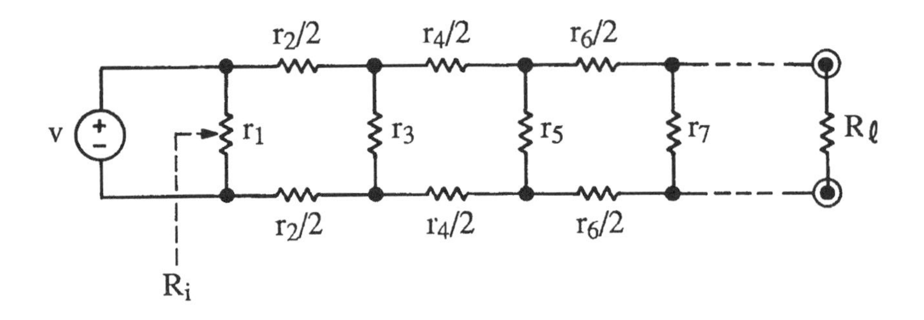

Figure 1.3. The infinite ladder loaded at infinity by the resistor R_l.

Moreover, if R_l can be connected to the infinite extremity of the ladder, so too can the input nodes of another infinite ladder. The process might be continued indefinitely to obtain an infinite cascade of infinite ladders. In order to connect a branch at the "far end" of this cascade will require another kind of node, say, a "2-node" as in Section 1.2 again. And so it goes. This gambit leads us on to a hierarchy of transfinite cascades that call for "ν-nodes," where ν might be any natural number and even any countable ordinal.

The absence of transfinite graphs and transfinite networks during 250 years of graph theory and 150 years of electrical network theory is puzzling — especially since Cantor introduced transfinite ideas over 100 years ago [4]. The examples that have led us on to a vision of transfiniteness are plain enough. Perhaps nothing more than simple oversight is the explanation. Whatever be the case, transfiniteness is a radically new concept in (undirected) graphs and networks. This book examines some of its implications, but it may well be that far more will yet be discovered than the results we presently have.

Chapter 2

Transfinite Graphs

Section 1.3 points toward our method for constructing transfinite graphs. Its most primitive concept is neither a node nor a branch but is instead an elementary tip. Nodes and branches are simply certain sets of such tips. Indeed, we may think of two elementary tips of a branch as being its two extremities. Then nodes are shorts between the extremities of different branches, and the result is a 0-graph.

An extension of these heuristic ideas will lead to transfinite graphs of higher ranks. The infinite extremities of 0-graphs will be represented by "0-tips," and shorts between them will connect 0-graphs together to yield a "1-graph." We continue recursively. For any countable ordinal rank ν the infinite extremities of a "ν-graph" will be viewed as "ν-tips," and shorts between them will yield a "$(\nu+1)$-graph."This is how we construct a transfinite graph whose rank is a countable successor ordinal $\nu+1$. However, an additional complication arises when the graph to be constructed has a countable limit ordinal λ as its rank. Our solutionis to introduce a rank $\vec{\lambda}$ — called an "arrow rank," which immediately precedes λ and whose corresponding "$\vec{\lambda}$-graphs" provide a transition to the "λ-graphs" by supplying the "$\vec{\lambda}$-tips" that serve for λ-graphs as do ν-tips for $(\nu+1)$-graphs. As compared to ν-tips, $\vec{\lambda}$-tips are constructed somewhat differently.

This very briefly sketches how we will generate in this chapter transfinite graphs for countable-ordinal ranks. With regard to graphs with uncountable-ordinal ranks, we have nothing to offer; this remains a puzzle.

2.1 1-Graphs

Our immediate objective is to render precise the idea of a "connection at infinity. " We use the device introduced in [33]. This will lead to our first kind of transfinite graph, the "1-graph. "

We start with a 0-graph $\mathcal{G}^0 = \{\mathcal{B}, \mathcal{N}^0\}$, where the branch set $\mathcal{B}$ has any cardinality and $\mathcal{N}^0$ is the set of 0-nodes. $\mathcal{G}^0$ need not contain any one-ended 0-path even when $\mathcal{B}$ is an infinite set, but, in the event that it does, we can partition the set of all one-ended 0-paths in $\mathcal{G}^0$ into equivalence classes by defining two one-ended 0-paths to be equivalent if they differ on no more than a finite number of nodes and branches (or, what amounts to the same thing, if there is a third one-ended 0-path that is "in" the first two 0-paths in the sense that all the branches of the third path are embraced by each of the first two paths). Each equivalence class will be called a 0-*tip*, and any one-ended 0-path in that class will be called a *representative* of the 0-tip. These 0-tips signify the infinite extremities of $\mathcal{G}^0$ and enable connections to those extremities in much the same way as elementary tips enabled connections between branches.

Example 2.1-1. How many 0-tips can an infinite 0-graph have? It need not have any, this being the case of the *star graph* all of whose 0-nodes are singletons except for its single central node which is incident to every branch and may be of any degree.

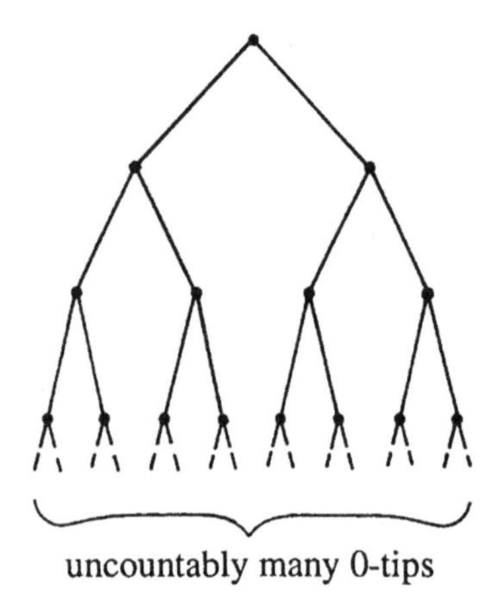

Figure 2.1. The infinite binary tree. The set of its 0-tips has the cardinality **c** of the continuum. The 0-tips are pairwise disconnectable.

On the other hand, a countable 0-graph may have uncountably many 0-tips. Consider the infinite binary tree indicated in Figure 2.1. Each one-ended 0-path is equivalent to a unique one-ended 0-path starting at the apex 0-node, but no two one-ended 0-paths starting at the apex node can be equivalent. Thus, each 0-tip can be identified with a unique 0-path of the latter kind. Moreover, a tracing of any such path involves making a binary choice at each 0-node to determine the next branch. Thus, the number of those 0-paths — and therefore the number of 0-tips — is $2^{\aleph_0} = \mathbf{c}$, where $\aleph_0$ is the cardinality of a denumerably infinite set and $\mathbf{c}$ is the cardinality of the continuum [1, page 387](see also Appendices A7 and A8).

Consider now a tree, every one of whose 0-nodes is of degree $\mathbf{c}$. Does it have more 0-tips than the binary tree? The answer is no. Indeed, choose any node as the apex node. Then, by the argument concerning the choices to be made in tracing any one-ended path starting at the apex node, the number of 0-tips is $\mathbf{c}^{\aleph_0} = \mathbf{c}$ [1, page 387]. ♣

Example 2.1-2. As another example consider the one-way infinite ladder $\mathcal{L}^0$ of Figure 2.2. Its set of 0-tips also has the cardinality $\mathbf{c}$ of the continuum. To see this, note first of all that any 0-tip of this graph can be uniquely identified with the set of just those of its representatives that start at the 0-node n^0. Any two such representatives for a given 0-tip differ on no more than finitely many branches. Moreover, every such representative for any 0-tip can be specified by a binary sequence $\{x_0, x_1, x_2, \ldots\}$, where $x_k = 0$ (resp. $x_k = 1$) if the representative passes through a_k (resp. b_k). Of course, within the representative a vertical branch c_k appears between the branches corresponding to x_k and x_{k+1} whenever $x_k \neq x_{k+1}$, and c_0 precedes b_0 when $x_0 = 1$. Now, the set of one-ended binary sequences has the cardinality $\mathbf{c}$ of the continuum as before. However, in order to determine the cardinality of the set of 0-tips of the ladder, we must examine a partitioning of those binary sequences, where two sequences are considered equivalent if they differ on no more than finitely many elements. Note that every set S of equivalent sequences is countable. Indeed, choose any binary sequence $s_0 = \{x_0, x_1, x_2, \ldots\}$. Then, count the one binary sequence s_1 that differs from s_0 only in x_0; then count the two additional binary sequences s_2 and s_3 that differ from s_0 or $s_1$only in x_1; then count the four additional binary sequences that differfrom s_0, s_1, or s_2 only in x_2; and so forth. Thus, the cardinality of S is $\aleph_0$. Consequently, the cardinal number of the set of 0-tips for the ladder is $\mathbf{c} \div \aleph_0 = \mathbf{c}$ [25, pages 277-278] (see also Appendix A9). ♣

As the next step, partition the set $\mathcal{T}^0$ of all 0-tips of $\mathcal{G}^0$ into subsets $\mathcal{T}_\tau^0$, where τ denotes the indices of the partition. Any partition is allowed. Thus, $\mathcal{T}^0 = \cup_\tau \mathcal{T}_\tau^0$, $\mathcal{T}_\tau^0 \neq \emptyset$ for every τ, and $\mathcal{T}_{\tau_1}^0 \cap \mathcal{T}_{\tau_2}^0 = \emptyset$ if $\tau_1 \neq \tau_2$. Furthermore, for each τ let $\mathcal{N}_\tau^0$ denote either the void set or a singleton whose sole element is a 0-node of $\mathcal{G}^0$; we also require that $\mathcal{N}_{\tau_1}^0 \cap \mathcal{N}_{\tau_2}^0 = \emptyset$ if $\tau_1 \neq \tau_2$.

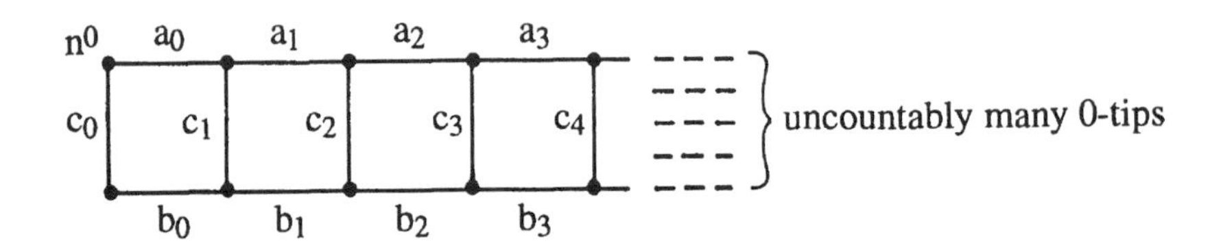

Figure 2.2. A ladder network $\mathcal{L}^0$. The a_k, b_k, and c_k denote branches, and n^0 is a 0-node. $\mathcal{L}^0$ has an infinite set of 0-tips, whose cardinality $\mathbf{c}$ is that of the continuum.

Definition of a 1-node: For each τ the set $\mathcal{N}_\tau^0 \cup \mathcal{T}_\tau^0$ is called a 1-*node*.

Since $\mathcal{T}_\tau^0$ is not void, neither is the 1-node. Also, if a 1-node is a singleton, its sole element will be a 0-tip, not a 0-node. The single 0-node in the 1-node, if it exists, is called the *exceptional element* of the 1-node. That exceptional element will not be contained in any other 1-node. A 1-node is said to *embrace* itself, each of its elements, and every one of the $\vec{0}$-tips embraced by its embraced 0-node if the latter exists. When n^1 is a nonsingleton, we say that all the entities embraced by n^1 are *shorted together* or that one such entity is *shorted* to all the others. However, if n^1 is a singleton, it and its only 0-tip are said to be *open*. Any 0-node that is not embraced by a 1-node will be called a *maximal* 0-node. (In [35] a maximal 0-node was called an "ordinary" 0-node.)

As we shall see later on, it is only the nonsingleton 1-nodes that play an essential role in transfinite electrical network theory, for these are the only 1-nodes that allow a current to pass from one 0-tip to another 0-tip or to a $\vec{0}$-tip. The singleton 1-nodes cannot serve this function. It is tempting therefore to define as 1-nodes only the nonsingleton sets $\mathcal{N}_\tau^0 \cup \mathcal{T}_\tau^0$. However, complications arise in other definitions if the singleton 1-nodes are ignored. So, we shall include them among the 1-nodes even though for most purposes they are superfluous.

We should also explain why we allow only one 0-node as the exceptional element of a 1-node. Had we allowed two or more 0-nodes to be members of a single 1-node, that 1-node would short together all the elementary tips of its several 0-nodes. In effect, we would be altering the partitioning of elementary tips that we had chosen when setting up the 0-nodes. That same alteration could have been accomplished when first defining the 0-nodes by combining all of the 0-nodes in the 1-node into a single 0-node. Thus, no generality is lost by restricting the number of 0-nodes in a 1-node to one or none.

We will use a natural terminology when discussing relationships between nodes, branches, and paths. Let us be explicit about meanings. A branch that is incident to the exceptional element of a 1-node is said to be *incident to* that 1-node. We say that a 0-path P^0 *traverses* a tip if it embraces a representative of that tip. (Traversing a tip is not the same as embracing a tip; see Example 2.1-6 below.) P^0 is *incident to* a tip, node, branch, or subgraph if P^0 traverses a tip that is shorted to a tip or a node of the said entity. Also, P^0*passes through* each of its embraced elements and every node that embraces one of its nodes — except for any terminal node if it has one. *To reach* will be synonymous with "to be incident to."

Definition of a 1-graph: A *1-graph* $\mathcal{G}^1$is a triplet $\{\mathcal{B}, \mathcal{N}^0, \mathcal{N}^1\}$, where $\mathcal{B}$ is a nonvoid set of branches, $\mathcal{N}^0$ is a set of 0-nodes obtained froma partitioning of the set $\mathcal{T}^{\vec{0}}$ of $\vec{0}$-tips of those branches, and $\mathcal{N}^1$ is a nonvoid set of 1-nodes constructed from the 0-nodes and 0-tips of the 0-graph $\{\mathcal{B}, \mathcal{N}^0\}$ as stated above. $\{\mathcal{B}, \mathcal{N}^0\}$ is the *0-graph of* $\mathcal{G}^1$.

This definition requires that the 0-graph$\{\mathcal{B}, \mathcal{N}^0\}$ have at least one 0-tip and thereby an infinity of branches and 0-nodes; otherwise, $\mathcal{N}^1$ will be void and $\mathcal{G}^1$will not exist. Thus, for any 1-graph $\mathcal{G}^1 = \{\mathcal{B}, \mathcal{N}^0, \mathcal{N}^1\}$, $\mathcal{B}$ is an infinite set. Examples 2.1-1 and 2.1-2 and others like them show that $\overline{\overline{\mathcal{N}^1}}$ can be less than, equal to, or larger than $\overline{\overline{\mathcal{B}}}$. On the other hand, for a 1-graph, $\overline{\overline{\mathcal{N}^0}}$ cannot exceed $\overline{\overline{\mathcal{B}}}$ since each branch has only two $\vec{0}$-tips.

A subgraph of a 1-graph $\mathcal{G}^1 = \{\mathcal{B}, \mathcal{N}^0, \mathcal{N}^1\}$ is defined as follows. As before, let $\mathcal{B}_s$ be a nonvoid subset of $\mathcal{B}$. We define the subset $\mathcal{N}_s^0$ of $\mathcal{N}^0$ as before; that is, a 0-node is in $\mathcal{N}_s^0$ if and only if it is incident to at least one branch in $\mathcal{B}_s$. Next, let $\mathcal{N}_s^1$ be the set of all 1-nodes in $\mathcal{N}^1$, each of which has at least one 0-tip with a representative whose branches are all in $\mathcal{B}_s$. If $\mathcal{N}_s^1$ is not void, set $\mathcal{G}_s^1 = \{\mathcal{B}_s, \mathcal{N}_s^0, \mathcal{N}_s^1\}$. If $\mathcal{N}_s^1$ is void, set $\mathcal{G}_s^0 = \{\mathcal{B}_s, \mathcal{N}_s^0\}$.

Definition of a subgraph of a 1-graph: $\mathcal{G}_s^1$ or $\mathcal{G}_s^0$ will be called the *subgraph of* $\mathcal{G}^1$ *induced by the branch set* $\mathcal{B}_s$.

A 0-path, as defined in Sections 1.2 and 1.3, is a subgraph of the 0-graph of $\mathcal{G}^1$; if the 0-path is two-ended, it is a subgraph of $\mathcal{G}^1$ as well. (A 1-path, which we shall define shortly, will be a subgraph of $\mathcal{G}^1$.)

Note that here too a subgraph is in general neither a 0-graph nor a 1-graph because there may be a 1-node in $\mathcal{N}_s^1$ having a 0-tip none of whose representatives has all its branches in $\mathcal{B}_s$ or there may be a 0-node with a $\vec{0}$-tip that does not belong to a branch in $\mathcal{B}_s$.

A collection of subgraphs is said to *partition* another subgraph or $\mathcal{G}^1$ itself if the branch sets of the former comprise a partition of the branch set of the latter.

Two subgraphs, a subgraph and a node, or two nodes are called *totally disjoint* if they do not embrace a common node. Whenever those two entities are not totally disjoint, they are said to *meet*, and they meet *with* the commonly embraced node or nodes — and also *with* any tip, branch, or subgraph that is commonly embraced. A subgraph considered here may be a subgraph of the 0-graph of $\mathcal{G}^1$ and therefore may have 0-tips but not their embracing 1-nodes. As a result, "meeting" is a stronger concept than "incidence" or "reaching"; meeting implies incidence, but not in general conversely. See Example 2.1-3 below.

We now present some examples of 1-graphs. They will also be used subsequently to illustrate ideas that have not as yet been introduced.

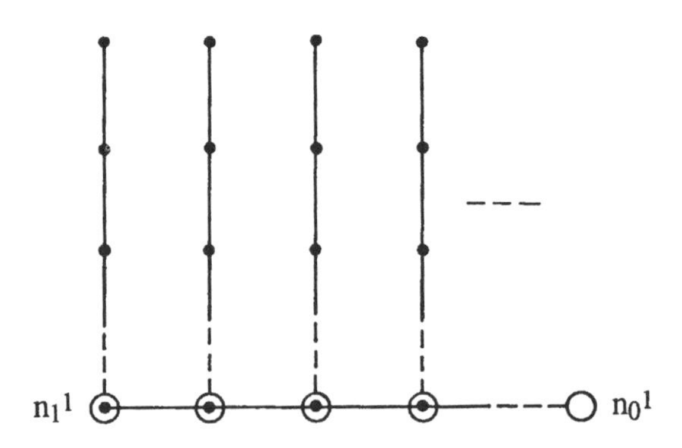

Figure 2.3. A 1-graph. The heavy dots are 0-nodes, the small circles are 1-nodes, and the lines between the 0-nodes are branches. Note that this graph does not contain a one-ended 1-path even though it has an infinity of 1-nodes.

Example 2.1-3. Figure 2.3 indicates a 1-graph. (The symbolism of that figure is defined in its caption; that symbolism will be used in many other figures as well.) Here we have an infinity of one-ended 0-paths (shown vertically) that reach 1-nodes with 0-tips. Those 1-nodes embrace 0-nodes, which along with their incident branches comprise another one-ended 0-path that reaches the singleton 1-node n_0^1. Note that each of the former 1-nodes has two incident branches — or just one incident branch in the case of n_1^1 — that cuts off that 1-node from all the other 1-nodes. However, there is no such branch for n_0^1. For certain kinds of analyses such as that for random walks in Chapter 7, this situation regarding n_0^1 leads to difficulties.

Note also that each vertical one-ended 0-path is a subgraph of the 0-graph of this 1-graph. As such, it does not meet the 1-node to which it is incident; however, it does reach that 1-node as well as the 0-node embraced by that 1-node. Furthermore, the horizontal 0-path is incident to the 0-tips of the vertical 0-paths, but it does not meet any of those vertical 0-paths; however, it does meet all the 1-nodes except for n_0^1, which it only reaches. ♣

Example 2.1-4. In Example 2.1-1 we noted that the infinite binary tree of Figure 2.1 has a set of 0-tips with the cardinality **c** of the continuum. We can short an elementary tip of a branch to each 0-tip — a different branch for each different 0-tip — to obtain the uncountable 1-graph of Figure 2.4. This is an uncountable 1-tree, a kind of graph that will be defined subsequently.

In the same way we can connect branches bijectively to the 0-tips of the ladder of Figure 2.2 to obtain another uncountable 1-graph (see Figure 2.5) whose branch set also has the cardinality of the continuum. ♣

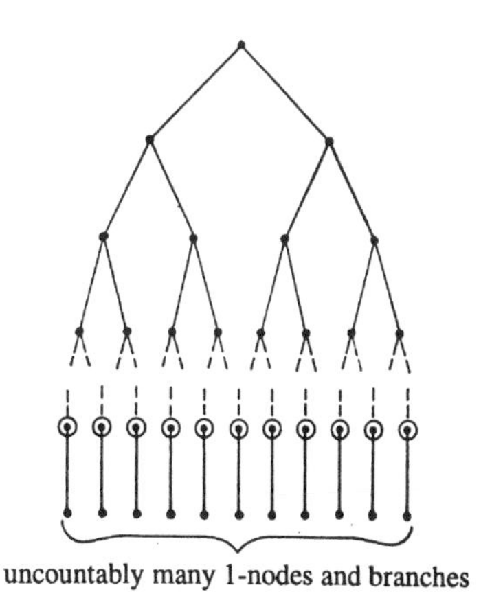

Figure 2.4. An uncountable 1-graph having a branch appended to each 0-tip of a binary tree.

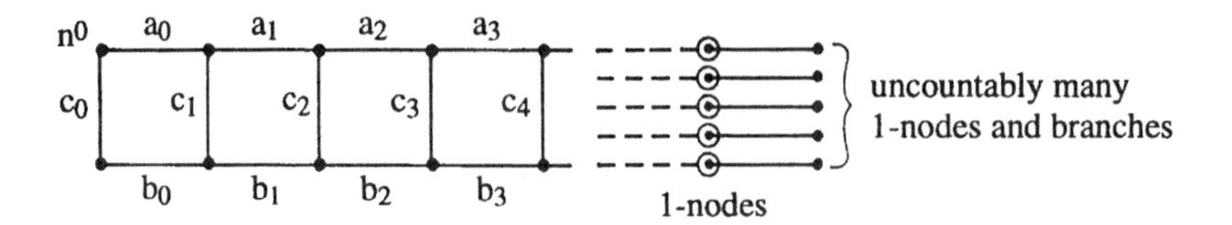

Figure 2.5. An uncountable 1-graph wherein a branch is appended to each 0-tip of a ladder.

Example 2.1-5. It may be worth noting at this point that every one of the examples of 1-graphs given so far possess maximal 0-nodes. This need not be the case. There are 1-graphs none of whose 0-nodes are maximal; that is, all 0-nodes are embraced by 1-nodes. Such a 1-graph can be constructed as follows: Start with a single branch b. Then, introduce two one-ended 0-paths P_1^0 andP_2^0, whose 0-tips are shorted to the two nodes of b through two 1-nodes; this is shown in Figure 2.6(a). Next, short the 0-nodes of P_1^0 and P_2^0 to the 0-tips of other one-ended 0-paths in a bijective fashion, as shown in Figure 2.6(b). Do the same thing to the 0-nodes of the newly introduced one-ended 0-paths. Continuing in this way indefinitely, we obtain the said 1-graph $\mathcal{G}^1$. The result of the first four steps of this construction is indicated in Figure 2.6(c).

(Actually, transfinite graphs of higher ranks ν also have peculiarities of this sort; that is, there are ν-graphs all of whose "maximal" nodes are either of rank ν or are singletons of lower ranks. See [39, Example 3.4].) ♣

Our next objective is to define the first kind of transfinite path, the "1-path." In order to do so, we have to be more specific about how 0-paths maybe "incident to" 1-nodes. If a 0-path P^0 terminates on one side at a 0-node n_t^0, it will traverse a $\vec{0}$-tip $t^{\vec{0}}$ that is a member of n_t^0. On the other hand, if P^0 extends infinitely on that side, it will traverse a 0-tip t^0 on that side. In either case, we call $t^{\vec{0}}$ or t^0 a *terminal tip of* P^0. Let n be a 0-node or 1-node that embraces a terminal tip of P^0. Then, we say that P^0 is *terminally incident to* or *terminally reaches* n *with* or *through* that terminal tip. We also say that P^0 is *terminally incident to* n *with* n_t^0 if P^0 terminates at n_t^0 and n embraces n_t^0 (e.g. , $n = n_t^0$ if n is a 0-node); in this case, P^0 *meets* n *with* n_t^0. Finally, we say that P^0 is *terminally incident to* n *but otherwise totally*

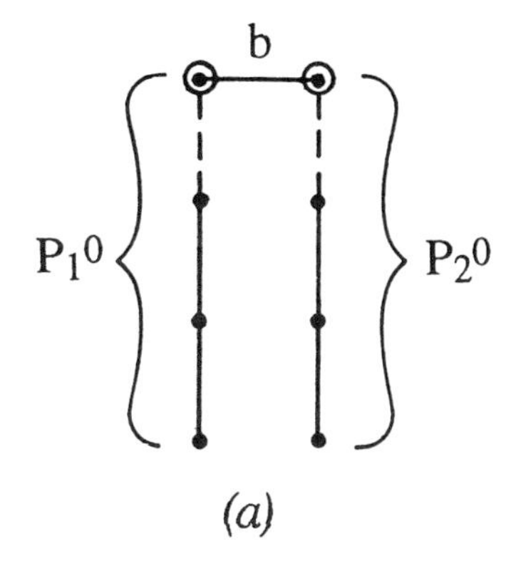

(a)

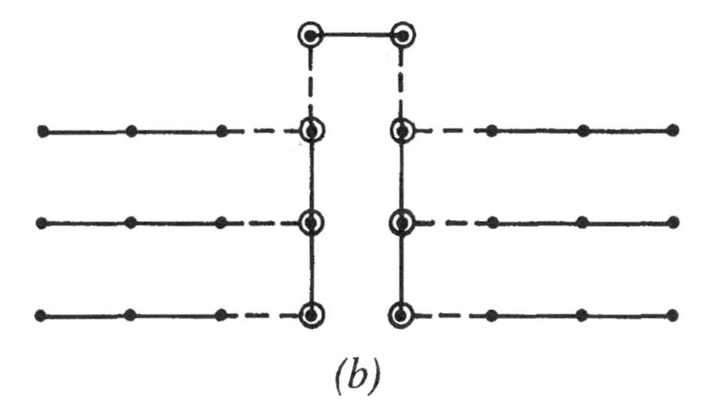

(b)

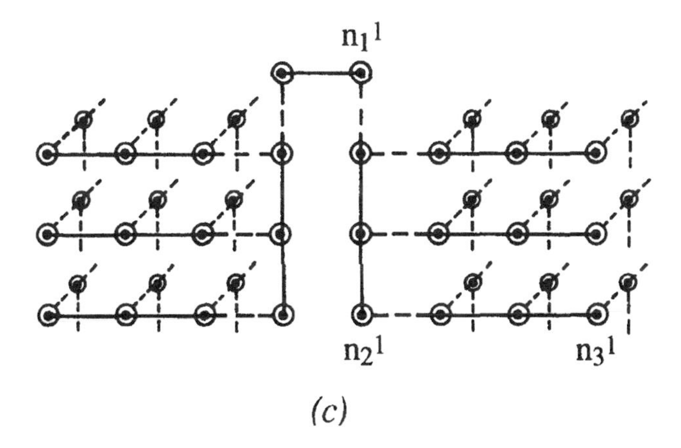

(c)

Figure 2.6. The first four steps in the construction of Example 2.1-5 for a 1-graph, none of whose 0-nodes is maximal.

disjoint if P^0reaches n with a terminal tip and if n does not embrace any other tip traversed by P^0; thus, P^0 reaches n with only one (not both) of its terminal tips.

Definition of a 1-path: The alternating sequence

$$P^1 = \{\ldots, n_m^1, P_m^0, n_{m+1}^1, P_{m+1}^0, \ldots\} \tag{2.1}$$

of 1-nodes n_m^1 and 0-paths P_m^0 of $\mathcal{G}^1$ is called a *nontrivial* 1*-path* if the following three conditions are satisfied:

(a) P^1 contains at least one 1-node and one 0-path. Every P_m^0 is a nontrivial 0-path. Moreover, every $n_m^1$is a 1-node except possibly when (2.1) terminates on the left and/or on the right. In the latter case, each terminal element is either a 0-node or a 1-node.

(b) Each P_m^0 is terminally incident to the two nodes adjacent to it in (2.1) but is otherwise totally disjoint from those nodes. Moreover, P_m^0 either reaches each of its adjacent nodes through a terminal 0-tip or meets it with a terminal 0-node. If that adjacent node is a terminal node n_t^δ ($\delta = 0, 1$) of (2.1), then P_m^0 reaches n_t^δ with a ($\delta - 1$)-tip, and, if in addition $\delta = 0$, then n_t^δ is also a terminal node of P_m^0.

(c) Every two elements in (2.1) that are not adjacent therein are totally disjoint.

Since the 0-paths in (2.1) are themselves sequences, a 1-path is in fact a sequence of sequences. Moreover, a 1-path is a subgraph.

As a particular case of (c), the requirement that P_m^0 and P_{m+1}^0 be totally disjoint implies that at least one of them must reach n_{m+1}^1 with a 0-tip. This is because 0-paths are subgraphs of the 0-graph of $\mathcal{G}^1$. Thus, were P_m^0 and P_{m+1}^0 to meet n_{m+1}^1 with 0-nodes, they would meet n_{m+1}^1 with the same 0-node — the 0-node embraced by n_{m+1}^1 — and could not therefore be totally disjoint.

A *trivial 1-path* is a singleton $\{n\}$ where n is either a 0-node or a 1-node.

When (2.1) is a finite sequence, we will call it a *two-ended* 1-path — but not a "finite" 1-path (in contrast to the terminology in [35] and [37]) so as to avoid the oxymoron, a "finite transfinite" path. Similarly, that 1-path is called *one-ended* or *endless* when the sequence (2.1) is respectively one-way infinite or two-way infinite.

A 1-*loop* is defined exactly as is a two-ended 1-path except that one of its terminal nodes is required to embrace the other.

As with a 0-path, a 1-path or a subgraph is said to *traverse* an elementary tip (resp. a 0-tip) if it embraces the branch for the elementary tip (resp. all the branches of a representative for the 0-tip).

Example 2.1-6. Traversing a tip is different from embracing a tip. Figure 2.7 exemplifies this. Let P_a^0 and P_b^0 be the two one-ended 0-paths consisting of the a-branches and the b-branches, respectively, and let t_a^0 and t_b^0 be their corresponding 0-tips. n_1^1 is the 1-node $\{n_c^0, t_a^0, t_b^0\}$, where $n_c^0 = \{t_c^{\vec{0}}\}$ is the singleton 0-node containing an elementary tip $t_c^{\vec{0}}$ of the branch c. Finally, let P_a^1 be the 1-path $\{n_1^0, P_a^0, n_1^1\}$. P_a^1 embraces the three tips t_a^0, t_b^0, and $t_c^{\vec{0}}$, but it only traverses t_a^0. On the other hand, the 0-path P_a^0 traverses t_a^0 but does not embrace it. (Actually, every two-ended 1-path will embrace every 0-tip it traverses. See Lemma 2.2-4 below.)

Figure 2.7 also illustrates how two 0-paths P_a^0 and P_b^0 may reach the same 1-node n_1^1 but nevertheless be totally disjoint. Indeed, there is no node embraced by both P_a^0 and P_b^0. Similarly, the 0-path consisting of the branch c and its two incident 0-nodes is totally disjoint from both P_a^0 and P_b^0. ♣

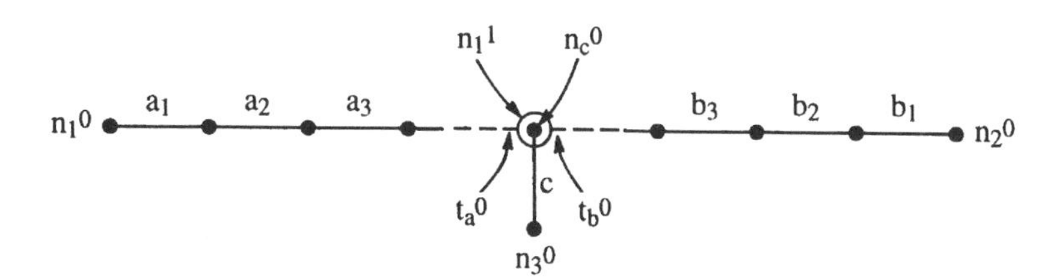

Figure 2.7. The 1-graph discussed in Example 2.1-6.

2.2 μ-Graphs

We have seen in the preceding section that with the invention of a new kind of node, the 1-node, we can connect together infinite 0-graphs at their infinite extremities to obtain a new kind of graph, a transfinite graph called a 1-graph. Those infinite extremities were defined as equivalence classes of one-ended 0-paths — the 0-tips. This process might be repeatable because a 1-graph might contain one-ended 1-paths. Indeed, we might define another kind of infinite extremity, a "1-tip," as an equivalence class of one-ended 1-paths and then define a "2-node" as a set of 1-tips. Those 2-nodes might serve to connect 1-graphs together at their infinite extremities. The result should be a "2-graph" — a transfinite graph of higher rank. All this turns out to be true. Moreover, the process can be repeated any number of times to obtain transfinite graphs whose ranks extend through all the natural numbers.

We shall define such graphs recursively. Let μ be any natural number. Let us assume that, for each $\gamma = 0, 1, \ldots, \mu - 1$, the γ-graphs $\mathcal{G}^\gamma = \{\mathcal{B}, \mathcal{N}^0, \ldots, \mathcal{N}^\gamma\}$ have already been defined along with their concomitant ideas such as a γ-node n^γ and a γ-path:

$$P^\gamma \;=\; \{\ldots, n_m^\gamma, P_m^{\alpha_m}, n_{m+1}^\gamma, P_{m+1}^{\alpha_{m+1}}, \ldots\} \tag{2.2}$$

where $\vec{0} \leq \alpha_m < \gamma$ for all m. ($P^{\vec{0}}$ denotes a branch.) These ideas were explicitly defined for $\gamma = 0$ and $\gamma = 1$ in Sections 1.3 and 2.1, respectively. The constructions of this section will extend them by recursion to all natural-number ranks μ.

Consider the $(\mu - 1)$-graph $\mathcal{G}^{\mu-1} = \{\mathcal{B}, \mathcal{N}^0, \ldots, \mathcal{N}^{\mu-1}\}$ and assume it has at least one one-ended $(\mu - 1)$-path. Partition the set of all its one-ended $(\mu - 1)$-paths into equivalence classes by treating two such paths as being *equivalent* if they differ on no more than finitely many $(\mu - 1)$-nodes and intervening paths, that is, if the members of those two sequences are all the same except for finitely many of them. Each equivalence class will be called a $(\mu - 1)$-*tip*. A *representative* of a $(\mu - 1)$-tip is any one of its members.

Next, partition the set $\mathcal{T}^{\mu-1}$ of all $(\mu - 1)$-tips into subsets $\mathcal{T}_\tau^{\mu-1}$, where τ is the index for the partitioning. Thus, $\mathcal{T}^{\mu-1} = \cup_\tau \mathcal{T}_\tau^{\mu-1}$, where $\mathcal{T}_\tau^{\mu-1} \neq \emptyset$ for every τ and $\mathcal{T}_{\tau_1}^{\mu-1} \cap \mathcal{T}_{\tau_2}^{\mu-1} = \emptyset$ if $\tau_1 \neq \tau_2$. Furthermore, for each τ, let $\mathcal{N}_\tau^{\mu-1}$ be either the void set or a singleton whose only member is an α-node n_τ^α, where $0 \leq \alpha \leq \mu - 1$.

Definition of a μ-node: For each τ the set

$$n_\tau^\mu \;=\; \mathcal{T}_\tau^{\mu-1} \cup \mathcal{N}_\tau^{\mu-1} \tag{2.3}$$

is called a μ-*node* so long as the following condition is satisfied: Whenever $\mathcal{N}_\tau^{\mu-1}$ is not void, its single member n_τ^α is not a member of any other β-node $(\alpha < \beta \leq \mu)$.

When $\mathcal{N}_\tau^{\mu-1}$ is not void so that $\mathcal{N}_\tau^{\mu-1} = \{n_\tau^\alpha\}$, we shall refer to n_τ^α as the *exceptional element of* n_τ^μ. According to our definition of a μ-node, each α-node can be the exceptional element of at most one node — necessarily of rank higher than α. The definition nonetheless allows n_τ^α to have another node of rank lower than α as its exceptional element, which in turn can have still another exceptional element, and so on. More generally, as was defined in Section 1.1, n_τ^μ *embraces* itself, all its elements, all elements of its exceptional element n_τ^α if it has one, all elements of the exceptional element of n_τ^α if that exists, and so on through exceptional elements of decreasing ranks. Thus, n_τ^μ can only contain $(\mu-1)$-tips and at most one node of rank lower than μ. However, n_τ^μ may embrace tips and nodes of many ranks lower than $\mu - 1$ — this being possible if n_τ^μ has an exceptional element. If n_τ^μ is a nonsingleton, all the embraced tips and nodes of n_τ^μ are said to be *shorted together*, and one such entity is said to be *shorted to* all the others. However, if n_τ^μ is a singleton, it and its sole $(\mu - 1)$-tip are said to be *open*.

Lemma 2.2-1. *If n_a^α and n_b^β are respectively an α-node and a β-node with $0 \leq \alpha \leq \beta$ and if n_a^α and n_b^β embrace a common node, then n_b^β embraces n_a^α. If in addition $\alpha = \beta$, then $n_a^\alpha = n_b^\beta$.*

Proof. Let n_c^γ denote a γ-node that is embraced by both n_a^α and n_b^β. If $\gamma = \alpha$, then $n_c^\gamma = n_a^\alpha$ because no node can embrace another node of the same rank. Therefore, n_b^β embraces n_a^α.

So, assume $\gamma < \alpha \leq \beta$. Corresponding to n_a^α (or n_b^β) there is a unique finite sequence S_a (resp. S_b) of all the nodes embraced by n_a^α (resp. n_b^β) such that the ranks of those nodes strictly decrease. Then, the node n_c^γ that is embraced by both n_a^α and n_b^β will be a member of both S_a and S_b but will not be the first member in either sequence. Suppose n_b^β does not embrace n_a^α. It follows that there will be a λ-node n_l^λ $(\gamma \leq \lambda < \alpha)$ in both S_a and S_b such that its immediate predecessor in S_a will differ from its immediate predecessor in S_b. This violates the condition stated in the definition of a μ-node.

Finally, if $\alpha = \beta$, we must have that $n_a^\alpha = n_b^\beta$ again because no node can embrace another node of the same rank. ♣

Definition of a μ-graph: A μ-graph is a $(\mu + 2)$-tuplet

$$\mathcal{G}^\mu = \{\mathcal{B}, \mathcal{N}^0, \ldots, \mathcal{N}^\mu\} \tag{2.4}$$

where $\mathcal{B}$ is a set of branches and, for each $\gamma = 0, \ldots, \mu$, $\mathcal{N}^\gamma$ is a nonvoid set of γ-nodes constructed out of all the $(\gamma - 1)$-tips and out of (none, some, or all of) the α-nodes $(0 \leq \alpha \leq \gamma - 1)$ of the $(\gamma - 1)$-graph $\mathcal{G}^{\gamma-1} = \{\mathcal{B}, \mathcal{N}^0, \ldots, \mathcal{N}^{\gamma-1}\}$. Furthermore, for each $\gamma = 0, \ldots, \mu - 1$, the subset $\mathcal{G}^\gamma = \{\mathcal{B}, \mathcal{N}^0, \ldots, \mathcal{N}^\gamma\}$ of (2.4) is called the *γ-graph of $\mathcal{G}^\mu$*.

When $\gamma = 0$, we have that $\gamma - 1 = \vec{0}$, the $(\gamma - 1)$-tips are elementary tips, there are no α-nodes, and $\mathcal{G}^{\gamma-1}$ simply denotes a set of branches.

Also note that, in order for $\mathcal{N}^\mu$ to be nonvoid, the $\mathcal{N}^\gamma$ for $\gamma = 0, \ldots, \mu - 1$ must be infinite sets. Indeed, each μ-node has at least one $(\mu - 1)$-tip, which in turn requires as a representative at least one one-ended $(\mu-1)$-path having an infinity of $(\mu - 1)$-nodes — and similarly, for the lower ranks.

An α-node n^α $(0 \leq \alpha < \mu)$ is called *maximal with respect to $\mathcal{G}^\mu$* or simply *maximal* when $\mathcal{G}^\mu$ is understood if n^α is not the exceptional element of any β-node $(\alpha < \beta \leq \mu)$. Otherwise, n^α is called *nonmaximal with respect to $\mathcal{G}^\mu$*. The μ-nodes of $\mathcal{G}^\mu$ are perforce *maximal with respect to $\mathcal{G}^\mu$*.

We turn now to the definition of a subgraph. Let $\mathcal{B}_s$ be a nonvoid subset of $\mathcal{B}$ again. With $\mathcal{G}^\mu$ being defined by (2.4) and for each $\lambda = 0, \ldots, \mu$, let $\mathcal{N}_s^\lambda$ be the subset of $\mathcal{N}^\lambda$ consisting of all λ-nodes n^λ such that n^λ contains at least one $(\lambda-1)$-tip having at least one representative all of whose branches are in $\mathcal{B}_s$. For $\lambda \geq 1$, $\mathcal{N}_s^\lambda$ may be void. ($\mathcal{N}_s^0$ will not be void because $\mathcal{B}_s$ is not void.) However, there will be some maximum rank γ $(0 \leq \gamma \leq \mu)$ for which all the $\mathcal{N}_s^\lambda$ $(\lambda = 0, \ldots, \gamma)$ are nonvoid; moreover, if $\gamma < \mu$, then all the $\mathcal{N}_s^\lambda$ $(\lambda = \gamma + 1, \ldots, \mu)$ will be void. This is because, if $\mathcal{N}_s^{\gamma+1}$ is void, there are no $(\gamma+1)$-nodes, no $(\gamma+1)$-paths, no $(\gamma+1)$-tips, and therefore no $(\gamma + 2)$-nodes that can be constructed out of $\mathcal{B}_s$ alone. Hence, $\mathcal{N}_s^{\gamma+2}$ is void too. Similarly, $\mathcal{N}_s^{\gamma+3}, \ldots, \mathcal{N}_s^\mu$ are all void too.

Definition of a subgraph of a μ-graph: $\mathcal{G}_s^\gamma = \{\mathcal{B}_s, \mathcal{N}_s^0, \ldots, \mathcal{N}_s^\gamma\}$ is called the *subgraph of $\mathcal{G}^\mu$ induced by $\mathcal{B}_s$*.

The subgraph $\mathcal{G}_s^\gamma$ need not be a γ-graph because there may be a $(\lambda-1)$-tip of a node n^λ in $\mathcal{N}_s^\lambda$ $(\lambda \leq \gamma)$ having no representative embracing only branches in $\mathcal{B}_s$.

(There is a similar but different concept called a "reduced graph" which is defined in [35] or [37]. Reduced graphs are always γ-graphs, but they introduce other nodes, the "reduced nodes," beyond those of $\mathcal{G}^\mu$. This complicates subsequent arguments. We can do without them.)

With $\mathcal{G}^\mu$ given by (2.4) and for $\gamma < \mu$, we have previously defined the "γ-graph $\mathcal{G}^\gamma$ of $\mathcal{G}^\mu$" as simply the subset $\{\mathcal{B}, \mathcal{N}^0, \ldots, \mathcal{N}^\gamma)$ of (2.4); this is not in general a subgraph of $\mathcal{G}^\mu$ for some choice of $\mathcal{B}_s$. Moreover, for a given $\mathcal{B}_s \subset \mathcal{B}$, the subgraphs of $\mathcal{G}^\gamma$ may be different from the subgraphs of

$\mathcal{G}^\mu$. For instance, a one-ended or endless γ-path P^γ $(\gamma < \mu)$ is a subgraph of $\mathcal{G}^\gamma$ induced by the set $\mathcal{B}_s$ of branches embraced by P^γ. However, with respect to $\mathcal{G}^\mu$, the subgraph induced by $\mathcal{B}_s$ is a two-ended $(\gamma+1)$-path or a $(\gamma+1)$-loop. For example, if P^γ is endless and if its two γ-tips are not shorted together, then the subgraph of $\mathcal{G}^\mu$ induced by $\mathcal{B}_s$ is the $(\gamma+1)$-path $\{n_1^{\gamma+1}, P^\gamma, n_2^{\gamma+1}\}$, where $n_1^{\gamma+1}$ and $n_2^{\gamma+1}$ are the two $(\gamma+1)$-nodes that contain the two γ-tips of P^γ.

A collection of subgraphs of another subgraph $\mathcal{G}^\gamma$ (possibly, $\mathcal{G}^\gamma = \mathcal{G}^\mu$) is said to *partition* $\mathcal{G}^\gamma$ if the branch sets of the former subgraphs comprise a partition of the branch set of the latter subgraph.

A subgraph $\mathcal{G}$ is said to *traverse* a tip if $\mathcal{G}$ embraces all the branches of a representative of the tip. As was indicated in Example 2.1-6, embracing and traversing a tip are different ideas. A subgraph $\mathcal{G}$ is said to be *incident to* or to *reach* a tip, a node, a branch, or another subgraph if $\mathcal{G}$ traverses a tip that is shorted to a tip or a node of the latter entity, and then we also say that the latter entity is *incident to* the subgraph.

Two subgraphs, a node and a subgraph, or two nodes are said to be *totally disjoint* if they do not embrace a common node; otherwise, they are said to *meet*, and they meet *with* that commonly embraced node or nodes — and also *with* any tip, branch, or subgraph that is commonly embraced. As before, meeting implies incidence, but the converse need not be true.

Next, let P^α be an α-path $(0 \leq \alpha \leq \mu)$. If P^α terminates on one side at a δ-node n_t^δ $(0 \leq \delta \leq \alpha)$, then P^α traverses a $(\delta-1)$-tip $t^{\delta-1}$ (e.g., a $\vec{0}$-tip if $\delta = 0$) that is a member of n_t^δ. (This follows from the definition of an α-path, which is explicated below and has already been stated for $\alpha = 0, 1$.) On the other hand, if P^α extends infinitely on that side, it will traverse an α-tip t^α on that side. In either case, we call $t^{\delta-1}$ or t^α a *terminal tip* of P^α. Furthermore, let n^γ be a γ-node that embraces that terminal tip (thus, $\gamma > \delta - 1$ or $\gamma > \alpha$); then we say that P^α is *terminally incident to* or *reaches* n^γ *with* or *through* that terminal tip. If P^α does terminate at n_t^δ and if n^γ embraces n_t^δ, then we also say that P^α is *terminally incident to* n^γ *with* n_t^δ and that P^α *meets* n^γ *with* n_t^δ. Finally, we say that P^α is *terminally incident to* n^γ *but otherwise totally disjoint from* n^γ if P^α reaches n^γ with a terminal tip and if n^γ does not embrace any other tip traversed by P^α; thus, P^α reaches n^γ with only one (not both) of its terminal tips. At times we will assign an orientation to a path — perhaps implicitly — and then will speak of that path as *starting at* or *stopping at* any node to which it is terminally incident.

Definition of a μ-path: For any natural number $\mu \geq 1$, consider the alternating sequence

$$P^{\mu} = \{\ldots, n_m^{\mu}, P_m^{\alpha_m}, n_{m+1}^{\mu}, P_{m+1}^{\alpha_{m+1}}, \ldots\} \tag{2.5}$$

of μ-nodes n_m^{μ} and α_m-paths $P_m^{\alpha_m}$, where $0 \leq \alpha_m < \mu$ and the α_m may differ. P^{μ} is called a *nontrivial μ-path* if the following three conditions are satisfied:

(a) Each $P_m^{\alpha_m}$ is a nontrivial α_m-path, and P^{μ} contains at least one such α_m-path and at least one μ-node. Moreover, every n_m^{μ} is a μ-node except possibly for terminal nodes when (2.5) terminates on the left and/or on the right. In the latter case, each terminal element is a δ-node n_t^{δ} $(0 \leq \delta \leq \mu)$, and its adjacent path is an α-path $(\alpha < \mu)$.

(b) Each $P_m^{\alpha_m}$ is terminally incident to the two nodes adjacent to it in (2.5) but is otherwise totally disjoint from those nodes. Moreover, $P_m^{\alpha_m}$ either reaches each of its adjacent nodes with a terminal $(\mu - 1)$-tip or it meets that adjacent node with a terminal δ-node where $\delta \leq \mu - 1$. If that adjacent node is a terminal node n_t^{δ} of (2.5), then $P_m^{\alpha_m}$ reaches n_t^{δ} with a $(\delta - 1)$-tip, and, if in addition $\delta < \mu$, then n_t^{δ} is also a terminal node of $P_m^{\alpha_m}$.

(c) Every two elements in (2.5) that are not adjacent therein are totally disjoint.

Lemma 2.2-2. *Let P^{μ} $(\mu \geq 1)$ be the nontrivial μ-path given by (2.5). When n_{m+1}^{μ} is not a terminal node, at least one of its adjacent paths $P_m^{\alpha_m}$ and $P_{m+1}^{\alpha_{m+1}}$ reaches n_{m+1}^{μ} with a $(\mu - 1)$-tip.*

Note. This implies that at least one of the ranks α_m and α_{m+1} must be $\mu - 1$.

Proof. Suppose the conclusion is false. Then, both $P_m^{\alpha_m}$ and $P_{m+1}^{\alpha_{m+1}}$ meet n_{m+1}^{μ} with nodes $n_m^{\delta_m}$ and $n_{m+1}^{\delta_{m+1}}$, respectively, whose ranks are no larger than $\mu - 1$. Thus, n_{m+1}^{μ} embraces both $n_m^{\delta_m}$ and $n_{m+1}^{\delta_{m+1}}$. But, by our recursive definition of a μ-node, the nodes embraced by n_{m+1}^{μ} form a finite sequence of nodes of decreasing ranks such that each node of the sequence embraces itself and all the nodes following it in that sequence. Thus, either

$n_m^{\delta_m}$ embraces $n_{m+1}^{\delta_{m+1}}$ or $n_{m+1}^{\delta_{m+1}}$ embraces $n_m^{\delta_m}$. Hence, $P_m^{\alpha_m}$ and $P_{m+1}^{\alpha_{m+1}}$ embrace a common node and therefore cannot be totally disjoint — in violation of condition (c). ♣

By the above definition of a μ-path, P^μ is a subgraph of $\mathcal{G}^\mu$. Moreover, for $\gamma < \mu$, each γ-path P^γ is a subgraph of the γ-graph of $\mathcal{G}^\mu$.

P^μ is called *two-ended, one-ended*, or *endless* when (2.5) is respectively a finite, one-way infinite, or two-way infinite sequence. A singleton containing just one α-node ($0 \le \alpha \le \mu$) will at times be called a *trivial μ-path.*

A *μ-loop* is defined exactly like a nontrivial two-ended μ-path except that one of its terminal nodes is required to embrace the other.

A γ-path or a γ-loop ($0 \le \gamma \le \mu$), all of whose branches are embraced by a subgraph, is said to be *in* that subgraph.

Example 2.2-3. Figure 2.8 illustrates a 2-graph $\mathcal{G}^2$. Let $\mathcal{B}_l$, $\mathcal{B}_u$, and $\mathcal{B}_d$ denote respectively the set of lower horizontal branches, the set of upper horizontal branches, and the set of branches that connect lower nodes to upper nodes. The subgraph of $\mathcal{G}^2$ induced by $\mathcal{B}_l$ is the two-ended 1-path $\{n_1^0, P_l^0, n_0^1\}$, where P_l^0 is the one-ended 0-path whose branches are the branches of $\mathcal{B}_l$; by stipulating that n_0^1 is a singleton, we get that $n_0^1 = \{t_l^0\}$, where t_l^0 is the 0-tip having P_l^0 as a representative. On the other hand, the subgraph of the 0-graph of $\mathcal{G}^2$ induced by $\mathcal{B}_l$ is simply P_l^0. n_0^1 is maximal with respect to the 1-graph of $\mathcal{G}^2$, but it is nonmaximal with respect to $\mathcal{G}^2$ because it is embraced by the 2-node $n_0^2 = \{n_0^1, t_u^1\}$, where t_u^1 is the 1-tip having as a representative the one-ended 1-path P_u^1 whose branches are the branches of $\mathcal{B}_u$:

$$P_u^1 = \{n_2^0, P_0^0, n_1^1, P_1^0, n_2^1, P_2^0, \ldots\}. \tag{2.6}$$

Here, each P_m^0 is a one-ended 0-path extending between the two nodes adjacent to it in (2.6). n_1^1 is a maximal 1-node, which can be written as $n_1^1 = \{n_3^0, t^0\}$, where t^0 is the 0-tip of P_0^0 and $n_3^0 = \{t_d^{\vec{0}}, t_u^{\vec{0}}\}$ where in turn $t_d^{\vec{0}}$ is a $\vec{0}$-tip of the branch of $\mathcal{B}_d$ that reaches n_1^1 and $t_u^{\vec{0}}$ is a $\vec{0}$-tip of the horizontal branch just to the right of n_1^1.

As is suggested in Figure 2.8, we take it that the branches of $\mathcal{B}_d$ define a bijection between the upper nodes and the lower nodes. As a result there are uncountably many 0-tips and uncountably many 1-tips whose representatives pass infinitely often between the upper and lower nodes. We assume that all these 0-tips and 1-tips comprise singleton nodes; they are not indicated in Figure 2.8. Moreover, these 0-tips and 1-tips do not exist in each of the subgraphs induced by each of $\mathcal{B}_l$, $\mathcal{B}_u$, and $\mathcal{B}_d$ alone.

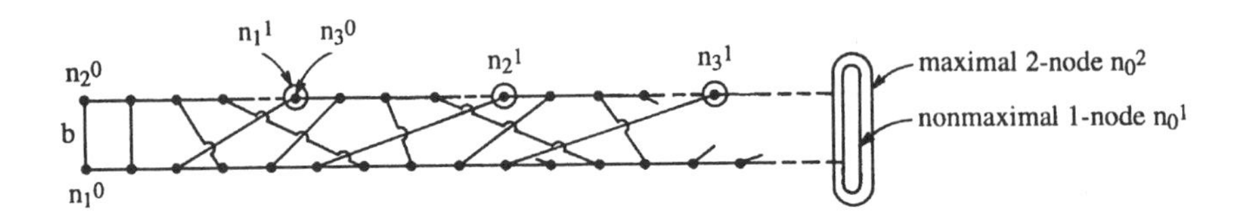

Figure 2.8. The 2-graph discussed in Example 2.2-3. The heavy dots denote 0-nodes, and the lines segments between them indicate branches. The small circles denote maximal 1-nodes, and the inner oval on the right represents a nonmaximal 1-node n_0^1. The outer oval denotes a 2-node n_0^2 that embraces n_0^1.

Let us also note that the subgraphs of $\mathcal{G}^2$ induced respectively by $\mathcal{B}_u$ and $\mathcal{B}_d$ meet (i.e., are not totally disjoint). Moreover, the subgraphs induced by $\mathcal{B}_l$, $\mathcal{B}_u$, and $\mathcal{B}_d$ partition $\mathcal{G}^2$. The subgraph of $\mathcal{G}^2$ induced by $\mathcal{B}_u$ is the two-ended 2-path $\{n_2^0, P_u^1, n_0^2\}$; it embraces t_l^0 but does not traverse it. On the other hand, the subgraph P_l^0 of the 0-graph of $\mathcal{G}^2$ induced by $\mathcal{B}_l$ traverses t_l^0 but does not embrace it. Finally, the subgraph of $\mathcal{G}^2$ induced by $\mathcal{B}_l \cup \mathcal{B}_u \cup \{b\}$, where b is the branch incident to n_1^0 and n_2^0, is a 2-loop.♣

Here is an easy result we shall be citing.

Lemma 2.2-4. *Every two-ended α-path P^α in any μ-graph $(\alpha \leq \mu)$ embraces every tip that P^α traverses.*

Proof. Since P^α is two-ended, it cannot traverse any λ-tip for $\alpha \leq \lambda \leq \mu$. Furthermore, if P^α traverses a δ-tip t^δ where $\vec{0} \leq \delta < \alpha$, P^α will embrace the $(\delta+1)$-node that contains t^δ because P^α is a subgraph. Hence, P^α embraces t^δ.♣

2.3 $\vec{\omega}$-Graphs

The next step of generalization in our construction of transfinite graphs is a special kind of graph denoted by $\mathcal{G}^{\vec{\omega}}$, where $\vec{\omega}$ represents the *rank* of $\mathcal{G}^{\vec{\omega}}$. ($\vec{\omega}$ is larger than every natural-number rank and smaller than every transfinite-ordinal rank. $\vec{\omega}$ is the first *arrow rank* for a transfinite graph; the lowest arrow rank $\vec{0}$ only has a meaning for elementary tips.) It will be possible to define an $\vec{\omega}$-graph so long as the constructions of the node sets $\mathcal{N}^0, \mathcal{N}^1, \ldots$ of progressively higher ranks can be continued without end throughout all the natural-number ranks μ. In other words, given the branch set $\mathcal{B}$, the sequence $\{\mathcal{G}^\mu\}_{\mu=1}^\infty$ of μ-graphs generated recursively by defining μ-nodes must be such that each $\mathcal{G}^\mu$ has one-ended μ-paths (and thereby an infinity of μ-nodes), for otherwise there would be no μ-tips with which to define $\mathcal{G}^{\mu+1}$. So, let us assume that, for each natural number μ, nonvoid sets $\mathcal{N}^\mu$ have been constructed.

We now introduce a new kind of node, the $\vec{\omega}$-node, something that was not done in [35] and [37].

Definition of an $\vec{\omega}$-node: An $\vec{\omega}$-*node*:

$$n^{\vec{\omega}} \;=\; \{n_0^{\mu_0}, n_1^{\mu_1}, n_2^{\mu_2}, \ldots\} \tag{2.7}$$

is an infinite sequence of μ_k-nodes $n_k^{\mu_k}$ $(k = 0, 1, 2, \ldots)$ having the following properties.

(a) Each rank μ_k is a natural number.

(b) Each node $n_k^{\mu_k}$ is the exceptional element of the next node $n_{k+1}^{\mu_{k+1}}$.

(c) $n_0^{\mu_0}$ does not have an exceptional element.

Thus, $\mu_0 < \mu_1 < \mu_2 \cdots$. The set of all $\vec{\omega}$-nodes will be denoted by $\mathcal{N}^{\vec{\omega}}$. As usual, we say that an $\vec{\omega}$-node $n^{\vec{\omega}}$ *embraces* itself, all its nodes, and all nodes and tips embraced by its nodes. Moreover, all those embraced elements are said to be *shorted together* or simply *shorted.* An $\vec{\omega}$-node cannot be "open" in the sense of a singleton node. It follows from this definition that two different $\vec{\omega}$-nodes are disjoint; in fact, as an easy consequence of Lemma 2.2-1, we have

Lemma 2.3-1. *If two $\vec{\omega}$-nodes embrace a common node, then the two $\vec{\omega}$-nodes are the same node. Moreover, if an $\vec{\omega}$-node $n^{\vec{\omega}}$ and a μ-node n^μ $(\mu < \vec{\omega})$ embrace a common node, then $n^{\vec{\omega}}$ embraces n^μ.*

Our definition of an $\vec{\omega}$-node differs substantially from that of a μ-node n^μ with a natural-number rank μ in two ways. First, $n^{\vec{\omega}}$ does not contain any tips of any ranks, whereas n^μ always contains μ-tips. (But, $n^{\vec{\omega}}$ does embrace all the $(\mu_k - 1)$-tips contained in every $n_k^{\mu_k}$.) Secondly, $n^{\vec{\omega}}$ embraces an infinity of nodes, whereas n^μ embraces only finitely many nodes — perhaps none at all.

Definition of an $\vec{\omega}$-graph: An *$\vec{\omega}$-graph* $\mathcal{G}^{\vec{\omega}}$ of *rank* $\vec{\omega}$ is the well-ordered, infinite set of sets

$$\mathcal{G}^{\vec{\omega}} = \{\mathcal{B}, \mathcal{N}^0, \mathcal{N}^1, \ldots, \mathcal{N}^{\vec{\omega}}\}, \tag{2.8}$$

where $\mathcal{B}$ is a set of branches and $\mathcal{N}^\nu$ $(\nu = 0, 1, \ldots, \vec{\omega})$ is a set of ν-nodes constructed as stated above. (For each natural number μ, the node set $\mathcal{N}^\mu$ is nonvoid. However, $\mathcal{N}^{\vec{\omega}}$ may be void.) For each $\mu = 0, 1, 2, \ldots$, $\{\mathcal{B}, \mathcal{N}^0, \ldots, \mathcal{N}^\mu\}$ is called the *μ-graph of* $\mathcal{G}^{\vec{\omega}}$.

This is a more general definition of an $\vec{\omega}$-graph $\mathcal{G}^{\vec{\omega}}$ than that given in [35] or [37] not only because the branch set $\mathcal{B}$ may now be uncountable but also because $\mathcal{G}^{\vec{\omega}}$ may now contain $\vec{\omega}$-nodes, something that was not included previously.

Example 2.3-2. Here are three simple examples illustrating countable $\vec{\omega}$-graphs. The first of these is a star graph, illustrated in Figure 2.9(a). It consists of a branch b (i.e., a $\vec{0}$-path) and the one-ended μ-paths P^μ $(\mu = 0, 1, 2, \ldots)$ all terminally incident to the 0-node m^0 through $\vec{0}$-tips and each terminally incident to a singleton $(\mu + 1)$-node (not shown). Thus, m^0 is of degree $\aleph_0$. Except for the meeting at m^0, these paths are totally disjoint. This is an $\vec{\omega}$-graph, but it has no $\vec{\omega}$-node; that is, $\mathcal{N}^{\vec{\omega}}$ is void.

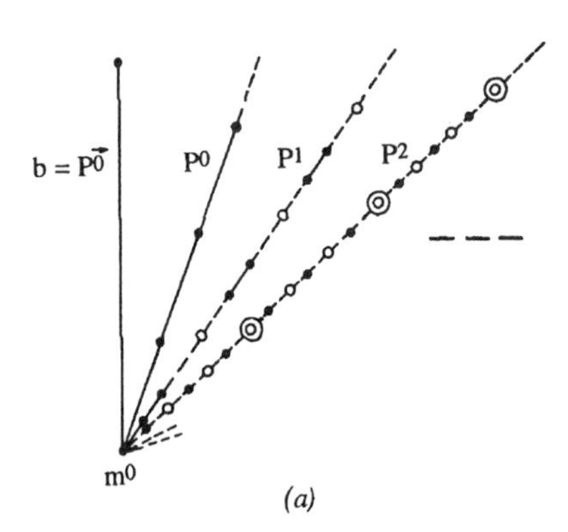

Figure 2.9. In this diagram heavy dots denote 0-nodes, single circles denote 1-nodes, and double circles denote 2-nodes. **(a)** A star graph consisting of a branch b and one-ended μ-paths P^μ $(\mu = 0, 1, 2, \ldots)$ terminally incident through $\vec{0}$-tips to the 0-node m^0 but otherwise totally disjoint. It is understood that, for each μ, there is a $(\mu + 1)$-node to which P^μ is terminally incident through a μ-tip. This is a $\vec{\omega}$-graph having no $\vec{\omega}$-node and no $\vec{\omega}$-path.

An $\vec{\omega}$-graph having an $\vec{\omega}$-node $n^{\vec{\omega}}$ is shown in Figure 2.9(b). The branch b and the μ-paths P^μ again meet through $\vec{0}$-tips at the 0-node m^0, but now they also meet at the $\vec{\omega}$-node $n^{\vec{\omega}}$ through tips of all ranks less than $\vec{\omega}$. More specifically, b meets the 0-node n^0, which is embraced by the 1-node n^1. Moreover, for each natural number μ, the μ-tip of P^μ is one member of the two-element $(\mu + 1)$-node $n^{\mu+1}$, whose other element is the μ-node n^μ. Thus, each node of the sequence $\{n^0, n^1, n^2, \ldots\}$ embraces all the nodes of lower ranks in the sequence, and the sequence itself is an $\vec{\omega}$-node $n^{\vec{\omega}}$.

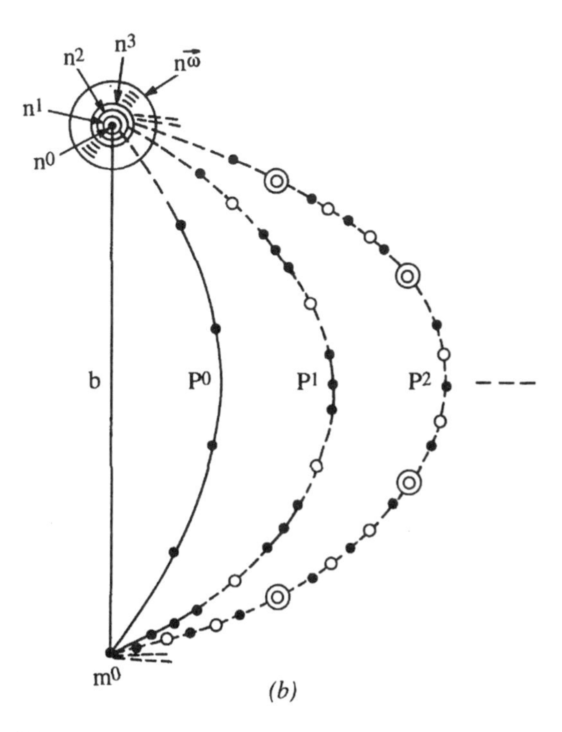

Figure 2.9. **(b)** A parallel connection of b and the $P^\mu (\mu = 0, 1, 2, \ldots)$ terminally incident to m^0 through $\vec{0}$-tips and terminally incident to an $\vec{\omega}$-node $n^{\vec{\omega}}$ through tips of all ranks less than $\vec{\omega}$. This is also a $\vec{\omega}$-graph. It has no $\vec{\omega}$-path.

Finally, Figure 2.9(c) shows a one-ended path starting with a 0-node m^0 incident to a branch b, followed by a one-ended 0-path P^0, which in turn is followed by a one-ended 1-path P^1, and so on through one-ended paths whose ranks increase through all the natural numbers. For each μ, P^μ and $P^{\mu+1}$ are both terminally incident to a $(\mu + 1)$-node through μ-tips. This too is an $\vec{\omega}$-graph, but like the graph of Figure 2.9(a), it has no $\vec{\omega}$-node. On the other hand, this graph is a particular example of a new kind of path, the $\vec{\omega}$-path, something we shall define shortly. Neither Figure 2.9(a) nor 2.9(b) have an $\vec{\omega}$-path. ♣

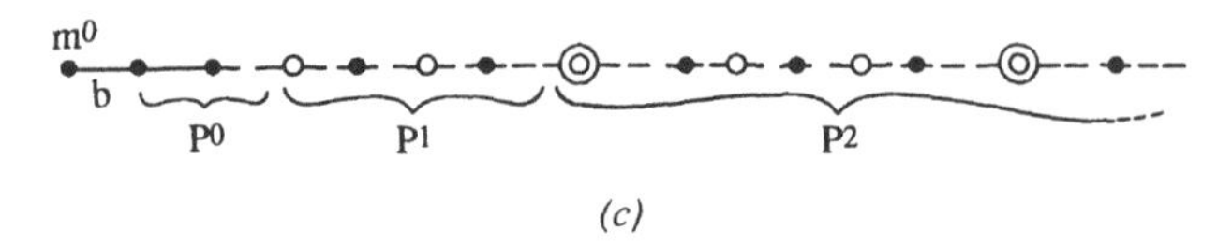

Figure 2.9. (c) An $\vec{\omega}$-path. This too is a $\vec{\omega}$-graph. It has an $\vec{\omega}$-tip but no $\vec{\omega}$-node.

A *subgraph* $\mathcal{G}_s$ of an $\vec{\omega}$-graph $\mathcal{G}^{\vec{\omega}}$ induced by a nonvoid subset $\mathcal{B}_s$ of the branch set $\mathcal{B}$ is defined much as before. With μ being a natural number, a μ-node of $\mathcal{G}^{\vec{\omega}}$ is a μ-node of $\mathcal{G}_s$ if and only if it has a $(\mu - 1)$-tip with a representative all of whose branches are in $\mathcal{B}_s$. Also, an $\vec{\omega}$-node $n^{\vec{\omega}}$ of $\mathcal{G}^{\vec{\omega}}$ is an $\vec{\omega}$-node of $\mathcal{G}_s$ if and only if an infinity of its members are nodes having the stated property. The rank of the resulting subgraph $\mathcal{G}_s$ may be either $\vec{\omega}$ or a natural number μ. In the former case, $\mathcal{N}^{\vec{\omega}}$ may be void.

The terminology that was defined for μ-graphs carries over to $\vec{\omega}$-graphs virtually word-for-word. Thus, just as before, we define *maximal nodes*, the *partitioning* of a graph or subgraph by a set of subgraphs, the *traversing* of a tip by a subgraph, *incidence* of a subgraph, *total disjointness* for subgraphs and/or nodes, and the *meeting* of subgraphs and/or nodes.

However, we now encounter a rather different kind of path.

Definition of a one-ended $\vec{\omega}$-path: A *one-ended $\vec{\omega}$-path* $P^{\vec{\omega}}$ is a one-way infinite sequence

$$P^{\vec{\omega}} = \{n_0^{\mu_0}, P_0^{\alpha_0}, n_1^{\mu_1}, P_1^{\alpha_1}, n_2^{\mu_2}, P_2^{\alpha_2}, \ldots\} \tag{2.9}$$

where the μ_m are natural numbers satisfying $\mu_0 < \mu_1 < \mu_2 < \cdots$, the α_m are natural numbers satisfying $0 \leq \alpha_m \leq \mu_{m+1} - 1$ for every m, each $n_m^{\mu_m}$ is a μ_m-node, each $P_m^{\alpha_m}$ is a nontrivial α_m-path, and the following conditions are satisfied.

(a) $P_0^{\alpha_0}$ meets $n_0^{\mu_0}$ with $n_0^{\mu_0}$ (that is, $P_0^{\alpha_0}$ terminates on the left with $n_0^{\mu_0}$).

(b) Each $P_m^{\alpha_m}$ is terminally incident to the two nodes adjacent to it in (2.9) but is otherwise totally disjoint from those nodes. Moreover, $P_m^{\alpha_m}$ either reaches $n_m^{\mu_m}$ (resp. $n_{m+1}^{\mu_{m+1}}$) through a terminal $(\mu_m - 1)$-tip (resp. a terminal $(\mu_{m+1}-1)$-tip) or meets it with a terminal δ-node where $\delta \leq \mu_m - 1$ (resp. $\delta \leq \mu_{m+1}-1$).

(c) Every two elements in (2.9) that are not adjacent therein are totally disjoint.

Here, too, $P^{\vec{\omega}}$ is a subgraph of $\mathcal{G}^{\vec{\omega}}$ induced by the branches embraced by $P^{\vec{\omega}}$. Actually, $\mathcal{G}^{\vec{\omega}}$ need not contain any $\vec{\omega}$-path. As was mentioned before, the $\vec{\omega}$-graphs of Figures 2.9(a) and (b) do not contain $\vec{\omega}$-paths, but that of Figure 2.9(c) does (in fact, is an $\vec{\omega}$-path).

Furthermore, with no changes at all, the proof of Lemma 2.2-2 holds and leads again to its conclusion; namely, for each $m > 0$, at least one of the paths $P_m^{\alpha_m}$ and $P_{m+1}^{\alpha_{m+1}}$ meets $n_{m+1}^{\mu_{m+1}}$ with a $(\mu_{m+1}-1)$-tip.

There is no such thing as a two-ended $\vec{\omega}$-path or a trivial $\vec{\omega}$-path, but an *endless $\vec{\omega}$-path* is meaningful. Such a path arises when two one-ended $\vec{\omega}$-paths start at the same initial node $n_0^{\mu_0}$ but are otherwise totally disjoint. It can be written as a two-way infinite sequence:

$$\{\ldots, n_{-2}^{\mu_{-2}}, P_{-2}^{\alpha_{-2}}, n_{-1}^{\mu_{-1}}, P_{-1}^{\alpha_{-1}}, n_0^{\mu_0}, P_0^{\alpha_0}, n_1^{\mu_1}, P_1^{\alpha_1}, n_2^{\mu_2}, \ldots\} \qquad (2.10)$$

where $n_0^{\mu_0}$ and the terms to the left (right) of $n_0^{\mu_0}$ comprise a one-ended $\vec{\omega}$-path, but those to the left are written in reverse order along with changes in its subscripts. Thus, for example, $\cdots > \mu_{-2} > \mu_{-1} > \mu_0 < \mu_1 < \mu_2 < \cdots$ and $0 \leq \alpha_{-m} \leq \mu_{-m-1} - 1$ for $-m < 0$.

Also, there is no such thing as an $\vec{\omega}$-loop, but there are loops of higher ranks as we shall see.

There is another difference between μ-paths and $\vec{\omega}$-paths that is worth mentioning. A μ-path can be written in only one way because the nodes it contains are all of a single rank μ. This is not so for $\vec{\omega}$-paths. For example, the first four terms of (2.9) can be encompassed within the path:

$$\{n_0^{\mu_0}, P_0^{\alpha_0}, n_1^{\mu_1}, P_1^{\alpha_1}, q^{\gamma_2}\} \;=\; \{n_0^{\mu_0}, Q_0^{\alpha}, q^{\gamma_2}\}$$

where q^{γ_2} is the node that contains the terminal tip on the right of $P_1^{\alpha_1}$. q^{γ_2} is embraced by $n_2^{\mu_2}$. Thus, (2.9) can be rewritten as

$$P^{\vec{\omega}} \;=\; \{n_0^{\mu_0}, Q_0^{\alpha}, n_2^{\mu_2}, P_2^{\alpha_2}, \ldots\}.$$

Nonetheless, both versions of $P^{\vec{\omega}}$ have the same branch set with the same total ordering as determined by a tracing of $P^{\vec{\omega}}$; in this sense we can consider both versions as being the "same" $\vec{\omega}$-path. In the next section we shall define an equivalence relationship between $\vec{\omega}$-paths that will encompass this idea of "sameness" as a special case.

2.4 ω-Graphs

Following the $\vec{\omega}$-graphs, the next kind of transfinite graph of higher rank is the "ω-graph." Such a graph can be constructed out of a given $\vec{\omega}$-graph $\mathcal{G}^{\vec{\omega}}$ so long as $\mathcal{G}^{\vec{\omega}}$ has at least one $\vec{\omega}$-path. As before, we start with equivalence classes of one-ended $\vec{\omega}$-paths. Now, however, we have to alter our definition of "equivalence" because one-ended $\vec{\omega}$-paths do not have unique representations as sequences, in contrast to one-ended μ-paths. We can use the fact that the branches embraced by a given one-ended $\vec{\omega}$-path $P^{\vec{\omega}}$ form a totally ordered set, the ordering being defined by a tracing of $P^{\vec{\omega}}$ starting at its terminal node. We shall say that two one-ended $\vec{\omega}$-paths $P_1^{\vec{\omega}}$ and $P_2^{\vec{\omega}}$ are *equivalent* if there is a node n_1 embraced by $P_1^{\vec{\omega}}$ and a node n_2 embraced by $P_2^{\vec{\omega}}$ such that the two totally ordered sets of branches following n_1 in $P_1^{\vec{\omega}}$ and following n_2 in $P_2^{\vec{\omega}}$ are identical. Thus, $P_1^{\vec{\omega}}$ and $P_2^{\vec{\omega}}$ are equivalent if and only if there is a third one-ended $\vec{\omega}$-path $P_3^{\vec{\omega}}$ that is "in" both $P_1^{\vec{\omega}}$ and $P_2^{\vec{\omega}}$ in the sense that all the branches of $P_3^{\vec{\omega}}$ are embraced by both $P_1^{\vec{\omega}}$ and $P_2^{\vec{\omega}}$.

This is truly an equivalence relationship, and it partitions the set of all one-ended $\vec{\omega}$-paths in $\mathcal{G}^{\vec{\omega}}$ into equivalence classes. Each such class will be called an *$\vec{\omega}$-tip*. A *representative* of a given $\vec{\omega}$-tip is any one of the paths in that equivalence class. Thus, for example, the endless $\vec{\omega}$-path (2.10) has exactly two $\vec{\omega}$-tips, and a representative of one tip or the other tip is the one-ended $\vec{\omega}$-path starting at $n_0^{\mu_0}$ and extending to the left or right. Each $\vec{\omega}$-tip of an $\vec{\omega}$-path $P^{\vec{\omega}}$ is called a *terminal tip* of $P^{\vec{\omega}}$. Also, if $P^{\vec{\omega}}$ is one-ended and therefore starts at a δ-node n_t^{δ} $(0 \le \delta < \vec{\omega})$, the $(\delta - 1)$-tip with which $P^{\vec{\omega}}$ meets n_t^{δ} is the other *terminal tip* of $P^{\vec{\omega}}$.

Partition the nonvoid set $\mathcal{T}^{\vec{\omega}}$ of all $\vec{\omega}$-tips of the $\vec{\omega}$-graph $\mathcal{G}^{\vec{\omega}}$ into subsets $\mathcal{T}_\tau^{\vec{\omega}}$, where τ denotes the index for the partitioning. Thus, $\mathcal{T}^{\vec{\omega}} = \cup_\tau \mathcal{T}_\tau^{\vec{\omega}}$, where each $\mathcal{T}_\tau^{\vec{\omega}}$ is nonvoid and $\mathcal{T}_{\tau_1}^{\vec{\omega}} \cap \mathcal{T}_{\tau_2}^{\vec{\omega}} = \emptyset$ if $\tau_1 \ne \tau_2$. Also, for each τ, let $\mathcal{N}_\tau^{\vec{\omega}}$ be either the void set or a singleton whose only member n_t^{α} is either an $\vec{\omega}$-node or a μ-node, μ being a natural number as always.

Definition of an ω-node: For each τ, the set

$$n_\tau^{\omega} \;=\; \mathcal{T}_\tau^{\vec{\omega}} \cup \mathcal{N}_\tau^{\vec{\omega}} \tag{2.11}$$

is called an *$\vec{\omega}$-node* so long as the following condition is satisfied: Whenever $\mathcal{N}_\tau^{\vec{\omega}}$ is not void, its single member n_τ^{α} $(0 \le \alpha \le \vec{\omega})$ is not a member of any other β-node $(\alpha < \beta \le \omega)$.

When n_τ^{α} exists, we call it the *exceptional element* of n_τ^{ω}. As usual, we say that n_τ^{ω} *embraces* itself, all its elements, and all the elements embraced by its exceptional element if the latter exists. When n_τ^ω is a nonsingleton, all the elements embraced by n_τ^ω are said to be *shorted together* or simply *shorted.* However, when n_τ^ω is a singleton, it and its sole $\vec{\omega}$-tip are called *open.*

Lemma 2.4-1. *If an α-node $(0 \leq \alpha \leq \omega)$ and an ω-node embrace a common node, then the ω-node embraces the α-node; in addition, if $\alpha = \omega$, then the ω-node is identical to the α-node.*

The proof of this lemma is exactly the same as that for Lemma 2.2-1, the only unessential difference being that the sequences S_a and S_b of embraced nodes may now be infinite.

Definition of an ω-graph: An *ω-graph* $\mathcal{G}^\omega$ is a well-ordered infinite set of sets:

$$\mathcal{G}^\omega = \{\mathcal{B}, \mathcal{N}^0, \mathcal{N}^1, \ldots, \mathcal{N}^{\vec{\omega}}, \mathcal{N}^\omega\} \tag{2.12}$$

where again $\mathcal{B}$ is a set of branches and the $\mathcal{N}^\nu$ $(\nu = 0, 1, \ldots, \vec{\omega}, \omega)$ is a set of ν-nodes constructed as stated above. (For every ν other than $\nu = \vec{\omega}$, $\mathcal{N}^\nu$ is nonvoid; $\mathcal{N}^{\vec{\omega}}$ may be void.) For each $\gamma = 0, 1, \ldots, \vec{\omega}$, $\{\mathcal{B}, \mathcal{N}^0, \ldots, \mathcal{N}^\gamma\}$ is called the *γ-graph of $\mathcal{G}^\omega$*.

Example 2.4-2. A simple example of an ω-graph can be obtained from the $\vec{\omega}$-graph of Figure 2.9(c) by appending another branch b_0, making b_0 incident to the initial 0-node m_0 in that figure and also incident to a new ω-node $n^\omega = \{n^0, t^{\vec{\omega}}\}$, where n^0 is a singleton 0-node containing an elementary tip of b_0 and $t^{\vec{\omega}}$ is the sole $\vec{\omega}$-tip of the $\vec{\omega}$-graph of Figure 2.9(c). The result is a particular case of an "ω-loop." ♣

Here, too, a *subgraph* $\mathcal{G}_s$ of an ω-graph $\mathcal{G}^\omega$ induced by a subset $\mathcal{B}_s$ of $\mathcal{B}$ is defined as before. The nodes of $\mathcal{G}_s$ with natural-number ranks and with rank $\vec{\omega}$ are specified as they were for a subgraph of an $\vec{\omega}$-graph. Also, an ω-node of $\mathcal{G}^\omega$ is an ω-node of $\mathcal{G}_s$ if and only if it has an $\vec{\omega}$-tip with a representative all of whose branches are in $\mathcal{B}_s$. The rank of the resulting subgraph $\mathcal{G}_s$ may be either μ, $\vec{\omega}$, or ω.

Moreover, the following terms are defined for ω-graphs in the same way as they were for μ-graphs (see Section 2.2): a *maximal node*, the *partitioning* of a graph or subgraph by a set of subgraphs, the *traversing* of a tip by a subgraph, *incidence* for a subgraph, *total disjointness* for subgraphs and/or nodes, and the *meeting* of subgraphs and/or nodes.

Our next objective is to define an "ω-path." Again our prior definitions require hardly any alterations, but there are some minor changes; so let us now repeat those ideas. We have already defined the terminal tips of a μ-path (resp. $\vec{\omega}$-path) in Section 2.2 (resp. Section 2.3). If n^γ is a γ-node that embraces a terminal tip of a ν-path P^ν $(0 \leq \nu \leq \vec{\omega})$, then we say as before that P^ν is *terminally incident to* or *reaches* n^γ *with* or *through* that terminal tip; if in addition P^ν terminates at a δ-node n_t^δ $(0 \leq \delta < \vec{\omega})$ and if n^γ embraces n_t^δ, then we also say that P^ν is *terminally incident to* n^γ *with* n_t^δ and that P^ν *meets* n^γ *with* n_t^δ. Finally, we say that P^ν is *terminally incident to* n^γ *but otherwise totally disjoint from* n^γ if P^ν reaches n^γ with a terminal tip and if n^γ does not embrace any other tip traversed by P^ν.

Definition of an ω-path: Consider the alternating sequence

$$P^\omega = \{\ldots, n_m^\omega, P_m^{\alpha_m}, n_{m+1}^\omega, P_{m+1}^{\alpha_{m+1}}, \ldots\} \tag{2.13}$$

of ω-nodes n_m^ω and α_m-paths $P_m^{\alpha_m}$, where $0 \leq \alpha_m \leq \vec{\omega}$ and the α_m may differ. P^ω is called a *nontrivial ω-path* if the following three conditions are satisfied:

(a) Each $P_m^{\alpha_m}$ is a nontrivial α_m-path, and P^ω contains at least one such α_m-path and at least one ω-node. Moreover, every n_m^ω is an ω-node except possibly for terminal nodes when (2.13) terminates on the left and/or on the right. In the latter case, each terminal element is a δ-node n_t^δ, where either $\delta = \omega$ or $0 \leq \delta < \vec{\omega}$, and its adjacent path is either an $\vec{\omega}$-path or an α-path $(\alpha \leq \vec{\omega})$, respectively.

(b) Each $P_m^{\alpha_m}$ is terminally incident to the two nodes adjacent to it in (2.13) but is otherwise totally disjoint from those nodes. Moreover, $P_m^{\alpha_m}$ either reaches each of its adjacent nodes with a terminal $\vec{\omega}$-tip or meets it with a terminal δ-node where $\delta < \vec{\omega}$. If that adjacent node is a terminal node n_t^δ of (2.13), then $P_m^{\alpha_m}$ reaches n_t^δ with a $(\delta - 1)$-tip; if in addition $\delta < \vec{\omega}$, then n_t^δ is also a terminal node of $P_m^{\alpha_m}$.

(c) Every two elements in (2.13) that are not adjacent therein are totally disjoint.

Here, too, an ω-path P^ω is a subgraph of $\mathcal{G}^\omega$ induced by the branches embraced by P^ω. Also, the proof of Lemma 2.2-2 still holds, and we thereby have the following consequence of condition (c): If n_{m+1}^ω is not a terminal node of (24.3), then at least one of its adjacent paths $P_m^{\alpha_m}$ and $P_{m+1}^{\alpha_{m+1}}$ meets n_{m+1}^ω with an $\vec{\omega}$-tip and must therefore be of rank $\vec{\omega}$.

P^ω is called *two-ended, one-ended,* or *endless* when (2.13) is respectively a finite, one-way infinite, or two-way infinite sequence. A *trivial ω-path* is any singleton containing an α-node ($0 \leq \alpha < \vec{\omega}$ or $\alpha = \omega$). An *ω-loop* is defined exactly like a nontrivial two-ended ω-path except that one of its terminal nodes is required to embrace the other. A γ-path ($0 \leq \gamma \leq \omega$) or γ-loop ($0 \leq \gamma < \vec{\omega}$ or $\gamma = \omega$), all of whose branches are embraced by a subgraph, is said to be *in* that subgraph. Finally, Lemma 2.2-4 extends directly to ω-paths; that is, a two-ended ω-path embraces every tip it traverses.

2.5 Graphs of Higher Ranks

We have come full circle: ω-graphs are similar to 0-graphs in certain ways. For instance, the ranks 0 and ω are the immediate successors of arrow ranks, which is not the case for any rank μ ($0 < \mu < \vec{\omega}$). To be sure, $\vec{0}$-tips are postulated, whereas $\vec{\omega}$-tips are constructed. Moreover, ω-nodes may have exceptional elements, in contrast to 0-nodes. Other than these differences, our procedure for defining an ω-graph is the same as that for a 0-graph, which in turn is the same as that for the μ-graphs. On the other hand, an $\vec{\omega}$-graph is set up rather differently.

With an ω-graph in hand, we can repeat the constructions of the μ-graphs to obtain transfinite graphs of ranks $\omega + \mu$ for all the natural numbers μ so long as the node sets $\mathcal{N}^{\omega+\mu}$ remain nonvoid. In particular, we define ω-tips as equivalence classes of one-ended ω-paths. Then, $(\omega+1)$-nodes are obtained by partitioning the set of all ω-tips and appending exceptional elements to none, some, or all of the subsets of the partition. This yields an $(\omega+1)$-graph:

$$\mathcal{G}^{\omega+1} = \{\mathcal{B}, \mathcal{N}^0, \mathcal{N}^1, \ldots, \mathcal{N}^{\vec{\omega}}, \mathcal{N}^{\omega}, \mathcal{N}^{\omega+1}\},$$

wherein all the sets are nonvoid except possibly $\mathcal{N}^{\vec{\omega}}$. Finally, $(\omega+1)$-paths are defined just as 1-paths are defined, and this allows us to proceed on to a construction of an $(\omega+2)$-graph. Continuing in this way, we can get all the $(\omega+\mu)$-graphs.

To go still further, we mimic our construction of an $\vec{\omega}$-graph. This involves the constuction of an $(\omega+\vec{\omega})$-node:

$$n^{\omega+\vec{\omega}} = \{n^\gamma\}_{\gamma\in\Gamma} \tag{2.14}$$

where Γ is a subset of the well-ordered set:

$$\{\vec{0}, 0, 1, \ldots, \vec{\omega}, \omega, \omega + 1, \ldots\} \tag{2.15}$$

of all ranks less than $\omega+\vec{\omega}$. Each n^γ in (2.14) is the exceptional element of its immediate successor in (2.14). Many such nodes might be set up, given a prior construction of the γ-graphs $\mathcal{G}^\gamma$ $(\gamma < \omega+\vec{\omega})$. With $\mathcal{N}^{\omega+\vec{\omega}}$ being the (possibly void) set of all such $(\omega+\vec{\omega})$-nodes, we define an $(\omega+\vec{\omega})$-graph as

$$\mathcal{G}^{\omega+\vec{\omega}} = \{\mathcal{B}, \mathcal{N}^0, \mathcal{N}^1, \ldots, \mathcal{N}^{\vec{\omega}}, \mathcal{N}^\omega, \mathcal{N}^{\omega+1}, \ldots, \mathcal{N}^{\omega+\vec{\omega}}\}.$$

We then define a one-ended $(\omega+\vec{\omega})$-path $P^{\omega+\vec{\omega}}$ in much the same way as we defined a one-ended $\vec{\omega}$-path (2.9), but now the ranks μ_k of the nodes $n_k^{\mu_k}$ in (2.9) approach but do not equal $\omega+\vec{\omega}$ (that is, the μ_k comprise a subset of (2.15) with $\mu_k < \mu_{k+1}$ for each k; moreover, for each member λ of (2.15), there is a μ_k with $\mu_k > \lambda$). This allows us to define $(\omega+\vec{\omega})$-tips and then $(\omega \cdot 2)$-nodes like $\vec{\omega}$-tips and ω-nodes. (Here, $\omega \cdot 2 = \omega + \omega$.)

We have now come full circle twice and can continue on to still higher ranks by repeating this cycle of constructions as often as we wish. To this end, we can use the procedure of Section 2.2 when ν is a successor-ordinal rank, the procedure of Section 2.3 when ν is an arrow rank, and the procedure of Section 2.4 when ν is a limit-ordinal rank.

But, can we get to any countable-ordinal rank ν in this way? In order to do so, we need to have defined β-graphs for all $\beta < \nu$ if $(\nu-1)$-tips or $\vec{\nu}$-tips (ν being a successor or limit ordinal, respectively) are to be available for the construction of ν-nodes. It is not clear that this is assured for all countable ν. What is clear is that we can proceed beyond the rank ω. Henceforth, it will be tacitly assumed that ν is restricted to those countable ordinals for which ν-graphs can be defined. Moreover, we will in general present detailed arguments only for the ranks up to and including ω since the extensions beyond ω can be obtained by repeating arguments already given.

Chapter 3

Connectedness

A key difference between conventional infinite graphs and transfinite graphs is that in conventional graphs two nodes are either connected through a finite path (i.e., a path of finitely many branches) or not connected at all, whereas in transfinite graphs two nodes may be connected through a transfinite path (i.e., one having infinitely many branches) but not through any finite path. In fact, for transfinite graphs there is a hierarchy of connectedness concepts, which is indexed by the countable ordinals. Thus, we speak of two nodes being "ρ-connected" when the two nodes are connected through a two-ended α-path for some α no larger than ρ. As ρ increases, ρ-connectedness weakens in the sense that, when $\rho_1 < \rho_2$, ρ_1-connectedness implies ρ_2-connectedness, but not the converse. Thus, ρ_2-connectedness is a more inclusive concept than ρ_1-connectedness.

Moreover, ρ-connectedness ($\rho \geq 1$) is peculiar in the following way. Although it is a reflexive and symmetric binary relation between branches, it may not be transitive. This has undesirable ramifications. For example, it may prevent node voltages from being unique in a transfinite electrical network, as we shall see in Section 5.5. Another consequence concerns "ρ-sections." A ρ-section in a ν-graph $\mathcal{G}^\nu$ ($\rho < \nu$) is a subgraph induced by a maximal set of branches that are pairwise ρ-connected. When ρ-connectedness is not transitive, ρ-sections may overlap and thereby fail to partition $\mathcal{G}^\nu$. Thus, we are led to a search for conditions insuring the transitivity of ρ-connectedness. Most of this chapter is devoted to that search.

Two sufficient sets of conditions for that transitivity are established. The first relates to the paths connecting the nodes of a $(\rho-1)$-section. It requires that, for every two (boundary or nonboundary) nodes of any given $(\rho-1)$-

section $S^{\rho-1}$, there must be a β-path ($\beta \leq \rho$) connecting those nodes that either is isolated except terminally or is in $S^{\rho-1}$ but does not meet any other boundary node of $S^{\rho-1}$. The other set of sufficient conditions requires that if one-ended paths meet infinitely often in a certain way, then either their infinite extremities are shorted together or at least one of them is open. A substantial part of the argument establishing the last result is devoted to a characterization of the totally ordered set of all the embraced nodes along a transfinite path, that characterization being a certain hierarchical structure of nested sequences, called "ν-sequences." ν-sequences generalize sequences in much the same way that transfinite paths generalize 0-paths.

The last principal result of this chapter is a sufficient set of conditions on all ρ-sections for all ranks $\rho < \nu$ that establishes a bound on the cardinality of the branch set $\mathcal{B}$ of $\mathcal{G}^\nu$. This is a generalization of the standard result that a conventional connected graph has a countable branch set if each of its nodes is of countable degree. The transfiniteness of the graphs considered herein complicates matters considerably.

3.1 Transfinite Connectedness

Given any ν-graph $\mathcal{G}^\nu$, let ρ be any rank such that $\rho \leq \nu$. Furthermore, let m and n be two totally disjoint nodes in $\mathcal{G}^\nu$ of possibly different ranks. Any of these ranks may be either ordinal or arrow ranks.

Definition of ρ-connectedness: Let $\rho \geq 0$. m and n are said to be *ρ-connected* if there is a two-ended α-path P^α with $\alpha \leq \rho$ if ρ is an ordinal rank and with $\alpha < \rho$ if ρ is an arrow rank such that P^α meets both m and n. Two branches b_1 and b_2 are said to be *ρ-connected* if a 0-node incident to b_1 and a 0-node incident to b_2 are ρ-connected. Every branch is taken to be ρ-connected to itself. $\mathcal{G}^\nu$ itself is called *ρ-connected* if every two of its branches are ρ-connected.

Thus, if m and n are ρ-connected, they are also θ-connected for every ordinal θ with $\rho \leq \theta \leq \nu$. Also, in the above definition, α has to be an ordinal rank since there is no two-ended path with an arrow rank.

(This use of the word "ρ-connected" conflicts with a different meaning which is standard in conventional graph theory [2, page 119], [3, page 50], but it conforms with our way of designating ranks. To avoid this disharmony, one might say "transfinitely ρ-connected" or perhaps "ρ-transconnected.")

It is important to note that in the definition of ρ-connectedness, P^α is required to meet the nodes m and n. Only requiring P^α to reach m and n may not be sufficient for ρ-connectedness. Actually, the condition that P^α be two-ended insures that P^α will meet — not just reach — m and n. Indeed, only the one-ended and endless α-paths can reach a node through a terminal α-tip without meeting that node. In this regard, see Lemma 2.2-4; that lemma extends to ω-graphs and graphs of still higher ranks.

Example 3.1-1. Consider for example the 1-graph of Figure 2.7. As before, let P_a^0 be the one-ended 0-path consisting of the a-branches. Then, P_a^0 meets the 0-node n_1^0 but only reaches the 1-node n_1^1. Thus, we cannot say that n_1^0 and n_1^1 are 0-connected. However, the 1-path $\{n_1^0, P_a^0, n_1^1\}$ meets both n_1^0 and n_1^1 and thereby 1-connects them. Similarly, the branches a_1 and c are 1-connected but not 0-connected. ♣

Definition of a ρ-section: Let $\rho \geq 0$. A *ρ-section* S^ρ is a subgraph of $\mathcal{G}^\nu$ induced by a maximal set of branches that are pairwise ρ-connected. We also define a *component* of $\mathcal{G}^\nu$ to be a ν-section of $\mathcal{G}^\nu$. It follows that an equivalent definition of a ρ-section is this: It is a component of the ρ-graph of $\mathcal{G}^\nu$. Finally, a *$\vec{0}$-section* is taken to be a single branch (and thus in this case alone it is not a subgraph).

As was defined in Chapter 2, a ρ-section S^ρ traverses a tip if S^ρ embraces every branch of some representative of that tip. It can happen that a ρ-section may traverse a tip of rank greater than ρ. An example of this occurs in Figure 2.8. There we have a 2-graph, which is itself a 0-section that traverses a 1-tip.

A node n is called a *boundary node of S^ρ* if n embraces a tip traversed by S^ρ and also embraces a tip not traversed by S^ρ. Thus, n is a *nonboundary node of S^ρ* if all the tips n embraces are traversed by S^ρ.

Lemma 3.1-2. *If n is a boundary node of a ρ-section $\mathcal{S}^\rho$, then n satisfies at least one of the following two conditions:*

(i) *All the tips embraced by n and traversed by $\mathcal{S}^\rho$ have ranks no less than ρ.*

(ii) *All the tips embraced by n and not traversed by $\mathcal{S}^\rho$ have ranks no less than ρ.*

Moreover, every boundary node of $\mathcal{S}^\rho$ has a rank larger than ρ, and any path P that reaches a branch in $\mathcal{S}^\rho$ and a branch not in $\mathcal{S}^\rho$ must have a rank larger than ρ.

Proof. If n satisfies neither (i) nor (ii), then n embraces an α-node n^α, where $\alpha \leq \rho$ if ρ is an ordinal rank and where $\alpha < \rho$ if ρ is an arrow rank, and there is an α-path passing through n^α and meeting a branch of $\mathcal{S}^\rho$ and a branch not in $\mathcal{S}^\rho$. This violates the definition of a ρ-section. The rest of the lemma follows directly. ♣

Example 3.1-3. Consider the $\vec{\omega}$-graph of Figure 2.9(a) and let $0 \leq \rho < \vec{\omega}$. For each μ with $\rho < \mu < \vec{\omega}$, let $m_\mu^{\rho+1}$ be the first $(\rho+1)$-node in the path P^μ with respect to a tracing from m^0, and let $n_\mu^{\rho+1}$ be the second $(\rho+1)$-node in P^μ. Then, the branches in the paths P^η $(\eta = \vec{0}, 0, \ldots, \rho)$ along with the branches between m^0 and $m_\mu^{\rho+1}$ in the paths P^μ $(\mu = \rho + 1, \rho+2, \ldots)$ induce a ρ-section $\mathcal{S}^\rho$. The boundary nodes of $\mathcal{S}^\rho$ are the $m_\mu^{\rho+1}$ $(\mu = \rho+1, \rho+2, \ldots)$. The $(\rho+1)$-node that P^ρ reaches is a nonboundary node of $\mathcal{S}^\rho$. Another ρ-section is induced by the branches between $m_\mu^{\rho+1}$ and $n_\mu^{\rho+1}$ in the path P^μ for any given $\mu \geq \rho+1$. There are infinitely many of these latter ρ-sections, one for each $\mu \geq \rho + 1$. (Later on, we shall refer to the set of these latter ρ-sections as the "$(\rho + 1)$-adjacency of $\mathcal{S}^\rho$.") Of course, there are still other ρ-sections. ♣

0-connectedness is obviously a reflexive and symmetric binary relation between the branches of a ν-graph $\mathcal{G}^\nu$. It is also transitive. Indeed, let b_1, b_2, and b_3 be three branches with b_1 and b_2 being 0-connected through the finite 0-path P_{12}^0 and with b_2 and b_3 being 0-connected through the finite 0-path P_{23}^0. Starting from a 0-node incident to b_1, we can trace along P_{12}^0 and, if need be, through b_2 as well until a 0-node m of P_{23}^0 is reached for the first time. We can then trace along P_{23}^0 starting from m to reach a 0-node of b_3. These tracings yield a finite 0-path connecting b_1 and b_3. Thus, 0-connectedness is an equivalence relation between branches. It partitions the ν-graph $\mathcal{G}^\nu$ into 0-sections.

Example 3.1-4. As an example of such a partitioning of a 1-graph, consider again the 1-graph $\mathcal{G}^1$ of Figure 2.7. The 0-section induced by the a-branches is the 1-path $\{n_1^0, P_a^0, n_1^1\}$, where P_a^0 is the one-ended 0-path containing the a-branches. Note that the rank of this 0-section as a subgraph of $\mathcal{G}^1$ is 1. On the other hand, 0 is the rank of the subgraph of the 0-graph of $\mathcal{G}^1$ induced by the a-branches; this latter subgraph is P_a^0.

Similarly, the b-branches induce the 0-section $\{n_2^0, P_b^0, n_1^1\}$, where P_b^0 is also a one-ended 0-path. Finally, the branch c by itself induces a third 0-section, namely, the finite 0-path $\{n_3^0, c, n_c^0\}$, where n_c^0 is the 0-node embraced by the 1-node n_1^1. n_1^1 is the one and only boundary node of the three 0-sections, but the 0-node n_c^0 is not a boundary node of any of those three 0-sections because it does not embrace any tips traversed by the first two 0-sections. ♣

In contrast to 0-connectedness, ρ-connectedness for $\rho \geq 1$ is not in general an equivalence relation between branches. Although reflexivity and symmetry obviously hold, transitivity may fail. As a result, different ρ-sections may overlap; moreover, when $\rho = \nu$, a ν-section of a ν-graph may have a smaller branch set than does the ν-graph, and different components of the ν-graph may not be totally disjoint. ρ-connecedness is transitive in the $\vec{\omega}$-graph of Figure 2.9(a). However, here is an example where it is not.

Example 3.1-5. The 1-graph of Figure 3.1 consists of the 0-nodes n_j^0, $j = 1, 2, 3, \ldots$; the parallel branches a_j and b_j incident to n_j^0 and n_{j+1}^0; the nonsingleton 1-node $n_a^1 = \{n_a^0, t_a^0\}$, where the 0-node n_a^0 is the exceptional element of n_a^1 and where t_a^0 is the 0-tip having as a representative the 0-path passing through all the a_j; the nonsingleton 1-node $n_b^1 = \{n_b^0, t_b^0\}$, where the 0-node n_b^0 is the exceptional element of n_b^1 and where t_b^0 is the 0-tip having as a representative the 0-path passing through all the b_j; and two more branches β_a and β_b — the first one incident to n_a^0 and the 0-node n_c^0 and the second one incident to n_b^0 and the 0-node n_d^0. All the other 1-nodes are singletons and are not shown.

In this 1-graph, the branch a_1 is 1-connected to the branch β_a and also to the branch β_b. However, β_a and β_b are not 1-connected; indeed, any tracing from β_a to β_b will perforce meet at least one of the 0-nodes n_j^0 at least twice — thereby preventing that tracing from being a 1-path. Thus, 1-connectedness is not transitive as a binary relation for the branches in this 1-graph.

Moreover, this 1-graph contains exactly three 0-sections: S_1^0 induced by all the branches a_j and b_j, S_2^0 induced by β_a alone, and S_3^0 induced by β_b alone. These 0-sections do not have overlapping branch sets, as expected. On the other hand, there are exactly two 1-sections: S_1^1 induced by β_a and all the a_j and b_j, and S_2^1 induced by β_b and all the a_j and b_j. Because of the nontransitivity of 1-connectedness in this case, these two 1-sections have overlapping but not identical branch sets.

(Before leaving this example, let us note that branches β_a and β_b are nonetheless connected in a certain weaker sense. In Section 7.2, we will define a "1-walk"; it extends the idea of a walk in a conventional graph.

For example, any tracing from β_a to, say, node n_1^0 and then back to β_b is a 1-walk. In this way, we might say that β_a and β_b are "1-walk-connected." This distinction between walk-connectedness and path-connectedness does not arise in conventional graph theory [2, Theorem 2.1].) ♣

Examples of nontransitivity can be constructed for $\rho > 1$ in a similar way. This peculiarity of ρ-connectedness leads to difficulties, which we shall avoid at times by restricting the kinds of ν-graphs under consideration. The question arises: What conditions on a ν-graph will insure that ρ-connectedness is transitive? With regard to the 1-graph of Figure 3.1, if a two-ended 1-path were to be appended incident to n_c^0 and n_d^0, the 1-nodes n_a^1 and n_b^1 would become 1-connected, and 1-connectedness would become transitive for all the branches (and for all the nodes as well) in the resulting 1-graph. It is tempting therefore to conjecture that the transitivity of 1-connectedness for branches will hold in any 1-graph that satisfies the following condition: If two nonsingleton 1-nodes are incident to the same 0-section, then they are 1-connected. (As in the case above where a two-ended 1-path is appended to n_c^0 and n_d^0, we are not requiring that the 1-connectedness be through that same 0-section.) However, this conjecture is not true.

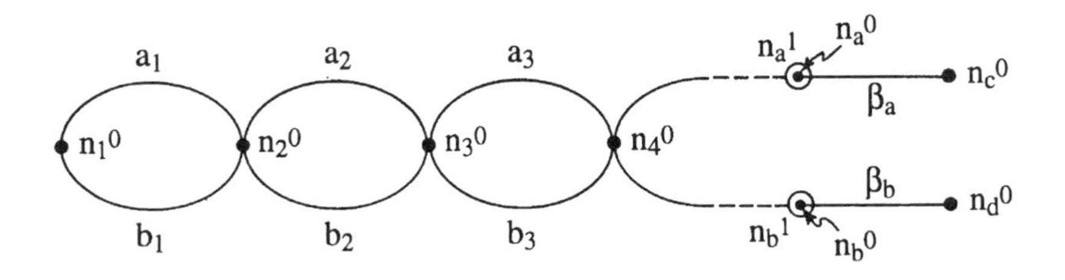

Figure 3.1. The 1-graph discussed in Example 3.1-5.

Example 3.1-6. Consider the 1-graph $\mathcal{G}^1$ of Figure 3.2. Each heavy dot therein represents a nonsingleton 1-node. Each S_j^0 is a 0-section like the 0-section of Figure 3.1 induced by the a_j and b_j branches. The 0-node of each S_j^0 corresponding to n_1^0 in Figure 3.1 is embraced by a 1-node of Figure 3.2 — except for that 0-node of S_1^0. Furthermore, the R_j^0 and T_j^0 are 0-sections, each consisting of a single endless 0-path along with its two incident 1-nodes. Finally, this 1-graph extends infinitely to the right.

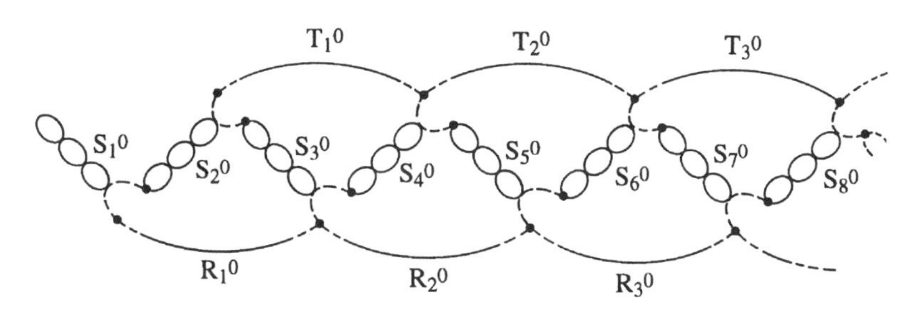

Figure 3.2. The 1-graph discussed in Example 3.1-6.

It can be seen that every two nonsingleton 1-nodes in Figure 3.2 that are incident to the same 0-section are 1-connected, and therefore the above conjectured condition is fulfilled. For example, consider the two nonsingleton 1-nodes incident to S_1^0; they are 1-connected by a 1-path that passes along R_1^0, then through S_3^0, and finally through S_2^0. Moreover, the branches of all S_j^0 and all R_j^0 induce a 1-section W^1, and the branches of all S_j^0 and all T_j^0 induce another 1-section Z^1. However, there is no 1-path 1-connecting any branch of any R_j^0 to any branch of any T_k^0; the "forked ends" of the S_j^0 block such 1-paths. Thus, 1-connectedness is not transitive for the branches in this 1-graph; moreover, W^1 and Z^1 overlap but are not the same. ♣

3.2 A Transitivity Criterion for Transfinite Connectedness

In this section we establish the first of our two criteria ensuring transitivity for transfinite connectedness. The other one will be presented in Section 3.5. Neither criterion implies the other. We will provide a detailed discussion for the natural-number ranks μ. Results for higher ranks can be obtained through obvious modifications.

Let $\mathcal{G}^\nu$ be a given ν-graph where ν is a countable ordinal. μ will denote a fixed natural number with $\mu < \nu$. Also, let n_1 and n_2 be two totally disjoint, nonsingleton nodes of any ranks (not necessarily the same rank).

Conditions 3.2-1. *$\mathcal{G}^\nu$ is such that the following holds for every rank γ with $\vec{0} \leq \gamma \leq \mu$ and for every two nodes of the sort just stated. If n_1 and n_2 are incident to the same γ-section S^γ of $\mathcal{G}^\nu$, then n_1 and n_2 are $(\gamma+1)$-connected through a two-ended β-path P^β $(0 \leq \beta \leq \gamma + 1)$ terminally* *incident to n_1 and n_2 and satisfying one or the other or both of the following two conditions:*

(a) *All the tips embraced by P^β are also traversed by P^β except possibly for some tips embraced by n_1 and n_2.*

(b) *All the branches of P^β are embraced by S^γ, and P^β does not meet any node of rank larger than γ other than possibly n_1 and n_2.*

Here, n_1 and n_2 may or may not be boundary nodes of S^γ. With regard to the two conditions (a) and (b), (a) does not require that P^β be in S^γ, whereas (b) does do so. Furthermore, (a) can be restated in more suggestive terms by saying that P^β is an isolated path totally disjoint from the rest of $\mathcal{G}^\nu$ except at its terminal nodes. On the other hand, since a boundary node of S^γ is necessarily of rank larger than γ, (b) implies that P^β meets no boundary node of S^γ except when n_1 or n_2 is a boundary node of S^γ.

Theorem 3.2-2. *Under Conditions 3.2-1, for each γ such that $\vec{0} \leq \gamma \leq \mu$, $(\gamma+1)$-connectedness is transitive in $\mathcal{G}^\nu$ (that is, $(\gamma+1)$-connectedness is a transitive binary relation among the branches of $\mathcal{G}^\nu$ — as well as among the nodes of $\mathcal{G}^\nu$).*

Proof. Transitivity with regard to branches will follow immediately from transitivity with regard to nodes. We prove the latter inductively. We know that 0-connectedness is always transitive. (Also, Conditions 3.2-1 are always satisfied when $\gamma = \vec{0}$, that is, when $\gamma + 1 = 0$, because a $\vec{0}$-section is a single branch along with its incident 0-node or 0-nodes.) So, assume the transitivity of γ-connectedness for some γ where now $0 \leq \gamma \leq \mu$. Consequently, γ-sections partition $\mathcal{G}^\nu$. We shall prove the transitivity of $(\gamma + 1)$-connectedness.

Let n_a, n_b, and n_c be three totally disjoint nodes of any ranks such that n_a and n_b are $(\gamma+1)$-connected and n_b and n_c are $(\gamma+1)$-connected. Let P_{ab} be a two-ended path of rank no larger than $\gamma+1$ connecting n_a and n_b and oriented from n_a to n_b, and let P_{ba} be the same path but with the reverse orientation. Also, let P_{bc} be a two-ended path of rank no larger than $\gamma+1$ connecting n_b to n_c and oriented from n_b to n_c. We now show that there exists a two-ended path of rank no larger than $\gamma+1$ connecting n_a and n_c. If P_{ab} meets n_c or if P_{bc} meets n_a, then n_a and n_c are $(\gamma+1)$-connected. So, assume henceforth that these meetings do not occur.

P_{ba} and P_{bc} each traverse only finitely many γ-tips, perhaps none at all. All other tips traversed by these paths will have ranks less than γ. If P_{ba} and P_{bc} meet at a node n, they will meet there with four tips (two for each), except when n is n_b, in which case they will meet at n_b with two tips (one for each). If all four tips (or both tips when $n = n_b$) have ranks less than γ, they will have representatives that are γ-connected to each other, that is, the branches of all those representatives will be γ-connected; in this case, we shall say that P_{ba} and P_{bc} *meet at* n *with only* $(\gamma-)$-tips. P_{ba} and P_{bc} can meet in this way infinitely many times because they can traverse infinitely many tips of ranks less than γ. On the other hand, if at least one of those four (or two) tips has a rank γ, we shall say that P_{ba} and P_{bc} *meet at* n *with at least one* γ-tip; P_{ba} and P_{bc} can meet in this latter way only finitely many times since they traverse only finitely many γ-tips.

We now consider four cases:

Case 1: P_{ba} and P_{bc} meet only at n_b. In this case, $P_{ab} \cup P_{bc}$ is a two-ended path of rank no larger than $\gamma + 1$. So, assume in the following three cases that P_{ba} and P_{bc} meet at at least one node totally disjoint from n_b.

Case 2: If n_a and n_c are incident to the same γ-section, we can invoke Conditions 3.2-1 to conclude that n_a and n_c are $(\gamma + 1)$-connected. So, assume in the following two cases that n_a and n_b are not incident to the same γ-section.

Case 3: Assume for this case that P_{ba} and P_{bc} meet at least once with only $(\gamma-)$-tips (possibly infinitely often). They will do so only within finitely many γ-sections because they can pass from one γ-section to another only through tips whose ranks are no less than γ (Lemma 3.1-2). Let S_l^γ be the last such γ-section that P_{ba} meets. Also, let n_d be n_a if n_a is incident to S_l^γ; otherwise, let n_d be the last boundary node of S_l^γ that P_{ba} meets. Let P_{ad} be the path obtained by tracing P_{ab} from n_a to n_d. The rank of P_{ad} is no larger than $\gamma + 1$. Also, P_{ad} will be the trivial path $\{n_a\}$ if $n_a = n_d$. If $n_a \neq n_d$, P_{ad} and P_{bc} will meet at most finitely many times because each time they meet they do so with at least one γ-tip and because P_{ad} and P_{bc} traverse only finitely many γ-tips.

Next, let n_e be n_c if n_c is incident to S_l^γ; otherwise, let n_e be the last boundary node of S_l^γ that P_{bc} meets. Also, let P_{ec} be the (possibly trivial) path obtained by tracing P_{bc} from n_e to n_c. The rank of P_{ec} is also no larger than $\gamma+1$. We can invoke Conditions 3.2-1 to assert the existence of a (possibly trivial) two-ended path P_{de} of rank no larger than $\gamma + 1$ that connects

n_d to n_e, does not meet P_{ad} except terminally at n_d, and does not meet P_{ec} except terminally to n_e. If P_{ad} and P_{ec} do not meet — except possibly terminally when P_{de} is trivial, then $P_{ad} \cup P_{de} \cup P_{ec}$ is a two-ended path of rank no larger than $\gamma + 1$ that connects n_a to n_c.

On the other hand, if P_{ad} and P_{ec} do meet at a node not incident to S_l^γ, they will do so with at least one γ-tip. In this case, there will be a first node n_f at which P_{ad} meets P_{ec} — "first" with respect to a tracing of P_{ad} from n_a to n_d. Now, let P_{af} (and P_{fc}) be the path obtained by tracing P_{ab} from n_a to n_f (resp. by tracing P_{bc} from n_f to n_c). Then, $P_{af} \cup P_{fc}$ is the two-ended path of rank no larger than $\gamma + 1$ that we seek.

Case 4: Finally, assume that P_{ab} and P_{bc} never meet with only $(\gamma-)$-tips but do meet one or more times with at least one γ-tip. Then, as in the preceding paragraph, there is a first node n_f — "first" with respect to a tracing of P_{ab} from n_a to n_b — at which P_{ab} and P_{bc} meet with at least one γ-tip. With P_{af} and P_{fc} defined as before, $P_{af} \cup P_{fc}$ is the path we seek.

This exhausts all possibilities and completes the proof. ♣

Corollary 3.2-3. *If Conditions 3.2-1 hold and if θ is any rank such that $1 \leq \gamma+1 < \theta \leq \nu$, where again γ is a natural number, then any θ-section is partitioned by the $(\gamma + 1)$-sections in it. Similarly, $\mathcal{G}^\nu$ is partitioned by its $(\gamma + 1)$-sections.*

This corollary follows directly from the transitivity of $(\gamma+1)$-connectedness. Note also that 0-sections always partition θ-sections and $\mathcal{G}^\nu$ too because 0-connectedness is always transitive.

Our results extend directly to the ranks $\vec{\omega}$ and ω as follows. The needed arguments are virtually the same as those already given.

Corollary 3.2-4. *Let $\nu \geq \vec{\omega}$. Also, let Conditions 3.2-1 hold for every natural number μ (thus, for every natural number γ). Then, $\vec{\omega}$-connectedness is transitive for both branches and nodes, and $\vec{\omega}$-sections partition any θ-section, whenever $\theta > \vec{\omega}$, and $\mathcal{G}^\nu$ as well.*

Corollary 3.2-5. *Let $\nu \geq \omega$. Also, assume that Conditions 3.2-1 continue to hold when μ is replaced by $\vec{\omega}$ (and thus $\nu > \vec{\omega}$). Then, ω-connectedness is transitive for both branches and nodes, and ω-sections partition any θ-section, whenever $\theta > \omega$, and $\mathcal{G}^\nu$ as well.*

Corollary 3.2-6. *If Conditions 3.2-1 hold for μ replaced by ν $(\nu \leq \omega)$, then the components of $\mathcal{G}^\nu$ are totally disjoint.*

Transitivity criteria for ranks higher than ω can be obtained by repeating the above with appropriate changes in notations.

3.3 About Nodes and Paths

The nontransitivity of 1-connectedness exhibited in Figures 3.1 and 3.2 arises because two totally disjoint 1-nodes, such as n_a^1 and n_b^1 in Figure 3.1, embrace 0-tips having representatives that meet infinitely often. Similar examples exhibit this kind of pathology for λ-connectedness ($\lambda > 1$). In the last section we restored transitivity for λ-connectedness by assuming that every two such nodes are λ-connected through certain paths. On the other hand, another way of restoring λ-connectedness suggests itself: Let us call two tips of any ranks "nondisconnectable " if they have representatives that meet infinitely often. (A more precise definition is given below.) Perhaps the requirement that every two nondisconnectable tips be shorted together will also restore transitivity for λ-connectedness. Actually, a weaker version of this is sufficient — as we shall show. We can infer from all this that the only cause of such nontransitivity is the presence of unrestricted nondisconnectable tips. (Conditions 3.2-1 provide another way of restoring the said transitivity.)

We shall establish this in Section 3.5, but to do so we shall need some preliminary results concerning the structure of the set of nodes at which two transfinite paths meet. These are obtained in this and the next section.

Recall that, for any natural number γ, a representative of a γ-tip is a one-ended γ-path of the form:

$$P^\gamma = \{n_0^\delta, P_0^{\alpha_0}, n_1^\gamma, P_1^{\alpha_1}, n_2^\gamma, P_2^{\alpha_2}, \ldots\} \qquad (3.1)$$

where $\delta \leq \gamma$, $0 \leq \alpha_m < \gamma$, and certain conditions are satisfied (see Section 2.2). Similarly, a representative of an $\vec{\omega}$-tip is a one-ended $\vec{\omega}$-path of the form:

$$P^{\vec{\omega}} = \{n_0^{\gamma_0}, P_0^{\alpha_0}, n_1^{\gamma_1}, P_1^{\alpha_1}, n_2^{\gamma_2}, P_2^{\alpha_2}, \ldots\} \qquad (3.2)$$

where the γ_k are natural numbers with $\gamma_0 < \gamma_1 < \gamma_2 < \cdots$, $0 \leq \alpha_m < \gamma_{m+1}$ and again certain conditions are satisfied (see Section 2.3).

Now consider an infinite sequence of nodes $\{m_1, m_2, m_3, \ldots\}$ of possibly differing ranks. We shall say that the m_l *approach* a γ-tip t^γ (resp. an $\vec{\omega}$-tip $t^{\vec{\omega}}$) if there is a representative (3.1) for t^γ (resp. (3.2) for $t^{\vec{\omega}}$) such that, for each natural number i, all but finitely many of the m_l are shorted to nodes embraced by the members of (3.1) (resp. (3.2)) lying to the right of n_i^γ (resp. $n_i^{\gamma_i}$). We also say that those nodes lie *beyond* n_i^γ (resp. $n_i^{\gamma_i}$) and that the m_l *approach* any node that is shorted to t^γ (resp. $t^{\vec{\omega}}$).

Let t_a and t_b be two tips, not necessarily of the same rank. We say that t_a and t_b are *nondisconnectable* if there is an infinite sequence of nodes that approach both t_a and t_b. On the other hand, t_a and t_b are called *disconnectable* if they are not nondisconnectable; the idea behind this is that t_a and t_b can be isolated from each other without destroying them by removing certain branches.

Example 3.3-1. Let us point to some instances of these two ideas. In Figure 3.1, the infinitely many 0-tips of the 0-section induced by all the a_j and b_j branches are pairwise nondisconnectable. In the ladder network of Figure 2.2, the two 0-tips, one having a representative along the a_j branches and the other having a representative along the b_j branches, are disconnectable; except for this one pair of 0-tips, all of the infinitely many 0-tips of that ladder are pairwise nondisconnectable. Similarly, in Figure 2.8 the 1-tip with a representative along the upper horizontal branches is disconnectable from the 0-tip with a representative along the lower horizontal branches; all other tips of any ranks are pairwise nondisconnectable. Finally, all the 0-tips of the infinite binary tree of Figure 2.1 are pairwise disconnectable.

As always, μ denotes a natural number.

Lemma 3.3-2. *Let*

$$\{m_1,\ m_2,\ m_3,\ \ldots\} \tag{3.3}$$

be an infinite sequence of nodes in a μ-graph $\mathcal{G}^\mu$, and let P^λ be an oriented two-ended λ-path in $\mathcal{G}^\mu$ ($\lambda \leq \mu$) that meets those nodes in the order given. (P^λ may meet other nodes as well.) Then, P^λ has in it a one-ended ρ-path R^ρ, where $\rho < \lambda$, such that R^ρ meets all of the m_l except possibly finitely many of them, R^ρ is a representative of a ρ-tip t^ρ traversed (and therefore embraced) by P^λ, and the m_l approach t^ρ.

Proof. For any two consecutive nodes m_l and m_{l+1}, there is in P^λ a two-ended ρ_l-path $Q_l^{\rho_l}$ ($\rho_l \leq \lambda$) that terminates at nodes shorted to m_l and m_{l+1}; that is, $Q_l^{\rho_l}$ starts at m_l and stops at m_{l+1}. ($Q_l^{\rho_l}$ need not embrace m_l and m_{l+1}.) Let ρ be the largest of the values ρ_l for which there is an infinity of $Q_l^{\rho_l}$ with that rank ρ. We must have that $\rho < \lambda$; indeed, since P^λ is a two-ended λ-path, it can embrace only finitely many nodes of rank λ, and therefore only finitely many of the $Q_l^{\rho_l}$ can be of rank λ. Furthermore, there will be only finitely many paths $Q_l^{\rho_l}$ whose ranks are larger than ρ. So, by choosing l_0 large enough, we can ensure that all $Q_l^{\rho_l}$ with $l \geq l_0$ have $\rho_l \leq \rho$. Infinitely many of the $Q_l^{\rho_l}$ with $l \geq l_0$ will have $\rho_l = \rho$.

The path R^ρ induced by all the branches embraced by all the $Q_l^{\rho_l}$ with $l \geq l_0$ must be of rank ρ. In fact, R^ρ is the one-ended ρ-path that we seek. Indeed, R^ρ clearly meets all except perhaps finitely many of the m_l. Moreover, since R^ρ is in P^λ, P^λ traverses the ρ-tip t^ρ that has R^ρ as a representative. Also, since P^λ is a two-ended path, we can invoke Lemma 2.2-4 to conclude that P^λ embraces t^ρ. Finally, the m_l obviously approach t^ρ. ♣

Lemma 3.3-3. *Let (3.3) be an infinite sequence of nodes in an ω-graph $\mathcal{G}^\omega$, and let P^λ be an oriented two-ended λ-path in $\mathcal{G}^\omega$ ($\lambda \leq \omega$, $\lambda \neq \vec{\omega}$) that meets those nodes in the order given. Then, P^λ has in it a one-ended ρ-path R^ρ, where $\rho < \lambda$ (possibly, $\rho = \vec{\omega}$) such that R^ρ meets all of the m_l except possibly finitely many of them, R^ρ is a representative of a ρ-tip t^ρ traversed (and therefore embraced) by P^λ, and the m_l approach t^ρ.*

Proof. Let $Q_l^{\rho_l}$ and ρ_l be as in the preceding proof. All except perhaps finitely many of the $Q_l^{\rho_l}$ will have natural numbers as their ranks ρ_l, for otherwise P^λ would embrace an infinity of ω-nodes and therefore would not be a two-ended λ-path with $\lambda \leq \omega$.

Now, if all but finitely many of the ρ_l are bounded by some fixed natural number, we can proceed as in the preceding proof to find a representative R^ρ of a ρ-tip, as asserted in the conclusion, where ρ is a natural number less than λ.

So, assume that the ranks ρ_l that are natural numbers are not bounded by any fixed natural number. We can now choose l_0 so large that all $Q_l^{\rho_l}$ with $l \geq l_0$ will have natural numbers as their ranks ρ_l. We can find a representative $R^{\vec{\omega}}$ of an $\vec{\omega}$-tip embraced by P^λ as follows: Let R_{l_1} be a two-ended path in P^λ such that R_{l_1} starts at m_{l_0}, proceeds toward the m_l of higher indices $l > l_0$ but meets only finitely many of them, and has in it a Q_{l_1} of rank $\rho_{l_1} > \rho_{l_0}$. R_{l_1} exists because of the unboundedness of the ρ_l. Inductively, for $i = 2, 3, 4, \ldots$, let R_{l_i} be a two-ended path in P^λ such that R_{l_i} starts at $m_{l_{i-1}}$, proceeds toward the m_l of higher indices $l > l_{i-1}$ but meets only finitely many of them, and has in it a Q_{l_i} of rank $\rho_{l_i} > \rho_{l_{i-1}}$. R_{l_i} exists for the same reason. $R = \bigcup_{i=1}^{\infty} R_{l_i}$ is a one-ended $\vec{\omega}$-path, which uniquely determines an $\vec{\omega}$-tip $t^{\vec{\omega}}$. R meets all of the m_l except possibly finitely many of them. P^λ traverses $t^{\vec{\omega}}$ and also embraces it because P^λ is two-ended (Lemma 2.2-4 extended to ω-paths — as noted in Section 2.4). Finally, the m_l obviously approach t^ρ, as before. ♣

3.4 ν-sequences

The nodes embraced by a path of rank 1 or higher are arranged in a particular structure — a hierarchy of sequences of sequences, which we need to explicate. A complication that arises at this point is that a γ-node of a γ-path may be hidden in a λ-path P^λ of higher rank ($\lambda > \gamma$) by being embraced by a λ-node of P^λ. This leads to a confusion of representations for the elements of the hierarchy we wish to explore. To avoid this difficulty, we can replace every node of a given path by the maximal node that embraces it. In this way, each path determines a unique totally ordered set of maximal nodes, and two different paths cannot meet the same such set of maximal nodes. In the following discussion of a hierarchy of sequences, we will refer to the elements at the lowest level of the hierarchy as "nodes" and will interpret them as maximal nodes in some ν-graph (but this interpretation is not at all essential).

0-sequences:

As was suggested in Section 1.1, a *nontrivial* 0*-sequence*

$$s^0 = \{\ldots, n_m, n_{m+1}, \ldots\} \tag{3.4}$$

is an ordinary sequence with at least two members, that is, a nonvoid set of two or more nodes that are indexed by some or all of the integers m and are ordered according to those integers.

A 0-sequence may be two-ended, one-ended, or endless. The *trivial* 0-sequence is a singleton $\{n\}$. A 0-sequence is said to *terminate on the left (right)* when there is a leftmost (resp. rightmost) node in (3.4), and it is said to *extend infinitely leftward (rightward)* when there is no such leftmost (resp. rightmost) node. We say that s^0 *embraces* itself and all its nodes.

For every nontrivial 0-sequence s^0, one can construct a nontrivial 0-path P^0 by inserting a branch between every pair of adjacent nodes in s^0. Conversely, the maximal nodes met by a nontrivial 0-path comprise a nontrivial 0-sequence.

As always, $\mathcal{X}\backslash\mathcal{Y}$ denotes the set of elements in the set $\mathcal{X}$ that are not in the set $\mathcal{Y}$. Let $\mathcal{S}$ be a totally ordered set of nodes. A subset $\mathcal{Z}$ of $\mathcal{S}$ will be called a *contiguous* subset of $\mathcal{S}$ if there is no node of $\mathcal{S}\backslash\mathcal{Z}$ that lies both to the right of some node of $\mathcal{Z}$ and to the left of some other node of $\mathcal{Z}$. Let s^0 be a 0-sequence of some of the nodes of $\mathcal{S}$ with a compatible ordering. s^0 is called *maximal with respect to* $\mathcal{S}$ (or simply *maximal* when $\mathcal{S}$ is understood) if s^0 is a contiguous subset of $\mathcal{S}$ and if there is no nonvoid subset $\mathcal{N}$ of $\mathcal{S}\backslash s^0$ such that $\mathcal{N}\cup s^0$ is a contiguous subset of $\mathcal{S}$ and is a 0-sequence. In more suggestive terms, s^0 is maximal with respect to $\mathcal{S}$ if s^0 is contiguous and cannot be expanded into a larger 0-sequence by appending parts of $\mathcal{S}$ "adjacent" to s^0.

1-sequences:

A 1-*sequence*

$$s^1 = \{\ldots, s^0_m, s^0_{m+1}, \ldots\} \tag{3.5}$$

is a 0-sequence of 0-sequences s^0_m such that the following holds: For every two adjacent members s^0_m and s^0_{m+1} in s^1, either s^0_m extends infinitely rightward and s^0_{m+1} terminates on the left, or s^0_m terminates on the right and s^0_{m+1} extends infinitely leftward. Moreover, if (3.5) terminates on the left (or right), then the terminal 0-sequence also terminates on the left (resp. on the right).

Here, too, a 1-sequence may be two-ended, one-ended, or endless. If (3.5) has at least two members, the 1-sequence s^1 is called *nontrivial*. The *minimum rank* of a nontrivial 1-sequence is 1. If (3.5) has only one member (i.e., $s^1 = \{s^0\}$), we say that the *minimum rank* of s^1 is 0, and if in addition s^0 is also a singleton $\{n\}$, we call the 1-sequence $\{\{n\}\}$ *trivial*. (According to this terminology, there are 1-sequences that are neither nontrivial nor trivial, namely those of the form $\{s^0\}$, where s^0 is a nontrivial 0-sequence.)

When the 1-sequence (3.5) is nontrivial, s^0_m and s^0_{m+1} do not extend infinitely toward each other, nor do they terminate next to each other. To save words, we shall say that *infinite extensions are separated by nodes*, that the terminal node n_0 between s^0_m and s^0_{m+1} *abuts an infinite extension*, and that n_0 *separates* s^0_m and s^0_{m+1}. We also say that s^1 *embraces* itself, all its 0-sequences, and all the nodes of its 0-sequences.

Let $\mathcal{E}(s^1)$ denote the set of all nodes in all the 0-sequences in s^1, and endow $\mathcal{E}(s^1)$ with the total ordering induced by the orderings of s^1 and its 0-sequences. $\mathcal{E}(s^1)$ will be called the *elementary set* of s^1. Note also that each 0-sequence s^0_m in (3.5) is maximal with respect to $\mathcal{E}(s^1)$. If s^0_m is not a first or last maximal 0-sequence in s^1 and is not a singleton, then there are exactly two nodes that separate s_m^0 from all other maximal 0-sequences in s^1; otherwise, there is exactly one node doing so.

Example 3.4-1. A two-ended 1-sequence having four members is

$$\begin{aligned} s^1 &= \{s_1^0, s_2^0, s_3^0, s_4^0\} \\ &= \{\{n_1, n_2, n_3, \ldots\}, \{n_a, n_b, n_c, \ldots\}, \{n_x\}, \{\ldots, n_\alpha, n_\beta, n_\gamma\}\}. \end{aligned}$$

Here $s_3^0 = \{n_x\}$ is a trivial 0-sequence. Note that n_a and n_x are the nodes that abut infinite extensions and separate s_2^0 from the other s_m^0. Also, n_x separates $\{n_x\}$ from the other s_m^0. The elementary set for s^1 is

$$\mathcal{E}(s^1) = \{n_1, n_2, n_3, \ldots, n_a, n_b, n_c, \ldots, n_x, \ldots, n_\alpha, n_\beta, n_\gamma\}.$$

Also note that each s_i^0 is maximal with respect to $\mathcal{E}(s^1)$; that is, we cannot contiguously extend any s_i^0 within $\mathcal{E}(s^1)$ as a 0-sequence. ♣

For any given nontrivial 1-sequence s^1, let us imagine that a branch has been inserted between every two adjacent nodes embraced by $\mathcal{E}(s^1)$. This yields a nontrivial 1-path P^1. Indeed, it is a routine matter to check that all the conditions in the definition of such a 1-path are fulfilled. The nodes of $\mathcal{E}(s^1)$ that abut infinite extensions take the roles of the 1-nodes in P^1, and all other nodes of $\mathcal{E}(s^1)$ become the 0-nodes embraced by P^1. Also, distinct nodes in $\mathcal{E}(s^1)$ are taken to be totally disjoint nodes in P^1.

Conversely, given any nontrivial 1-path P^1 in a ν-graph $\mathcal{G}^\nu$, the maximal nodes in $\mathcal{G}^\nu$ that P^1 meets comprise the elementary set $\mathcal{E}(s^1)$ of a 1-sequence s^1. This fact follows from Lemma 2.2-2. Moreover, we can uniquely specify P^1 by specifying the said maximal nodes in $\mathcal{G}^\nu$ along with their total ordering — so long as a 1-path is truly obtained thereby.

Lemma 3.4-2. *Let $\mathcal{A}$ be an infinite subset of the elementary set $\mathcal{E}(s^1)$ of a 1-sequence s^1 given by (3.5) and endow $\mathcal{A}$ with the ordering induced by $\mathcal{E}(s^1)$. Assume that for every strictly increasing (resp. strictly decreasing) infinite sequence $\{a_i\}_{i=1}^\infty$ that is contained in the union of at most finitely many of the 0-sequences in (3.5), the set $\mathcal{A}^* = \{s \in \mathcal{A} \colon s > a_i \ \forall i\}$ is not void and has a minimum member a (resp. $\mathcal{A}^* = \{s \in \mathcal{A} \colon s < a_i \ \forall i\}$ is not void and has a maximum member a). Then, $\mathcal{A}$ is the elementary set of a 1-sequence (whose minimum rank may be 0).*

Note. The hypothesis concerning a can be restated as follows: Given $\{a_i\}_{i=1}^{\infty}$ as stated, there exists an $a \in \mathcal{A}$ with $a_i < a \leq s$ for all i and for all $s \in \mathcal{A}$ such that $s > a_i$ for all i (resp. there exists an $a \in \mathcal{A}$ with $s \leq a < a_i$ for all i and for all $s \in \mathcal{A}$ such that $s < a_i$ for all i). Note also that $\{a_i\}_{i=1}^{\infty}$ need not be one of the members of s^1.

Proof. $\mathcal{A}$ has the structure of a 0-sequence of 0-sequences (perhaps just a single 0-sequence alone) because $\mathcal{E}(s^1)$ has that structure. We first argue that in $\mathcal{A}$ infinite extensions are separated by nodes. Let A_m^0 and A_{m+1}^0 be any two adjacent maximal 0-sequences in $\mathcal{A}$. Assume that A_m^0 extends infinitely rightward. Let $\{a_i\}_{i=1}^{\infty}$ be a strictly increasing, infinite subsequence of A_m^0. By hypothesis, there exists an $a \in \mathcal{A}$ such that $a_i < a \leq s$ for all i and for all $s \in A_{m+1}^0$. It follows that $a \notin A_m^0$ and that a is a member of A_{m+1}^0 lying to the left of all other members of A_{m+1}^0. Hence, A_{m+1}^0 does not extend infinitely leftward. Its leftmost node a is the node we seek. A similar argument works when A_{m+1}^0 extends infinitely leftward.

Next, assume that $\mathcal{A}$ has a last (or first) 0-sequence. Then, the fact that $\mathcal{A}^*$ is not void whenever $\{a_i\}_{i=1}^{\infty}$ is chosen as stated insures that there is a last (resp. first) node in that 0-sequence. ♣

Let $\mathcal{S}$ be a totally ordered set of nodes and let s^1 be a 1-sequence such that $\mathcal{E}(s^1) \subset \mathcal{S}$ and the ordering of $\mathcal{E}(s^1)$ is compatible with that of $\mathcal{S}$. s^1 is called *maximal with respect to* $\mathcal{S}$ (or simply *maximal* when it is clear what $\mathcal{S}$ is) if there does not exist any nonvoid subset $\mathcal{N}$ of $\mathcal{S}\backslash\mathcal{E}(s^1)$ such that $\mathcal{N} \cup \mathcal{E}(s^1)$ with the ordering induced by $\mathcal{S}$ is a contiguous subset of $\mathcal{S}$ and is the elementary set of a 1-sequence.

μ-sequences:

A "2-sequence" can be defined as a 0-sequence of 1-sequences such that infinite extensions are separated by nodes. In fact, our definitions can be extended recursively to obtain a "μ-sequence" for any natural number μ. To this end, let us now assume that η-sequences have been defined for $\eta = 0, 1, \ldots, \mu - 1$, where $\mu \geq 2$. Consider a 0-sequence of $(\mu-1)$-sequences $s_m^{\mu-1}$.

$$s^{\mu} = \{\ldots, s_m^{\mu-1}, s_{m+1}^{\mu-1}, \ldots\} \tag{3.6}$$

(We allow s^{μ} to be a singleton.) By recursion each $s_m^{\mu-1}$ is a 0-sequence of $(\mu-2)$-sequences, which in turn are 0-sequences of $(\mu-3)$-sequences, and so forth down to 0-sequences of nodes. We shall say that s^{μ} *embraces* itself, all its members, all members of its members, and so on down to the said nodes. Let $\mathcal{E}(s^\mu)$ be the set of all nodes embraced by s^μ. We call $\mathcal{E}(s^\mu)$ the *elementary set* of s^μ. We assign to $\mathcal{E}(s^\mu)$ the total ordering induced by this recursive sequences-of-sequences structure.

Let $\mathcal{S}$ be a superset of $\mathcal{E}(s_m^{\mu-1})$, where $\mathcal{S}$ has a total ordering that is compatible with that of $\mathcal{E}(s_m^{\mu-1})$. For example, $\mathcal{S}$ may be $\mathcal{E}(s^\mu)$. $s_m^{\mu-1}$ is called *maximal with respect to* $\mathcal{S}$ if there does not exist any nonvoid subset $\mathcal{N}$ of $\mathcal{S}\backslash\mathcal{E}(s^{\mu-1})$ such that $\mathcal{N}\cup\mathcal{E}(s^{\mu-1})$ with the ordering induced by $\mathcal{S}$ is a contiguous subset of $\mathcal{S}$ and is the elementary set of a $(\mu-1)$-sequence. (So far, this definition has been explicated for $\mu-1=0$ or 1, and it will become explicitly defined for $\mu-1>1$ when we complete our recursive definitions.)

Furthermore, we shall say that $s_m^{\mu-1}$ *extends infinitely leftward (rightward)* if $s_m^{\mu-1}$ extends in that direction through an infinity of $(\mu-2)$-sequences that are maximal with respect to $\mathcal{E}(s_m^{\mu-1})$. On the other hand, we shall say that $s_m^{\mu-1}$ *terminates on the left (right) at a node* n_0 if there exists a node $n_0 \in \mathcal{E}(s^\mu)$ such that n_0 is embraced by $s_m^{\mu-1}$ and no other node embraced by $s_m^{\mu-1}$ lies to the left (right) of n_0. This occurs when and only when $s_m^{\mu-1}$ contains a leftmost (rightmost) maximal $(\mu-2)$-sequence, which in turn contains a leftmost (rightmost) maximal $(\mu-3)$-sequence, and so on down to a leftmost (rightmost) maximal 0-sequence, which terminates on the left (right) at n_0. n_0 is called a *terminal node* of $s_m^{\mu-1}$. As a particular case, all these leftmost (rightmost) sequences may be trivial sequences of the form $\{\cdots\{n_0\}\cdots\}$.

Here now is our definition of a "μ-sequence." For $\mu\geq 1$, a *μ-sequence* s^μ is a 0-sequence of $(\mu-1)$-sequences (as indicated in (3.6)) such that the following conditions hold:

Conditions 3.4-3. *For every two adjacent members $s_m^{\mu-1}$ and $s_{m+1}^{\mu-1}$ of s^μ, either $s_m^{\mu-1}$ extends infinitely rightward and $s_{m+1}^{\mu-1}$ terminates on the left at a node, or $s_m^{\mu-1}$ terminates on the right at a node and $s_{m+1}^{\mu-1}$ extends infinitely leftward. Moreover, if s^μ terminates on the left (or right), then, for each $\eta=0,\ldots,\mu-1$, s^μ embraces a leftmost (resp. rightmost) η-sequence which also terminates on the left (resp. on the right).*

This definition insures that each $s_m^{\mu-1}$ is truly maximal as a $(\mu-1)$-sequence with respect to $\mathcal{E}(s^\mu)$, which in turn insures that the representation (3.6) of s^μ is unique under Conditions 3.4-3. μ is the *rank* of s^μ. If (3.6) has two or more members, we call s^μ *nontrivial.*

As with paths, a μ-sequence is either two-ended or one-ended or endless. The statement that *infinite extensions embraced by s^μ are separated* *by nodes* will mean that Conditions 3.4-3 hold not only for s^μ but also for all maximal η-sequences embraced by s^μ, where $\eta = 1, \ldots, \mu - 1$. Moreover, any node that separates an infinite extension from its adjacent sequence of whatever rank will be said to *abut an infinite extension* and to *separate* maximal sequences at or to the left of it from maximal sequences at or to the right of it.

Example 3.4-4. An illustration of this structure for a 3-sequence is indicted in Figure 3.3. So as not to clutter the diagram too much, we have deleted many of the subscripts. In that diagram, s_1^2 is a singleton 2-sequence $\{s_1^1\}$. Both s_1^2 and s_1^1 terminate on the right at the node n_0, which is the sole member of the singleton 0-sequence s_1^0; n_0 separates s_1^2 and s_2^2, as well as other maximal sequences of lower ranks. For instance, according to our terminology, n_0 separates $\{n_0\}$ from all the other maximal 0-sequences. On the other hand, s_2^2 extends infinitely leftward. $\mathcal{E}(s^3)$ consists of the nodes at the lowest level of this diagram. Note that every infinite extension at that level has an abutting node. ♣

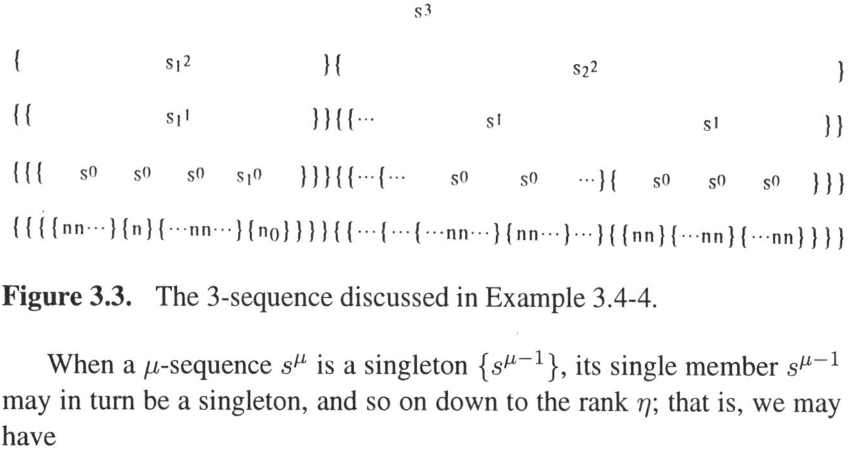

s3

{ s12 }{ s22 }

{{ s11 }}{{··· s1 s1 }}

{{{ s0 s0 s0 s10 }}}{{···{··· s0 s0 ···}{ s0 s0 s0 }}}

{{{{nn···}{n}{···nn···}{n0}}}}{{···{···{···nn···}{nn···}···}{{nn}{···nn}{···nn}}}}

Figure 3.3. The 3-sequence discussed in Example 3.4-4.

When a μ-sequence s^μ is a singleton $\{s^{\mu-1}\}$, its single member $s^{\mu-1}$ may in turn be a singleton, and so on down to the rank η; that is, we may have

$$\begin{aligned} s^\mu &= \{s^{\mu-1}\} = \{\{s^{\mu-2}\}\} = \cdots = \{\cdots\{s^\eta\}\cdots\} \\ &= \{\cdots\{\cdots, s_m^{\eta-1}, s_{m+1}^{\eta-1}, \cdots\}\cdots\}. \end{aligned}$$

The maximum natural number η, for which s^η has two or more members, will be called the *minimum rank* of s^μ. When s^μ embraces just one node (i.e., $s^\mu = \{\cdots\{n_0\}\cdots\}$), we have the *trivial* μ-sequence; we take its minimum rank to be 0. (Thus, there are μ-sequences that are neither trivial nor nontrivial.)

Given any nontrivial μ-sequence s^μ, let us imagine again that a branch has been inserted between every two adjacent nodes embraced by s^μ. It is easy to check that the result is a nontrivial μ-path P^μ. The embraced nodes of s^μ that abut infinite extensions take the role of the embraced α-nodes ($0 < \alpha \leq \mu$) of P^μ. For example, in Figure 3.3 the node n_0 plays the role of a 3-node between the 2-paths corresponding to s_1^2 and s_2^2; n_0 embraces a 0-tip on its left and a 2-tip on its right. Conversely, given any nontrivial μ-path P^μ in a ν-graph $\mathcal{G}^\nu$, the maximal nodes in $\mathcal{G}^\nu$ that P^μ meets comprise the elementary set $\mathcal{E}(s^\mu)$ of a nontrivial μ-sequence s^μ; this too is a consequence of Lemma 2.2-2. Moreover, a nontrivial μ-path in $\mathcal{G}^\mu$ can be uniquely specified by identifying the nontrivial μ-sequence of maximal nodes that the μ-path meets so long as a μ-path is in fact obtained that way.

Example 3.4-5. In some ν-graph let P^3 be a two-ended 3-path as follows:

$$P^3 = \{n_0^2, P_0^2, n_1^3, P_1^1, n_2^3, P_2^2, n_3^3, P_3^2, n_4^3\}.$$

Here,

$$P_0^2 = \{n_0^2, P_0^1, n_1^2, P_1^1, \ldots\}$$

is a one-ended 2-path with n_0^2 embracing a 1-tip of P_0^1 and n_1^3 embracing the 2-tip of P_0^2.

$$P_1^1 = \{n_a^1, P_a^0, n_b^1\}$$

is a two-ended 1-path with P_a^0 being an endless 0-path, n_1^3 embracing n_a^1 and n_2^3 embracing n_b^1.

$$P_2^2 = \{\ldots, n_\alpha^2, P_\alpha^1, n_\beta^2, P_\beta^1, \ldots\}$$

is an endless 2-path with n_2^3 embracing a 2-tip of P_2^2 and n_3^3 embracing the other 2-tip of P_2^2. Finally,

$$P_3^2 = \{\ldots, n_x^2, P_x^1, n_y^2, P_y^2, \ldots\}$$

is an endless 2-path whose 2-tips are embraced by n_3^3 and n_4^3. The set of maximal nodes that P^3 meets is the elementary set of a 3-sequence

$$s^3 = \{s_0^2, s_1^2, s_2^2, s_3^2, s_4^2, s_5^2\}$$

where the s_i^2 have elementary sets $\mathcal{E}(s_i^2)$ consisting of the maximal nodes met by the following paths: For s_0^2 we have P_0^2. For s_1^2 we have P_1^1; note that n_1^3 and n_a^1 are embraced by the same maximal node, and similarly for n_2^3

and n_b^1. For s_2^2 we have P_2^2. For s_3^2 we have the trivial sequence consisting only of the maximal node that embraces n_3^3. For s_4^2 we have P_3^2. Finally, for s_5^2 we have a trivial sequence again embracing n_4^3 alone.

Note also that we can reverse this discussion. Starting with s^3 we can insert branches between adjacent nodes in $\mathcal{E}(s^3)$ to obtain P^3. ♣

Lemma 3.4-6. *Let $\mathcal{A}$ be an infinite subset of the elementary set $\mathcal{E}(s^\mu)$ of a given μ-sequence s^μ (as in (3.6)), and endow $\mathcal{A}$ with the ordering induced by $\mathcal{E}(s^\mu)$. Assume that for every strictly increasing (resp. strictly decreasing) infinite sequence $\{a_i\}_{i=1}^\infty$, all of whose elements are embraced by at most finitely many of the $(\mu - 1)$-sequences in (3.6), the set $\mathcal{A}^* = \{s \in \mathcal{A} : s > a_i \ \forall i\}$ is nonvoid and has a minimum member a (resp. $\mathcal{A}^* = \{s \in \mathcal{A} : s < a_i \ \forall i\}$ is nonvoid and has a maximum member a). Then, $\mathcal{A}$ is the elementary set of a μ-sequence, whose minimum rank may be μ or less than μ.*

Proof. $\mathcal{A}$ will have the structure of a hierarchy of embraced sequences because $\mathcal{E}(s^\mu)$ has that structure. The rank of that hierarchy (that is, the number of levels within it minus one — see Figure 3.3) cannot be larger than μ. We first show that in $\mathcal{A}$ infinite extensions are separated by nodes. We can do so inductively. Arguing as in the proof of Lemma 3.4-2, we show that this is true for the maximal 0-sequences in $\mathcal{A}$. Next, for $\eta \leq \mu$, assume that this is true for all maximal ξ-sequences where $\xi = 0, \ldots, \eta - 2$. Let $A_m^{\eta-1}$ and $A_{m+1}^{\eta-1}$ be two adjacent maximal $(\eta - 1)$-sequences in $\mathcal{A}$. For definiteness, assume that $A_m^{\eta-1}$ extends infinitely rightward. Thus,

$$A_m^{\eta-1} = \{\ldots, A_i^{\eta-2}, A_{i+1}^{\eta-2}, \ldots\},$$

where the integers $i, i+1, \ldots$ extend infinitely rightward. Choose an $a_i \in \mathcal{E}(A_i^{\eta-2})$ for each $i \geq 1$. Thus, $\{a_i\}_{i=1}^\infty$ is a strictly increasing 0-sequence in $\mathcal{A}$. Let a be the node specified in the hypothesis. We can conclude that a is not embraced by $A_m^{\eta-1}$ and therefore must be embraced by $A_{m+1}^{\eta-1}$. Moreover, since $a \leq s$ for all $s \in A_{m+1}^{\eta-1}$, $A_{m+1}^{\eta-1}$ terminates on the left at a. This shows that the infinite extensions of the $(\eta - 1)$-sequences in $\mathcal{A}$ are separated by nodes. By induction this is so for all $\eta = 0, \ldots, \mu$.

Next, assume that $\mathcal{A}$ has a last (or first) $(\mu - 1)$-sequence; its minimum rank may be less than $\mu - 1$. Then, the fact that $\mathcal{A}^*$ is not void whenever $\{a_i\}_{i=1}^\infty$ is chosen as stated insures that s^μ embraces a last (resp. first) η-sequence for each $\eta = 0, \ldots, \mu - 2$, and in addition the last (resp. first) 0-sequence has a last (resp. first) node. ♣

$\vec{\omega}$-sequences:

Consider the following one-ended 0-sequence

$$s^{\vec{\omega}} = \{s_0^{\mu_0}, s_1^{\mu_1}, s_2^{\mu_2}, \ldots\} \tag{3.7}$$

of nontrivial μ_m-sequences $s_m^{\mu_m}$ of varying minimum ranks μ_m. The words and notations: "embraces," "elementary set $\mathcal{E}(s^{\vec{\omega}})$," "$s_m^{\mu_m}$ extends infinitely leftward (rightward)," "$s_m^{\mu_m}$ terminates on the left (right)," and "a node abuts an infinite extension and separates maximal sequences" are defined exactly as they were for μ-sequences except that now μ is replaced by $\vec{\omega}$ and $s_m^{\mu-1}$ by $s_m^{\mu_m}$. In the same way, we speak of $s_m^{\mu_m}$ being "maximal with respect to some superset $\mathcal{S}$ of $\mathcal{E}(s_m^{\mu_m})$," with the understanding that $\mathcal{S}$ has a compatible ordering; for example, $\mathcal{S}$ may be $\mathcal{E}(s^{\vec{\omega}})$.

A *rightward $\vec{\omega}$-sequence* is a one-ended 0-sequence of the form (3.7), wherein the μ_k are minimum ranks satisfying $\max(\mu_0, \ldots, \mu_m) \to \infty$ as $m \to \infty$, and moreover where every two adjacent members $s_m^{\mu_m}$ and $s_{m+1}^{\mu_{m+1}}$ $(m \geq 0)$ satisfy Conditions 3.4-3 with $s_m^{\mu-1}$ replaced by $s_m^{\mu_m}$ and $s_{m+1}^{\mu-1}$ by $s_{m+1}^{\mu_{m+1}}$.

A *leftward $\vec{\omega}$-sequence* is a one-ended 0-sequence:

$$s^{\vec{\omega}} = \{\ldots, s_{-3}^{\mu_{-3}}, s_{-2}^{\mu_{-2}}, s_{-1}^{\mu_{-1}}\} \tag{3.8}$$

of μ_{-m}-sequences, where now the μ_{-m} are minimum ranks satisfying $\max(\mu_{-1}, \ldots, \mu_{-m}) \to \infty$ as $m \to \infty$, and moreover where every two adjacent members $s_{-m}^{\mu_{-m}}$ and $s_{-m+1}^{\mu_{-m+1}}$ $(m \geq 2)$ satisfy Conditions 3.4-3 with $s_m^{\mu-1}$ replaced by $s_{-m}^{\mu_{-m}}$ and $s_{m+1}^{\mu-1}$ by $s_{-m+1}^{\mu_{-m+1}}$.

Finally, an *endless $\vec{\omega}$-sequence* is the conjunction of a leftward $\vec{\omega}$-sequence and a rightward $\vec{\omega}$-sequence:

$$s^{\vec{\omega}} = \{\ldots, s_{-1}^{\mu_{-1}}, s_0^{\mu_0}, s_1^{\mu_1}, \ldots\}.$$

Here, the leftward part (3.7) and rightward part (3.8) of this endless 0-sequence satisfy the corresponding conditions given above. Moreover, $s_{-1}^{\mu_{-1}}$ and $s_0^{\mu_0}$ satisfy Conditions 3.4-3 with $s_m^{\mu-1}$ replaced by $s_{-1}^{\mu_{-1}}$ and $s_{m+1}^{\mu-1}$ by $s_0^{\mu_0}$.

Altogether then, an *$\vec{\omega}$-sequence* is one of these three kinds of sequences. Note that an $\vec{\omega}$-sequence is always an infinite 0-sequence of sequences — never a finite one. Thus, every $\vec{\omega}$-sequence is *nontrivial*. Moreover, its *minimum rank* is taken to be $\vec{\omega}$.

A lemma just like that of Lemma 3.4-6 can be stated for the elementary set $\mathcal{E}(\vec{\omega})$ of an $\vec{\omega}$-sequence, but we won't need it.

Here, too, we can relate $\vec{\omega}$-sequences to the maximal nodes in a ν-graph that an $\vec{\omega}$-path meets. For instance, an $\vec{\omega}$-sequence becomes an $\vec{\omega}$-path when branches are connected between adjacent nodes in the $\vec{\omega}$-sequence. (There is an unimportant variation between the definitions of $\vec{\omega}$-sequences and $\vec{\omega}$-paths: For $\vec{\omega}$-paths, the ranks μ_m are required to be strictly monotone for $m \geq 0$ and also for $m < 0$. However, by combining adjacent sequences in $s^{\vec{\omega}}$ appropriately, we can get the needed monotonicities in the ranks.)

ω-sequences:

Finally, consider a (finite, one-ended, or endless) 0-sequence of the form

$$s^{\omega} = \{\ldots, s_m^{\rho_m}, s_{m+1}^{\rho_{m+1}}, \ldots\} \tag{3.9}$$

where each $s_m^{\rho_m}$ is a ρ_m-sequence of minimum rank ρ_m which is either a natural number or $\vec{\omega}$. Again the definitions of "embrace," "elementary set $\mathcal{E}(s^{\omega})$," "maximal member $s_m^{\rho_m}$ with respect to $\mathcal{E}(s^{\omega})$," "$s_m^{\rho_m}$ terminates on the left (right) at a node," and "a node abuts an infinite extension and separates maximal sequences" read exactly as they do for μ-sequences except for changes in notation. For instance, μ is replaced by ω and $\mu - 1$ by ρ_m. On the other hand, when $\rho_m = \vec{\omega}$, the phrase "$s_m^{\rho_m}$ extends infinitely leftward (rightward)" will now mean that $s_m^{\rho_m}$ is either a leftward (rightward) $\vec{\omega}$-sequence or an endless $\vec{\omega}$-sequence.

An *ω-sequence* is a 0-sequence of the form (3.9) such that every two adjacent members $s_m^{\rho_m}$ and $s_{m+1}^{\rho_{m+1}}$ satisfy Conditions 3.4-3 with $s_m^{\mu-1}$ replaced by $s_m^{\rho_m}$ and $s_{m+1}^{\mu-1}$ by $s_{m+1}^{\rho_{m+1}}$ and with the just-mentioned interpretation when either ρ_m or ρ_{m+1} is $\vec{\omega}$.

As a special case, an ω-sequence may be a singleton whose minimum rank is a natural number. An ω-sequence is *nontrivial* if it contains at least one $\vec{\omega}$-sequence. Also, as before, a nontrivial ω-sequence can be related to the set of maximal nodes in a ν-graph that a nontrivial ω-path meets.

The same proof as that for Lemmas 3.4-6 yields the following:

Lemma 3.4-7. *Let $\mathcal{A}$ be an infinite subset of the elementary set $\mathcal{E}(s^{\omega})$ of a given ω-sequence s^{ω}, and endow $\mathcal{A}$ with the ordering induced by $\mathcal{E}(s^{\omega})$. Assume that for every strictly increasing (resp. strictly decreasing) infinite sequence $\{a_i\}_{i=1}^{\infty}$, all of whose elements are embraced by at most finitely many of the members of s^{ω}, the set $\{s \in \mathcal{A}: s > a_i \ \forall i\}$ is nonvoid and has a minimum member a (resp. the set $\{s \in \mathcal{A}: s < a_i \ \forall i\}$ is nonvoid and* *has a maximum member a). Then, $\mathcal{A}$ is the elementary set of a ρ-sequence, where ρ is either a natural number or $\vec{\omega}$ or ω.*

The next lemma is the result we have been aiming for in this and the last section. In this lemma each of the ranks ρ and ζ is either a natural number or ω.

Lemma 3.4-8. *Let P^ρ be a two-ended ρ-path and let Q^ζ be a two-ended ζ-path in a ν-graph $\mathcal{G}^\nu$. Let t^{ρ_1} denote any one of the tips traversed by P^ρ (thus, $\rho_1 < \rho$ and possibly $\rho_1 = \vec{\omega}$), and let t^{ζ_1} denote any any one of the tips traversed by Q^ζ (thus, $\zeta_1 < \zeta$ and possibly $\zeta = \vec{\omega}$). Assume that t^{ρ_1} and t^{ζ_1} are shorted together whenever they are nondisconnectable. Let $\{n_i\}_{i\in I}$ be the set of maximal nodes that P^ρ and Q^ζ both meet, and let $\{n_i\}_{i\in I}$ have the total ordering induced by P^ρ. Then, $\{n_i\}_{i\in I}$ is the elementary set of a ξ-sequence, where $\xi \leq \rho$.*

Proof. If $\{n_i\}_{i\in I}$ is a finite set, it is the elementary set of a 0-sequence. So, assume that $\{n_i\}_{i\in I}$ is an infinite set. By appropriately choosing the orientation of P^ρ, we can make at least a part of $\{n_i\}_{i\in I}$ extend infinitely rightward. Choose any strictly increasing sequence $\{n_{i_j}\}_{j=1}^\infty$ in $\{n_i\}_{i\in I}$. Set $m_1 = n_{i_1}$. Now, starting at n_{i_1}, trace along Q^ζ. In at least one of the two possible directions of tracing Q^ζ from n_{i_1}, Q^ζ will meet an infinity of the n_{i_j}. Choose such a direction. In accordance with that direction of tracing, let m_2 be the first node in $\{n_{i_j}\}_{j=1}^\infty$ after m_1 that Q^ζ meets. (See Figure 3.4.) More generally, for each integer $l > 1$, let m_l be the first node in

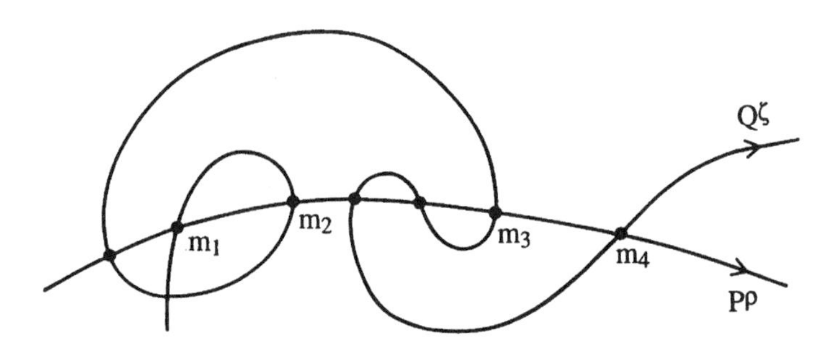

Figure 3.4. Illustration for the proof of Lemma 3.4-8.

$\{n_{i_j}\}_{j=1}^\infty$ after m_{l-1} that Q^ζ meets when tracing along Q^ζ. Then, $\{m_l\}_{l=1}^\infty$ is a strictly increasing sequence in $\{n_i\}_{i\in I}$ such that both P^ρ and Q^ζ meet the m_l in the order given. Also, since P^ρ and Q^ζ are two-ended, neither ρ nor ζ can be $\vec{\omega}$. By Lemmas 3.3-2 and 3.3-3, P^ρ traverses and embraces a ρ_1-tip t^{ρ_1} ($\rho_1 < \rho$) with a representative that meets all but possibly finitely many of the m_l; moreover, the m_l approach t^{ρ_1}. By the same lemmas, Q^ζ traverses and embraces a ζ_1-tip τ^{ζ_1} ($\zeta_1 < \zeta$) with a representative that meets all but possibly finitely many of the m_l; also, the m_l approach τ^{ζ_1}. Thus, t^{ρ_1} and τ^{ζ_1} are nondisconnectable. Hence, they are shorted together, and the maximal node n_x that shorts them is met by both P^ρ and Q^ζ according to Lemma 2.2-4 and its extension to two-ended ω-paths. Thus, we have that $n_x = n_i$ for some i, $n_{i_j} < n_x$ for all j, and $n_x \leq s$ for all s in $\{n_i\}_{i \in I}$ such that $s > n_{i_j}$ for all j. That is, the set $\{s \in \{n_i\}_{i \in I} : s > n_{i_j} \ \forall j\}$ has a minimum member n_x. (The analogous conclusion would hold had $\{n_{i_j}\}_{j=1}^{\infty}$ been chosen strictly decreasing.) Finally, recall that the set of all maximal nodes that P^ρ meets is the elementary set of a ρ-sequence ($\rho = \mu$ or ω). Thus, the hypothesis of Lemma 3.4-2 or 3.4-6 or 3.4-7 holds with resp. $\mathcal{E}(s^1)$ or $\mathcal{E}(s^\mu)$ or $\mathcal{E}(s^\omega)$ being the set of all maximal nodes that P^ρ meets, with $\mathcal{A}$ being the set $\{n_i\}_{i \in I}$, with $\{a_i\}_{i=1}^{\infty}$ being $\{n_{i_j}\}_{j=1}^{\infty}$, and with n_x being a. (The m_l were only used to find n_x.) By those lemmas, we can conclude that $\{n_i\}_{i \in I}$ is a ξ-sequence with $\xi \leq \rho$. ♣

3.5 Another Transitivity Criterion for Transfinite Connectedness

In this section we shall show that, under the forthcoming Condition 3.5-1, the ρ-sections (for a given rank $\rho \leq \omega$) partition a transfinite graph $\mathcal{G}^\nu$ ($\rho \leq \nu$) because ρ-connectedness is then transitive. Recall that two tips are shorted together if and only if they are embraced by the same node. Moreover, a tip is open if and only if it is embraced by only one node and that node is a singleton.

Condition 3.5-1. *If two tips (of ranks less than ν and possibly differing) are nondisconnectable, then either the two tips are shorted together or at least one of them is open.*

Recall also that μ denotes a natural number.

Theorem 3.5-2. *Let $\mathcal{G}^\nu$ be a ν-graph for which Condition 3.5-1 is satisfied. Let n_a, n_b, and n_c be totally disjoint nonsingleton nodes (possibly with different ranks) in $\mathcal{G}^\nu$ such that n_a and n_b are μ-connected and that n_b*

and n_c are μ-connected. Then, n_a and n_c are μ-connected.

Proof. There is a two-ended ρ-path P^ρ that terminates at n_a and n_b, and there is a two-ended ζ-path Q^ζ that terminates at n_b and n_c, where $\rho \leq \mu$ and $\zeta \leq \mu$. Let $\{n_i\}_{i\in I}$ be the maximal nodes met by both P^ρ and Q^ζ (that is, each n_i is embraced by both paths), but let $\{n_i\}_{i\in I}$ be totally ordered in accordance with a tracing of P^ρ from n_b to n_a. The node n_b is embraced by a maximal node in $\{n_i\}_{i\in I}$, which we shall also denote by n_b.

If $\{n_i\}_{i\in I}$ is a finite set, there will be a last node n_x in it. Then, a tracing of P^ρ from n_a to n_x followed by a tracing of Q^ζ from n_x to n_c will yield a two-ended λ-path terminating at n_a and n_c, where $\lambda \leq \max\{\rho, \zeta\} \leq \mu$. Hence, n_a and n_c are μ-connected.

Now, assume that $\{n_i\}_{i\in I}$ is an infinite set. No tip traversed by P^ρ (or Q^ζ) can be open. Indeed, that tip will also be embraced by P^ρ (or Q^ζ) according to Lemma 2.2-4, and every nonterminal node embraced by P^ρ (or Q^ζ) will be a nonsingleton. Moreover, the terminal nodes of P^ρ (or Q^ζ) are nonsingletons by hypothesis. Hence, by Condition 3.5-1, any tip traversed by P^ρ and any tip traversed by Q^ζ that are nondisconnectable will be shorted together. Therefore, by Lemma 3.4-8, $\{n_i\}_{i\in I}$ is the elementary set of an η-sequence ($\eta \leq \mu$).

Thus, either $\{n_i\}_{i\in I}$ has a last node n_x or it (that is, the said η-sequence) extends infinitely rightward through an infinite sequence $\{s_l^{\eta-1}\}_{l=1}^{\infty}$ of maximal $(\eta - 1)$-sequences $s_l^{\eta-1}$. (Here, $s_l^{\eta-1}$ is a node if $\eta = 0$.) Suppose $\{n_i\}_{i\in I}$ does not have a last member, that is, it extends infinitely rightward as stated. For each natural number l, choose a node m_l that is embraced by $s_l^{\eta-1}$. Thus, $\{m_l\}_{l=1}^{\infty}$ is a sequence in $\{n_i\}_{i\in I}$ such that no node of $\{n_i\}_{i\in I}$ lies to the right of all the m_l. As in the proof of Lemma 3.4-8, we can choose a subsequence $\{m_{l_k}\}_{k=1}^{\infty}$ of $\{m_l\}_{l=1}^{\infty}$ such that Q^ζ meets the m_{l_k} in sequence, that is, in the same order that P^ρ meets the m_{l_k}. So, by Lemma 3.3-2, P^ρ traverses a tip t^ρ and Q^ζ traverses a tip τ^ζ such that the m_{l_k} approach both tips. Hence, t^ρ and τ^ζ are nondisconnectable. We have already noted that neither of them is open. By Condition 3.5-1, there is a node n_x that embraces both of them. Moreover, $n_x \in \{n_i\}_{i\in I}$. Indeed, P^ρ, being a two-ended path, embraces every tip it traverses (Lemma 2.2-4) and thus embraces a node that embraces such a tip; the same is true for Q^ζ.

Finally, n_x lies to the right of all the m_{l_k}, therefore to the right of all the m_l, and therefore to the right of all the nodes in $\{n_i\}_{i\in I}$, according to our supposition that $\{n_i\}_{i\in I}$ extends infinitely rightward. This is a contradiction, and thus our supposition is invalid. It follows that $\{n_i\}_{i \in I}$ does have a last member n_x. We can now conclude as before that n_a and n_c are μ-connected. ♣

Corollary 3.5-3. *Theorem 3.5-2 remains true when μ is replaced either by $\vec{\omega}$ or ω.*

Proof. By the definition of $\vec{\omega}$-connectedness, there is a μ such that n_a and n_b are μ-connected and also that n_b and n_c are μ-connected. By Theorem 3.5-2, n_a and n_c are μ-connected and therefore $\vec{\omega}$-connected.

When μ is replaced by ω, the proof that n_a and n_c are ω-connected is the same as that of Theorem 3.5-2 except for some obvious modifications in wording and notations, the use of Lemma 3.3-3 in place of Lemma 3.3-2, and the following additional alteration: When $\{n_i\}_{i \in I}$ is taken to extend infinitely rightward, it may do so either through an infinite sequence $\{s_l^{\eta-1}\}_{l=1}^{\infty}$ of $(\eta-1)$-sequences $s_l^{\eta-1}$ as before or through an $\vec{\omega}$-sequence $s^{\vec{\omega}}$ such as (3.7). In the latter case, we choose each m_l to be a node embraced by $s_l^{\mu_l}$. ♣

Our results include the following, a version that we will cite later on.

Corollary 3.5-4. *Under Condition 3.5-1, let n_a, n_b, and n_c be totally disjoint nonsingleton nodes in a ν-graph $\mathcal{G}^\nu$. Let P_{ab}^ρ be a two-ended ρ-path connecting n_a and n_b and let P_{bc}^ζ be a two-ended ζ-path connecting n_b and n_c, where $\rho \le \omega$, $\zeta \le \omega$, $\rho \ne \vec{\omega}$, and $\zeta \ne \vec{\omega}$. Then, there is in $P_{ab}^\rho \cup P_{bc}^\zeta$ a two-ended λ-path P_{ac}^λ connecting n_a and n_c, where $\lambda \le \max\{\rho, \zeta\}$.*

Still another version is

Corollary 3.5-5. *Under Condition 3.5-1, let $\{n_i\}_{i \in I}$ be the set of all maximal nodes met by both of two two-ended paths (i.e., each n_i is embraced by both paths). Assume that the terminal nodes of both paths are nonsingletons. Assign to $\{n_i\}_{i \in I}$ the total ordering induced by an orientation of one of those two paths. Then, $\{n_i\}_{i \in I}$ has both a first node and a last node.*

Finally, we can conclude that ρ-sections do not overlap whenever Condition 3.5-1 is satisfied. This is subsumed under

Corollary 3.5-6. *Let $\mathcal{G}^\nu$ be a ν-graph that satisfies Condition 3.5-1. Let ρ denote either μ or $\vec{\omega}$ or ω. Then, the ρ-sections of $\mathcal{G}^\nu$ comprise a partition of $\mathcal{G}^\nu$. Also, for each $\lambda < \rho$, each ρ-section of $\mathcal{G}^\nu$ is partitioned by the λ-*

sections contained in that ρ-section. Finally, for $\nu \leq \omega$, the components of $\mathcal{G}^\nu$ are totally disjoint.

Proof. We need merely show that ρ-connectedness is an equivalence relation between the branches of $\mathcal{G}^\nu$. This will be so if ρ-connectedness is an equivalence relation for 0-nodes. Reflexivity and symmetry are obvious. As for transitivity, we can assume that all the 0-nodes are nonsingletons. Indeed, the appending of a self-loop to any 0-node will not alter the ρ-connectedness — or lack of it — between branches; hence, we may render all 0-nodes into nonsingletons in this way. It follows that we may invoke Theorem 3.5-2 or Corollary 3.5-3 to establish the transitivity of ρ-connectedness between branches. ♣

3.6 The Cardinality of the Branch Set

We turn now to a rather different result, one which emerges nevertheless from some ideas we have already discussed, namely sections, boundary nodes, and nondisconnectable tips. We shall derive a bound on the cardinality of the set of branches of a transfinite graph by limiting the number of sections that share boundary nodes with any given section. We will in fact extend the idea of the degree of a 0-node to a more general concept called the "adjacency degree" of a section, and, by restricting the latter, we will extend to transfinite graphs the standard result that a conventional connected graph has a countable branch set if each of its nodes is of countable degree [31, pages 39-40].

We assume throughout this section that Condition 3.5-1 holds. Hence, λ-connectedness in a ν-graph $\mathcal{G}^\nu$ ($\lambda \leq \nu \leq \omega$) is transitive, and λ-sections do not overlap; in fact, λ-sections partition $\mathcal{G}^\nu$. The condition $\nu \leq \omega$ is imposed only to conform with what we have specifically established, but, as was indicated in Section 2.5, our results also hold for ranks ν higher than ω.

Two ρ-sections $\mathcal{S}_1^\rho$ and $\mathcal{S}_2^\rho$ will be called $(\rho+1)$-*adjacent* if there is a $(\rho+1)$-node $n^{\rho+1}$ that embraces a tip of $\mathcal{S}_1^\rho$ and a tip of $\mathcal{S}_2^\rho$ with one of those tips having a rank of ρ and the other tip having a rank no larger than ρ.

Since $\mathcal{S}_1^\rho$ and $\mathcal{S}_2^\rho$ are different ρ-sections, $n^{\rho+1}$ must be a boundary node for both $\mathcal{S}_1^\rho$ and $\mathcal{S}_2^\rho$, with either condition (i) or (ii) of Lemma 3.1-2 holding – that is, $n^{\rho+1}$ is incident to at least one of $\mathcal{S}_1^\rho$ and $\mathcal{S}_2^\rho$ only through tips of ranks no less than ρ. It follows that $\mathcal{S}_1^\rho$ and $\mathcal{S}_2^\rho$ are both $(\rho+1)$-connected by a $(\rho+1)$-path passing through $n^{\rho+1}$ and are both in some single $(\rho+1)$-section. Furthermore, we have the possibility that, for $\gamma < \rho$, two γ-sections may be $(\rho+1)$-adjacent but not $(\lambda+1)$-adjacent for any λ such that $\gamma \leq \lambda < \rho$; indeed, since γ-connectedness implies ρ-connectedness, those γ-sections are also ρ-sections and may be $(\rho+1)$-adjacent but with no shared boundary $(\lambda+1)$-node fulfilling the conditions needed for them to be $(\lambda+1)$-adjacent. Thus, it is possible for a γ-section to have a void $(\gamma+1)$-adjacency.

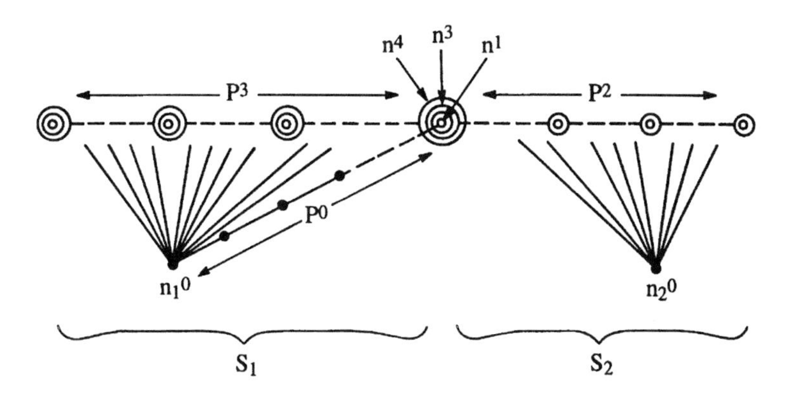

Figure 3.5. A 4-graph. The dots denote 0-nodes, the smallest circles are 1-nodes, the double circles are 2-nodes, the triple circles are 3-nodes, and the quadruple circle is a 4-node. That 4-node embraces a 3-node and a 1-node but neither a 2-node nor a 0-node. The P^μ $(\mu = 0, 2, 3)$ denote one-ended μ-paths. S_1 and S_2 are ρ-sections for each $\rho = 0, 1, 2$. This graph is discussed in Examples 3.6-1 and 4.1-2.

Example 3.6-1. Figure 3.5 illustrates a 4-graph. n^4 is the 4-node $\{n^3, t^3\}$, where t^3 is the 3-tip of the one-ended 3-path P^3. Also, n^3 is the 3-node $\{n^1, t^2\}$, where t^2 is the 2-tip of the one-ended 2-path P^2. Finally, n^1 is the singleton 1-node $\{t^0\}$, where t^0 is the 0-tip of the one-ended 0-path P^0. n^4 does not embrace a 2-node. Every 0-node embraced by P^3 (or P^2) is 0-adjacent to the 0-node n_1^0 (resp. n_2^0); that is, for each 0-node in the said path there is a branch that is incident to it and to n_1^0 (resp. n_2^0) as well. The subgraphs S_1 and S_2 to the left and to the right of n^4 are 0-connected in themselves and are connected to each other by a 3-path such as the one passing along P^0, then through n^3, and then along P^2. No path of rank lower than 3 connects $\mathcal{S}_1$ and $\mathcal{S}_2$. Therefore, $\mathcal{S}_1$ and $\mathcal{S}_2$ are ρ-sections for each of $\rho = 0, 1, 2$, but not for $\rho = 3, 4$. On the other hand, $\mathcal{S}_1 \cup \mathcal{S}_2$ is the entire 4-graph and is a 3-section as well. Note that even though $\mathcal{S}_1$ is a 0-section, it traverses a 3-tip. n^3 is a boundary node for both $\mathcal{S}_1$ and $\mathcal{S}_2$ as ρ-sections ($\rho = 0, 1, 2$), and so too is n^4; but, n^1 is not for either $\mathcal{S}_1$ or $\mathcal{S}_2$. Finally, as 2-sections, $\mathcal{S}_1$ and $\mathcal{S}_2$ are 3-adjacent because of the 3-node n^3. However, as 0-sections (or as 1-sections), $\mathcal{S}_1$ and $\mathcal{S}_2$ are not 1-adjacent (resp. not 2-adjacent). ♣

Given a ρ-section $\mathcal{S}^\rho$, the $(\rho+1)$*-adjacency of* $\mathcal{S}^\rho$ is the set $\mathcal{J}$ of all ρ-sections that are $(\rho+1)$-adjacent to $\mathcal{S}^\rho$. We allow $\rho = \vec{0}$; in this case, we have that $\rho + 1 = 0$, a $\vec{0}$-section is a single branch b, and its 0-adjacency is the set of all other branches that are 0-adjacent to b. The cardinality $\overline{\overline{\mathcal{J}}}$ of the $(\rho+1)$-adjacency $\mathcal{J}$ of $\mathcal{S}^\rho$ will be called the $(\rho+1)$*-adjacency degree of* $\mathcal{S}^\rho$. This can be viewed as a generalization of the degree of a 0-node whereby the 0-node and its incident branches are replaced by ρ-sections.

On the other hand, a generalization of "degree" that replaces the 0-node by a $(\rho+1)$-node $n^{\rho+1}$ $(\rho \geq 0)$ can be conceived as follows. The cardinality of the set of ρ-sections to which $n^{\rho+1}$ is incident through γ-tips, where $\gamma \leq \rho$, will be called the ρ*-sectional degree of* $n^{\rho+1}$. With $\mathcal{S}^\rho$ being a given ρ-section as before, let $\mathcal{M}$ be the set of boundary nodes for $\mathcal{S}^\rho$ and assume that the ρ-sectional degrees of all members of $\mathcal{M}$ are bounded by the cardinal number d. Then, the $(\rho+1)$-adjacency degree $\overline{\overline{\mathcal{J}}}$ of $\mathcal{S}^\rho$ can be bounded as follows:

$$\overline{\overline{\mathcal{J}}} \leq \overline{\overline{\mathcal{M}}} \cdot d.$$

Moreover, if either $\overline{\overline{\mathcal{M}}}$ or d is an infinite cardinal number, then $\overline{\overline{\mathcal{J}}} \leq \max\{\overline{\overline{\mathcal{M}}}, d\}$ [1, page 376] (see also Appendix A9). Thus, as a special case we have that if each boundary node of $\mathcal{S}^\rho$ has a countable ρ-sectional degree and if $\mathcal{M}$ itself is countable, then the adjacency $\mathcal{J}$ of $\mathcal{S}^\rho$ is countable too; but these two conditions are only sufficient — not necessary — for the countability of $\mathcal{J}$.

The principal result of this section is

Theorem 3.6-2. *Let the* ν*-graph* $\mathcal{G}^\nu$ $(\nu \leq \omega)$ *satisfy the following conditions:*

(a) $\mathcal{G}^\nu$ *is* ν*-connected.*

(b) *$\mathcal{G}^\nu$ satisfies Condition 3.5-1.*

(c) *There is an infinite cardinal number a that bounds the $(\rho+1)$-adjacency degrees of all ρ-sections for all ranks ρ such that $\vec{0} \le \rho < \nu$.*

Then, the branch set $\mathcal{B}$ of $\mathcal{G}^\nu$ has a cardinality no larger than a.

Proof. We shall use transfinite induction, working with ranks rather than ordinal numbers. First, choose any 0-section $\mathcal{S}^0$ and let b be any branch in $\mathcal{S}^0$. Since a $\vec{0}$-section is the same thing as a branch, hypothesis (c) asserts that the set of branches 0-adjacent to b has a cardinality no larger than a. Let $\mathcal{H}_0$ be the singleton branch set $\{b\}$. For each positive natural number k, let $\mathcal{H}_k$ be the set of branches each of which is 0-adjacent to a branch of $\mathcal{H}_{k-1}$. By hypothesis (c) and induction, $\overline{\overline{\mathcal{H}_k}} \le a^2 = a$ for every k. (For $a^2 = a$, see [1, page 376].) Since $\mathcal{S}^0$ is 0-connected, every two branches of $\mathcal{S}^0$ is connected by a two-ended 0-path in $\mathcal{S}^0$. Furthermore, no 0-path starting at b will leave $\mathcal{S}^0$. It follows that the branch set $\mathcal{B}(\mathcal{S}^0)$ of $\mathcal{S}^0$ is $\bigcup_{k=0}^{\infty} \mathcal{H}_k$. (The $\mathcal{H}_k$ may be void for all sufficiently large k.) Thus, $\overline{\overline{\mathcal{B}(\mathcal{S}^0)}} \le \aleph_0 \cdot a = a$.

Next, let μ be any natural number larger than 0 and let $\mathcal{S}^\mu$ be any μ-section. By hypothesis (b) and Corollary 3.5-6, $\mathcal{S}^\mu$ is partitioned by some or all of the $(\mu-1)$-sections of $\mathcal{G}^\nu$. For the inductive hypothesis, assume that the branch set of every $(\mu-1)$-section in $\mathcal{S}^\mu$ has a cardinality no larger than a. Let $\mathcal{S}^{\mu-1}$ be one of the $(\mu-1)$-sections in $\mathcal{S}^\mu$, but now let $\mathcal{H}_0$ be the singleton $\{\mathcal{S}^{\mu-1}\}$. For each positive natural number k, let $\mathcal{H}_k$ be the set of $(\mu-1)$-sections, each of which is μ-adjacent to a $(\mu-1)$-section of $\mathcal{H}_{k-1}$. By hypothesis (c), $\overline{\overline{\mathcal{H}_1}} \le a$. Since $\mathcal{S}^\mu$ is μ-connected, every two $(\mu-1)$-sections in $\mathcal{S}^\mu$ are connected through a two-ended μ-path in $\mathcal{S}^\mu$. Moreover, every two-ended μ-path meets only finitely many $(\mu-1)$-sections, and consecutively met $(\mu-1)$-sections are μ-adjacent. Also, no μ-path that meets $\mathcal{S}^\mu$ will ever leave $\mathcal{S}^\mu$. It follows that $\bigcup_{k=0}^{\infty} \mathcal{H}_k$ is the set of all $(\mu-1)$-sections in $\mathcal{S}^\mu$. (Again the $\mathcal{H}_k$ may be void for all sufficiently large k.) By hypothesis (c) and induction, $\overline{\overline{\mathcal{H}_k}} \le a^2 = a$ for every k. Hence, the set of all $(\mu-1)$-sections in $\mathcal{S}^\mu$ has a cardinality no larger than $\aleph_0 \cdot a = a$. This result coupled with our aforementioned inductive hypothesis implies that the branch set $\mathcal{B}(\mathcal{S}^\mu)$ of $\mathcal{S}^\mu$ satisfies $\overline{\overline{\mathcal{B}(\mathcal{S}^\mu)}} \le a^2 = a$ for every natural number $\mu \le \nu$. So, if $\mathcal{G}^\nu$ is a μ-graph, we have our conclusion.

Next, let $\mathcal{S}^{\vec{\omega}}$ be any $\vec{\omega}$-section in $\mathcal{G}^\nu$. Given any two branches b_1 and b_2 of $\mathcal{S}^{\vec{\omega}}$, there is for some natural number λ a two-ended λ-path that meets them. Moreover, for any two natural numbers γ and μ with $\gamma < \mu$, any $\mathcal{S}^\gamma$

in $\mathcal{S}^{\vec{\omega}}$ is contained in some $\mathcal{S}^{\mu}$ in $\mathcal{S}^{\vec{\omega}}$. Consequently,

$$\mathcal{S}^{\vec{\omega}} = \bigcup_{\gamma \leq \mu < \vec{\omega}} \mathcal{S}^{\mu}, \tag{3.10}$$

where $\mathcal{S}^{\gamma}$ is any arbitrarily chosen but then fixed γ-section in $\mathcal{S}^{\vec{\omega}}$ and where $\mathcal{S}^{\gamma} \subset \mathcal{S}^{\mu}$. By the preceding paragraph, $\overline{\overline{\mathcal{B}(\mathcal{S}^{\mu})}} \leq a$ for every $\mu \leq \vec{\omega}$. Hence, by (3.10), $\overline{\overline{\mathcal{B}(\mathcal{S}^{\vec{\omega}})}} \leq \aleph_0 \cdot a = a$. Thus, our conclusion holds for any $\vec{\omega}$-graph.

We can now repeat the argument of the penultimate paragraph with $\mathcal{S}^{\vec{\omega}}$ taking the role of $\mathcal{S}^{\mu-1}$ to conclude that, when $\nu = \omega$, the branch set $\mathcal{B}$ of $\mathcal{G}^{\nu}$ satisfies $\overline{\overline{\mathcal{B}}} \leq a$. ♣

As a special case we have

Corollary 3.6-3. *Let the ν-graph $\mathcal{G}^{\nu}$ $(\nu \leq \omega)$ be ν-connected and satisfy Condition 3.5-1. Also, assume that, for each ρ with $\vec{0} \leq \rho < \nu$, every ρ-section has a countable $(\rho + 1)$-adjacency. Then, $\mathcal{G}^{\nu}$ has a countable branch set.*

Chapter 4

Finitely Structured Transfinite Graphs

One of the objectives of this book is to develop a theory for transfinite random walks. More specifically, we wish to construct a random walk that may wander over a conventional infinite graph, then "pass through infinity," and wander over another conventional infinite graph, the "passage through infinity" occurring at a 1-node. This we do in Chapter 7 and in fact construct passages through nodes of higher ranks as well.

However, such results are achieved only for a limited class of transfinite graphs, those with structures that mimic local finiteness in conventional graphs. We shall refer to these as "finitely structured" graphs. They satisfy several restrictions such as the shorting together of nondisconnectable tips to obtain transitivity for transfinite connectedness, local finiteness for 0-nodes, and a generalization of local finiteness for nodes of higher ranks. The aim of this chapter is to develop the needed structure and to establish some preparatory lemmas for our subsequent theory of transfinite random walks.

One of those lemmas is an extension of Konig's classical lemma concerning the existence of a one-ended 0-path as a consequence of connectedness and local finiteness for a conventional infinite graph. Our result combines transfinite connectedness with our generalized form of local finiteness to obtain the existence of a one-ended transfinite path. Another idea we shall introduce is that of a "subsection," which is like a section but more restricted; it provides a finer partitioning of a transfinite graph. Still another idea is that of a "cut" for nodes of higher ranks; this is a certain set of branches that isolates a β-node from all other nodes of ranks no less than β. Our cuts will consist of only finitely many branches, and we will use them in Chapter 5 to apply Kirchhoff's current law to nodes of higher ranks, something that was not achieved in [35]. We also introduce the idea of a "contraction" of a sequence of cuts toward a node. These will be used when we develop our theory for transfinite random walks. Finally, we show that all finitely structured graphs have spanning trees, a property that is not possessed by transfinite graphs in general.

Example 4.0-1. It may be helpful at this point to consider heuristically one of the simplest kinds of finitely structured graphs. Such a 1-graph $\mathcal{G}^1$ can be obtained from any finite 0-graph $\mathcal{G}^0$ by replacing each branch b of $\mathcal{G}^0$ by an endless 0-path P^0. Thus, the elementary tips of b are replaced by the 0-tips of P^0, and every 0-node of $\mathcal{G}^0$ is replaced by a 1-node of $\mathcal{G}^1$. In this simple case, every 0-path of $\mathcal{G}^1$ is a "subsection" of "internal rank" 0. Moreover, given any 1-node n^1 of $\mathcal{G}^1$, we can select a branch from each 0-path incident to n^1 to obtain a "cut" for n^1 as the set of those selected branches. Finally, an infinite sequence of cuts for n^1 that move steadily closer to n^1 is essentially a "contraction" to n^1; actually, the "contraction" will be a sequence of sets of 0-nodes where each 0-node is incident to a branch of the said branch sets.

Note also that this process of generating simple finitely structured graphs can be continued: Take any finitely structured 1-graph of the simple type just described and replace each branch of it by an endless 0-path; then each 0-node becomes a 1-node, and each 1-node becomes a 2-node. The result will be a finitely structured 2-graph.

However, in the general case our finite structuring will be considerably more complicated than this. For example, the endless 0-paths will be replaced by finite or infinite 0-graphs, the 1-nodes may embrace 0-nodes, and a variety of restrictions will have to be imposed in order to prohibit various intractable complications. ♣

Perhaps we should also mention that several of our various assumptions for the general case could be simplified or eliminated by working with subsections that are connected together only through ladders or grids of finite width and moreover by disallowing embraced nodes. Something like this was in fact done in earlier versions of this work [36], [38]. Here, however, we aim for greater generality. Starting with transfinite graphs in general, we assume away the various structures that violate the idea of local finiteness — hence the several assumptions listed in Conditions 4.2-2 and 4.3-1.

As always, we take it that we are working with a particular ν-graph $\mathcal{G}^\nu$ whose rank ν is no larger than ω. (Nevertheless, the recursive techniques we are using do extend to ranks larger than ω; see Section 2.5.) Let us recall a notation regarding ranks that was introduced in Section 1.1. For any given rank β with $1 \leq \beta \leq \nu$, $\beta-$ and $\beta+$ will denote arbitrary and unspecified ranks such that $0 \leq (\beta-) < \beta \leq (\beta+) \leq \nu$. For example, $\omega-$ is any natural number or $\vec{\omega}$, and $\omega+$ is ω since ν is ω in this case. On the other hand, $\vec{\omega}-$ is any natural number, whereas $\vec{\omega}+$ is either $\vec{\omega}$ or ω. Two $(\beta-)$-nodes need not have the same rank, and similarly for two $(\beta+)$-nodes.

4.1 Subsections and Cores

Throughout this chapter we shall assume that Condition 3.5-1 (regarding the shorting of nondisconnectable tips) is satisfied. Our theory of transfinite random walks will be based upon the idea of a "subsection." To define it, choose and fix some rank β with $1 \leq \beta \leq \nu$, and then partition the branch set $\mathcal{B}$ by placing two branches in the same subset if they are $(\beta-)$-connected by a two-ended $(\beta-)$-path that does not meet any $(\beta+)$-node. Also, if a self-loop is incident to a $(\beta+)$-node, make that branch the sole member of a singleton subset. Corollary 3.5-4 implies that we can truly partition the branch set $\mathcal{B}$ in this way. A $(\beta-)$-*subsection* $\mathcal{S}_b^{\beta-}$ is defined as the subgraph of $\mathcal{G}^\nu$ induced by the branches in one of the partitioning subsets of $\mathcal{B}$. An immediate result of these definitions and Lemma 3.1-2 is

Lemma 4.1-1. *For $1 \leq \beta < \gamma \leq \nu$, every $(\gamma-)$-subsection is partitioned by $(\beta-)$-subsections. Moreover, for $1 \leq \beta \leq \gamma + 1 \leq \nu$, $\mathcal{G}^\nu$ and every γ-section is partitioned by $(\beta-)$-subsections.*

Let n be a node met by $\mathcal{S}_b^{\beta-}$; n is called a *bordering node* of $\mathcal{S}_b^{\beta-}$ if n is embraced by a $(\beta+)$-node and is called an *internal node* of $\mathcal{S}_b^{\beta-}$ otherwise. Thus, if n is a maximal node met by $\mathcal{S}_b^{\beta-}$, then n is an internal node of $\mathcal{S}_b^{\beta-}$ if its rank is less than β, and n is a bordering node of $\mathcal{S}_b^{\beta-}$ if its rank is no less than β.

The *rank* of a $(\beta-)$-subsection $\mathcal{S}_b^{\beta-}$ is simply its rank as a subgraph of $\mathcal{G}^\nu$; that rank may be either larger, smaller, or equal to β. On the other hand, let α be the least rank that is no less than every rank of every internal node of $\mathcal{S}_b^{\beta-}$; α will be called the *internal rank* of $\mathcal{S}_b^{\beta-}$. Thus, if β is a natural number or if $\beta = \omega$, then $\alpha < \beta$; but, if $\beta = \vec{\omega}$, then $\alpha \leq \beta$. When $\alpha \neq \vec{\omega}$, it will be more convenient at times to refer to a subsection by its internal rank; thus, we may call $\mathcal{S}_b^{\beta-}$ an *α-subsection* and denote it by $\mathcal{S}_b^{\alpha}$. In this case, $\mathcal{S}_b^{\alpha}$ will be a $(\beta-)$-subsection for every value of β from $\alpha + 1$ up to the least rank among all the bordering nodes of $\mathcal{S}_b^{\beta-}$ whose ranks are larger than α.

Example 4.1-2. Consider at first a single branch b with its two incident 0-nodes n_1^0 and n_2^0. Assume that b is also incident to two 2-nodes that do not embrace 1-nodes, that is, n_1^0 and n_2^0 are the exceptional elements of those 2-nodes. The 2-nodes may themselves be embraced by nodes of higher ranks. Then, b along with n_1^0 and n_2^0 comprise a 0-subsection (and a 0-section too); moreover, that 0-subsection is a $(1-)$-subsection and also a $(2-)$-subsection.

The 4-graph of Figure 3.5 has two $(1-)$-subsections, namely, $\mathcal{S}_{b1}^{1-} = \mathcal{S}_1$ and $\mathcal{S}_{b2}^{1-} = \mathcal{S}_2$. Their internal ranks are both 0. The rank of $\mathcal{S}_1$ is 4, and the rank of $\mathcal{S}_2$ is 3. The bordering nodes of $\mathcal{S}_1$ are n^4, n^3, n^1, and all the $(1+)$-nodes embraced by the 3-path P^3. The bordering nodes of $\mathcal{S}_2$ are n^4, n^3, n^1, and all the $(1+)$-nodes embraced by P^2. In this case, the $(1-)$-subsections coincide with the 0-sections; see Example 3.6-1. On the other hand, if we append a one-ended 0-path that meets n_2^0 and reaches n^1 but is otherwise totally disjoint from the graph of Figure 3.5, then the resulting graph will be a 1-section, but it will have two $(2-)$-subsections.

The $\vec{\omega}$-graph of Figure 2.9(b) has just one $(\vec{\omega}-)$-subsection, the entire graph itself; both its rank and its internal rank are $\vec{\omega}$. On the other hand, for each natural number $\mu \geq 1$ it has an infinity of $(\mu-)$-subsections; they coincide with the $(\mu - 1)$-sections of that graph.

The 1-graph of Figure 2.3 has an infinity of $(1-)$-subsections. The branches of each vertical 0-path induce a $(1-)$-subsection. Also, each horizontal branch induces a $(1-)$-subsection since any branch is 0-connected to itself. Every 1-node is a bordering node of two or three of these $(1-)$-subsections — except for n_0^1; n_0^1 is not a bordering node of any $(1-)$-subsection for it is not incident to any $(1-)$-subsection. ♣

We now define the "core" of a $(\beta-)$-subsection $\mathcal{S}_b^{\beta-}$. If $\mathcal{S}_b^{\beta-}$ has no internal node, the *core* of $\mathcal{S}_b^{\beta-}$ is void. If $\mathcal{S}_b^{\beta-}$ has exactly one internal maximal node n, the *core* of $\mathcal{S}_b^{\beta-}$ is defined to be n. If $\mathcal{S}_b^{\beta-}$ has two or more internal maximal nodes, the *core* of $\mathcal{S}_b^{\beta-}$ is the subgraph of $\mathcal{G}^\nu$ induced by all branches in $\mathcal{S}_b^{\beta-}$ that are not incident to $(\beta+)$-nodes.

Some simple consequences of these definitions are given by

Lemma 4.1-3. *Let $\mathcal{S}_b^{\beta-}$ be a $(\beta-)$-subsection in a ν-connected ν-graph $\mathcal{G}^\nu$ that satisfies Condition 3.5-1. Then, the following hold:*

(i) *$\mathcal{S}_b^{\beta-}$ has at least one bordering node — except in the following circumstance: When $\beta = \nu = \vec{\omega}$ and $\mathcal{S}_b^{\beta-}$ coincides with $\mathcal{G}^\nu$, $\mathcal{S}_b^{\beta-}$ may — but need not have — a bordering node.*

(ii) *If $\mathcal{S}_b^{\beta-}$ has no internal node, then $\mathcal{S}_b^{\beta-}$ has only one branch and is in fact a $(1-)$-subsection.*

(iii) *If $\mathcal{S}_b^{\beta-}$ has exactly one internal maximal node n, then n is a maximal 0-node, every branch of $\mathcal{S}_b^{\beta-}$ is incident to n, and $\mathcal{S}_b^{\beta-}$ is also a $(1-)$-subsection.*

(iv) *If $\mathcal{S}_b^{\beta-}$ has two or more internal maximal nodes, then its core has at least one branch and is $(\beta-)$-connected through itself as follows: If the core has two or more branches , every two branches in the core are $(\beta-)$-connected through a two-ended $(\beta-)$-path that does not meet any bordering node.*

(v) *If $\mathcal{S}_b^{\beta-}$ has at least two branches , then the core of $\mathcal{S}_b^{\beta-}$ is not void and $\mathcal{S}_b^{\beta-}$ has at least one internal node.*

(vi) *If $\mathcal{S}_b^{\beta-}$ has an internal α-node n^α with $\alpha \geq 1$ (whence $\beta \geq 2$) and if $\mathcal{S}_b^{\beta-}$ has only finitely many bordering nodes, then the core of $\mathcal{S}_b^{\beta-}$ has an infinity of branches and an infinity of internal maximal nodes.*

Proof. (i) In the exceptional case when $\beta = \nu = \vec{\omega}$ and $\mathcal{S}_b^{\beta-}$ coincides with $\mathcal{G}^{\vec{\omega}}$, $\mathcal{S}_b^{\beta-}$ need not have a bordering node. Examples of this are provided by the $\vec{\omega}$-graphs of Figures 2.9(a) and (c).

If $\nu \neq \vec{\omega}$ and if $\mathcal{S}_b^{\beta-}$ coincides with $\mathcal{G}^\nu$, then each ν-node of $\mathcal{G}^\nu$ is a bordering node of $\mathcal{S}_b^{\beta-}$.

Finally, if $\mathcal{S}_b^{\beta-}$ does not coincide with $\mathcal{G}^\nu$, there is a branch b_1 in $\mathcal{S}_b^{\beta-}$ and a branch b_2 not in $\mathcal{S}_b^{\beta-}$. By the ν-connectedness of $\mathcal{G}^\nu$, there is a path that connects b_1 and b_2. That path must meet a bordering node of $\mathcal{S}_b^{\beta-}$.

(ii) The only way $\mathcal{S}_b^{\beta-}$ can have no internal node is if each of its branches is incident through each of its elementary tips to a bordering node of $\mathcal{S}_b^{\beta-}$. But then, $\mathcal{S}_b^{\beta-}$ will have only one branch b, which means by definition that $\mathcal{S}_b^{\beta-}$ is a $(1-)$-subsection too.

(iii) As in (ii), if a branch is incident to two bordering nodes of $\mathcal{S}_b^{\beta-}$, then by itself it is a $(1-)$-subsection having no internal node. Thus, every branch of $\mathcal{S}_b^{\beta-}$ must either be a self-loop incident to n or be incident to n and to a bordering node. Hence, n is a maximal 0-node. Since any bordering node must be of rank greater than the rank of n, $\mathcal{S}_b^{\beta-}$ is also a $(1-)$-subsection.

(iv) If a branch is incident to two internal nodes of $\mathcal{S}_b^{\beta-}$, it will lie in the core of $\mathcal{S}_b^{\beta-}$. Now, suppose that the core has no branch. Then, every branch of $\mathcal{S}_b^{\beta-}$ must be incident to a bordering node. This implies that all branches of $\mathcal{S}_b^{\beta-}$ must be incident to a single internal maximal node of $\mathcal{S}_b^{\beta-}$, for otherwise at least two such branches would not be in the same $(\beta-)$-subsection. Thus, $\mathcal{S}_b^{\beta-}$ has only one internal node, in contradiction to the hypothesis. Consequently, the core of $\mathcal{S}_b^{\beta-}$ has at least one branch.

Furthermore, suppose two branches of the core are not $(\beta-)$-connected through the core as stated. Then, they can be connected only by paths that meet $(\beta+)$-nodes and therefore must lie in different $(\beta-)$-subsections — another contradiction.

(v) $\mathcal{S}_b^{\beta-}$ has no internal node if and only if its core is void. So, suppose the core is void. By (ii) and the definition of the core, $\mathcal{S}_b^{\beta-}$ has only one branch — in violation of our hypothesis.

(vi) Every non-elementary tip embraced by n^α must have a representative that meets no bordering node of $\mathcal{S}_b^{\beta-}$, for otherwise every such representative would meet at least one bordering node and therefore infinitely many bordering nodes; the latter result would contradict the hypothesis. Thus, the core contains such a representative. That representative has an infinity of branches and an infinity of maximal nodes, all of which will be internal. ♣

Example 4.1-4. The hypothesis in (vi) of Lemma 4.1-3 stating that $\mathcal{S}_b^{\beta-}$ has only finitely many bordering nodes is truly needed. To see this, examine the 2-graph of Figure 4.1. The branches $b_1, b_2, b_3, \ldots, b_\omega$ induce a $(2-)$-subsection $\mathcal{S}_0^{2-}$ having an infinity of bordering 2-nodes $n_1^2, n_2^2, n_3^2, \ldots$ but only a finite core consisting of the branch b_ω along with the 0-node n^0 and the 0-node embraced by the 1-node n^1. (n^1 itself is an internal 1-node.) On the other hand, each vertical 1-path P_k^1 ($k = 1, 2, 3, \ldots$) along with the 2-node n_k^2 that it reaches is a $(2-)$-subsection $\mathcal{S}_{b,k}^{2-}$, which has only one bordering 2-node n_k^2 and an infinite core, namely, $\mathcal{S}_{b,k}^{2-}$ itself.

Another peculiarity that can arise is that a two-ended α-path may meet an infinity of $(\beta+)$-nodes ($\beta > \alpha$). This is illustrated in Figure 4.1 by the three-element 1-path $\{n_1^0, P_a^0, n^1\}$, where P_a^0 is the one-ended 0-path induced by the branches $a_1, a_2, a_3, \ldots$. That path meets the 2-nodes $n_1^2, n_2^2, n_3^2, \ldots$. We show through the next lemma that this possibility can be eliminated by imposing the Condition (c), a restriction that will be in force throughout the rest of this chapter. ♣

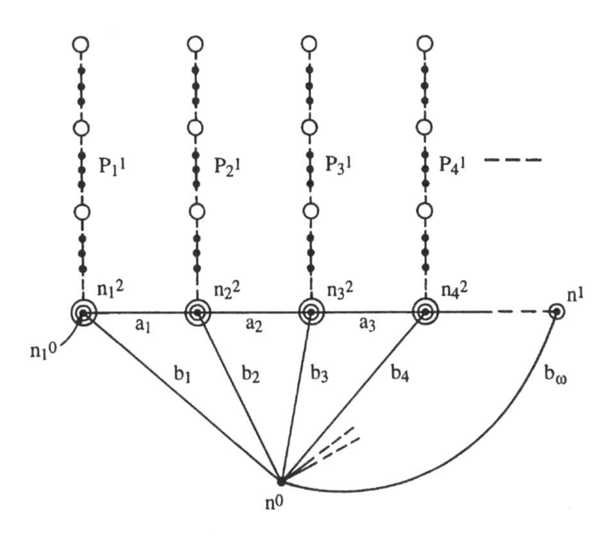

Figure 4.1. The 2-graph discussed in Example 4.1-4.

Lemma 4.1-5. *Let $\mathcal{G}^\nu$ be a ν-connected ν-graph ($\nu \leq \omega$) that satisfies Condition 3.5-1 and the following.*

Condition (c): *For each $\beta = 1, \ldots, \nu$, every nonelementary $(\beta-)$-tip $t^{\beta-}$ embraced by a maximal $(\beta+)$-node n^β has a representative that reaches exactly one maximal $(\beta+)$-node, namely, n^β.*

Then, the following hold.

(i) *Every two-ended α-path P^α ($0 \leq \alpha \leq \nu$) meets only finitely many maximal $(\alpha+)$-nodes.*

(ii) *Any one-ended path Q that meets infinitely many $(\alpha+)$-nodes is a one-ended $(\alpha+)$-path.*

Note. To say that $t^{\beta-}$ has a representative that reaches exactly one maximal $(\beta+)$-node is the same as saying that $t^{\beta-}$ has a representative that reaches no maximal $(\beta+)$-node other than the maximal $(\beta+)$-node that embraces $t^{\beta-}$. The negation of this condition is the assertion that every representative of $t^{\beta-}$ reaches two or more maximal $(\beta+)$-nodes — or equivalently every representative of $t^{\beta-}$ reaches an infinity of maximal $(\beta+)$-nodes. The last condition follows from the facts that any representative is a one-ended path and removing any two-ended part of it to eliminate any node will still leave a representative reaching two or more $(\beta+)$-nodes.

Note also another immediate consequence of Condition (c): $t^{\beta-}$ has a representative lying entirely within a single $(\beta-)$-subsection.

Proof. (i) Since P^α is two-sided, α is either a natural number or ω. Our conclusion is obvious if $\alpha = 0$. So, let $\alpha > 0$. Suppose P^α reaches an infinity of maximal $(\alpha+)$-nodes. Then, exactly two cases arise: Either P^α traverses an $(\alpha-1)$-tip every representative of which reaches two or more maximal $(\alpha+)$-nodes, or every one of the finitely many $(\alpha-1)$-tips traversed by P^α has a representative that reaches exactly one maximal $(\alpha+)$-node. In the first case, Condition (c) is violated. In the second case, we can remove those finitely many representatives to find a two-ended λ-path Q^λ $(\lambda < \alpha)$ such that Q^λ is in P^α and Q^λ reaches an infinity of maximal $(\alpha+)$-nodes. Note in particular that, if $\alpha = \omega$, the removal of those representatives will eliminate all $\vec{\omega}$-tips traversed by P^α, and what remains from P^α will be finitely many two-ended paths whose ranks are natural numbers.

We can now treat Q^λ the way we did P^α to find either a violation of Condition (c) or a two-ended γ-path R^γ $(\gamma < \lambda)$ in Q^λ that reaches an infinity of maximal $(\alpha+)$-nodes. We can continue repeating this argument as needed to find a sequence of two-ended paths with decreasing ranks and with each path reaching an infinity of maximal $(\alpha+)$-nodes. This process must stop before the rank 0 is attained, for no two-ended 0-path can reach an infinity of maximal nodes. Thus, we will eventually find a violation of Condition (c).

Thus, P^α reaches only finitely many maximal $(\alpha+)$-nodes. Finally, since P^α is two-ended, it meets every node that it reaches.

(ii) Suppose the rank γ of the said one-ended path Q is less than α. Then, Q is the representative of a γ-tip t^γ where $0 \leq \gamma < \alpha$. Moreover, t^γ will be embraced by a maximal λ-node where $\gamma < \lambda$. Setting $\beta = \gamma+1$, we have that γ is a $\beta-$ rank and λ is a $\beta+$ rank. Since $\alpha > \gamma$, every $(\alpha+)$-node is a $(\beta+)$-node. Thus, Q is a representative of the $(\beta-)$-tip t^γ that is embraced by a maximal $(\beta+)$-node, and Q does not fulfill the requirement of Condition (c) because it meets infinitely many maximal $(\beta+)$-nodes. Moreover, by our first conclusion (i), no two-ended portion of Q can meet infinitely many $(\alpha+)$-nodes. It follows that every representative of the γ-tip t^γ violates Condition (c). Hence, our supposition is wrong. ♣

4.2 A Generalization of Konig's Lemma

Here is a version of Konig's lemma for a conventional infinite graph as defined in Section 1.2. It reads somewhat differently from — but is in fact equivalent to — his "Unendlichkeitslemma" [15, page 81].

Lemma 4.2-1. *If a conventional infinite graph is connected and locally finite (i.e., each node has only finitely many incident branches), then, given any node, there is at least one one-ended path starting from that node.*

This result is easily established and is moreover quite useful [20], [29]. We shall now extend it to transfinite graphs and will use that extension in Sections 4.5 and 6.3.

As before, β is a rank no less than 1. Two nodes of any ranks are said to be *β-adjacent* (or *0-adjacent*) if they are totally disjoint and are incident to the same $(\beta-)$-subsection (resp. same branch).

The following conditions on a given ν-graph $\mathcal{G}^\nu$ $(\nu \leq \omega)$ will be part of the hypothesis of our generalization of Konig's lemma.

Conditions 4.2-2.

(a) *$\mathcal{G}^\nu$ is ν-connected.*

(b) *If two tips are nondisconnectable, they are shorted together.*

(c) *For each $\beta = 1, \ldots, \nu$, every nonelementary $(\beta-)$-tip $t^{\beta-}$ embraced by a maximal $(\beta+)$-node has a representative that reaches exactly one maximal $(\beta+)$-node.*

(d) *For each $\beta = 1, \ldots, \nu$, every $(\beta+)$-node is β-adjacent to only finitely many maximal $(\beta+)$-nodes.*

Condition (b) is a strengthened version of Condition 3.5-1. Condition (c) is the same as Condition (c) in Lemma 4.1-5. Condition (d) is our version of local finiteness as extended to transfinite graphs; it implies that every $(\beta-)$-subsection has only finitely many bordering maximal $(\beta+)$-nodes.

Lemma 4.2-3. *Assume Conditions 4.2-2(b) and (c). Let n_1 and n_2 be two β-adjacent maximal $(\beta+)$-nodes, and let $\mathcal{S}_b^{\beta-}$ be a $(\beta-)$-subsection to which they are both incident. Assume $\mathcal{S}_b^{\beta-}$ has at least one internal node. Then, there is a $(\beta-)$-path $P^{\beta-}$ that reaches n_1 and n_2, lies in $\mathcal{S}_b^{\beta-}$, and does not reach any $(\beta+)$-node other than n_1 and n_2. The same conclusion holds when n_1 is an internal node of $\mathcal{S}_b^{\beta-}$ and n_2 is a maximal bordering node of $\mathcal{S}_b^{\beta-}$ (now, of course, n_1 is not a $(\beta+)$-node).*

Proof. If the core of $\mathcal{S}_b^{\beta-}$ is void or has exactly one internal node, the conclusion follows directly from Lemma 4.1-3(ii) and (iii). So, assume that the core of $\mathcal{S}_b^{\beta-}$ has two or more internal nodes. By Condition 4.2-2(c), since n_1 is incident to $\mathcal{S}_b^{\beta-}$, n_1 embraces an α-tip t_1^α ($\vec{0} \leq \alpha < \beta$) with a representative lying in $\mathcal{S}_b^{\beta-}$ and reaching no maximal $(\beta+)$-node other than n_1. (Condition 4.2-2(c) is invoked when $\alpha \geq 0$; this assertion is clearly true when $\alpha = \vec{0}$.) Likewise, n_2 embraces a γ-tip t_2^γ ($\vec{0} \leq \gamma < \beta$) with similar properties. By Condition 4.2-2(b) and the fact that n_1 and n_2 are not shorted together, t_1^α and t_2^γ are disconnectable. If $\alpha = \gamma = \vec{0}$, that is, if n_1 and n_2 are incident to branches in $\mathcal{S}_b^{\beta-}$, we can obtain the conclusion by choosing a path in the core of $\mathcal{S}_b^{\beta-}$ connecting those two branches (Lemma 4.1-3(iv)).

Next, assume that $\alpha \neq \vec{0}$ and $\gamma \neq \vec{0}$. This case is illustrated in Figure 4.2. We can shorten the said representatives as needed to get a representative α-path P_1^α for t_1^α and a representative γ-path P_2^γ for t_2^γ such that they are totally disjoint except possibly terminally, lie in $\mathcal{S}_b^{\beta-}$, do not meet any $(\beta+)$-nodes other than n_1 and n_2, respectively, and terminate at $(\beta-)$-nodes n_b for P_1^α and n_c for P_2^γ. If n_b and n_c meet, we have our conclusion again. So, assume n_b and n_c are totally disjoint (this is the case illustrated in Figure 4.2). Then, there is a two-ended $(\beta-)$-path $P_{bc}^{\beta-}$ in the core of $\mathcal{S}_b^{\beta-}$ that terminates at n_b and n_c and does not reach any $(\beta+)$-node (invoke Condition 4.2-2(c) and Lemma 4.1-3(iv)). Furthermore, there is a nonsingleton node n_a in P_1^α such that the path in P_1^α from n_1 to n_a is totally disjoint from $P_{bc}^{\beta-}$, for otherwise $P_{bc}^{\beta-}$ would traverse a tip that is nondisconnectable from a tip embraced by n_1 (Lemma 3.3-2) and would therefore meet the $(\beta+)$-

node n_1. We can now invoke Corollary 3.5-5 as follows: Upon tracing P_1^α from n_b to n_a, we will find a last maximal node n_x that is met by both P_1^α and $P_{bc}^{\beta-}$. The same argument yields the following: Upon tracing P_2^γ from n_c to n_2, we will find a last maximal node n_y that is met by both P_2^γ and $P_{bc}^{\beta-}$. Now the part P_{1x}^α of P_1^α from n_1 to n_x does not meet any maximal node of $P_{bc}^{\beta-}$ other than n_x, and similarly the part $P_{y2}^{\beta-}$ of P_2^γ from n_y to n_2 also does not meet any maximal node of $P_{bc}^{\beta-}$ other than n_y. So, the part $P_{xy}^{\beta-}$ of $P_{bc}^{\beta-}$ between n_x and n_y is totally disjoint from P_{1x}^α and P_{y2}^γ except terminally. In fact, the branches of P_{1x}^α, $P_{xy}^{\beta-}$, and P_{y2}^γ induce the two-ended $(\beta-)$-path $P^{\beta-}$ we seek.

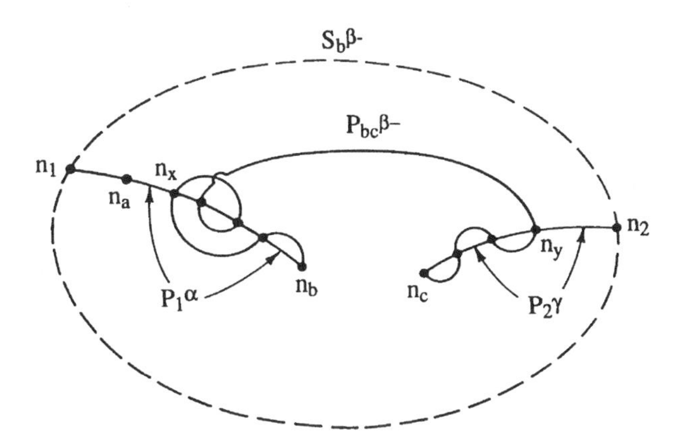

Figure 4.2. The paths constructed in the proof of Lemma 4.2-3.

There is one more case that should be considered — the case where one but not both of n_1 and n_2 are incident to $\mathcal{S}_b^{\beta-}$ through $\vec{0}$-tips. This can be handled by combining our foregoing arguments.

Finally, our proof has already established the second conclusion. ♣

Theorem 4.2-4. *Let $\mathcal{G}^\nu$ be a ν-graph with $1 \le \nu \le \omega$. Assume $\mathcal{G}^\nu$ has infinitely many ν-nodes and satisfies Conditions 4.2-2. Then, given any ν-node n_0^ν, there is at least one one-ended ν-path starting at n_0^ν.*

Proof. Corresponding to $\mathcal{G}^\nu$ we set up a "surrogate" 0-graph $\mathcal{G}^0$ by setting up one and only one 0-node m_a^0 in $\mathcal{G}^0$ for each ν-node n_a^ν in $\mathcal{G}^\nu$ and inserting branches as follows: Insert a branch between two 0-nodes m_a^0 and m_b^0 of $\mathcal{G}^0$ when and only when the corresponding ν-nodes n_a^ν and n_b^ν in $\mathcal{G}^\nu$

are ν-adjacent. (We will identify corresponding nodes n_a^ν and m_a^0 by using the same subscripts.) By Conditions 4.2-2(a) and (d), $\mathcal{G}^0$ is a locally finite 0-connected 0-graph with infinitely many 0-nodes. Therefore, we can invoke Konig's lemma (Lemma 4.1-2) to conclude that there is a one-ended 0-path P^0 in $\mathcal{G}^0$ starting at the 0-node m_0^0 corresponding to n_0^ν. Orient P^0 from m_0^0 onward.

Let $\mathcal{M}_0^0$ be the singleton set $\{m_0^0\}$. Also, let $\mathcal{M}_1^0$ be the finite set of all 0-nodes in $\mathcal{G}^0$ that are 0-adjacent to m_0^0. Let m_1^0 be the last node in $\mathcal{M}_1^0$ that P^0 meets. No node of P^0 beyond m_1^0 will be in $\mathcal{M}_0^0 \cup \mathcal{M}_1^0$. Let $\mathcal{M}_2^0$ be the finite set of all 0-nodes in $\mathcal{G}^0$ that are 0-adjacent to m_1^0. Let m_2^0 be the last node in $\mathcal{M}_2^0$ that P^0 meets. No node of P^0 beyond m_2^0 will be in $\mathcal{M}_0^0 \cup \mathcal{M}_1^0 \cup \mathcal{M}_2^0$. We can continue recursively this way to get an infinite alternating sequence

$$\{\mathcal{M}_0^0, m_0^0, \mathcal{M}_1^0, m_1^0, \mathcal{M}_2^0, m_2^0, \ldots\}$$

where, for each $k \geq 1$, $\mathcal{M}_k^0$ is the finite set of 0-nodes in $\mathcal{G}^\nu$ that are 0-adjacent to m_{k-1}^0 and where m_k^0 is the last node in $\mathcal{M}_k^0$ that P^0 meets. Again, no node of P^0 beyond m_k^0 will be in $\bigcup_{i=0}^k \mathcal{M}_i^0$.

Now let $\{n_0^\nu, n_1^\nu, n_2^\nu, \ldots\}$ be the sequence of ν-nodes in $\mathcal{G}^\nu$ corresponding bijectively to the sequence $\{m_0^0, m_1^0, m_2^0, \ldots\}$ of 0-nodes in $\mathcal{G}^0$ as stated above. As a result of how the m_k^0 were chosen and how 0-adjacency in $\mathcal{G}^0$ corresponds to ν-adjacency in $\mathcal{G}^\nu$, we have that each n_k^ν $(k \geq 1)$ is ν-adjacent to n_{k-1}^ν but not ν-adjacent to any n_i^ν for $0 \leq i \leq k-1$. Thus, for each $k \geq 1$, there is a $(\nu-)$-subsection $\mathcal{S}_{b,k}^{\nu-}$ to which n_{k-1}^ν and n_k^ν are both incident and to which no other n_i^ν $(i \neq k-1, i \neq k)$ is incident. By Lemma 4.2-3, we can choose in each $\mathcal{S}_{b,k}^{\nu-}$ a $(\nu-)$-path $P_k^{\nu-}$ that reaches n_{k-1}^ν and n_k^ν and is totally disjoint from all other $P_i^{\nu-}$ $(i \neq k)$ except terminally for $P_{k-1}^{\nu-}$ and $P_{k+1}^{\nu-}$ at n_{k-1}^ν and n_k^ν, respectively. Since there are infinitely many n_k^ν, we can conclude that the branches of all the $P_k^{\nu-}$ $(k = 1, 2, 3, \ldots)$ induce a one-ended path Q that meets infinitely many ν-nodes. By Lemma 4.1-5(ii), Q is a one-ended ν-path. It starts at the arbitrarily chosen ν-node n_0^ν. ♣

Corollary 4.2-5. *Under the hypothesis of Theorem 4.2-4, given any node of any rank in $\mathcal{G}^\nu$, there is at least one one-ended ν-path starting at that node.*

Proof. If $\delta < \nu$ and if n^δ is a δ-node not embraced by any ν-node, we can choose a two-ended path P_1 in the $(\nu-)$-subsection that embraces n^δ

such that P_1 terminally meets n^δ and a bordering ν-node n^ν. Let P_2^ν be a one-ended ν-path starting at n^ν; the existence of P_2^ν is assured by Theorem 4.2-4. Since $\mathcal{S}_b^{\nu-}$ has only finitely many bordering ν-nodes (a consequence of Condition 4.2-2(d)), P_2^ν will eventually be totally disjoint from $\mathcal{S}_b^{\nu-}$. It therefore follows from Corollary 3.5-4 that there is in $P_1 \cup P_2^\nu$ a one-ended ν-path that starts at n^δ. ♣

Our results can be extended to subsections of $\mathcal{G}^\nu$ as follows. Given any rank β with $1 \leq \beta \leq \nu$, let $\mathcal{S}_b^{\beta-}$ be a $(\beta-)$-subsection of $\mathcal{G}^\nu$ of internal rank $\alpha \geq 0$. Thus, $\alpha < \beta$, and there will be at least one internal α-node in $\mathcal{S}_b^{\beta-}$. Consider the subgraph $\mathcal{H}$ of the α-graph of $\mathcal{G}^\nu$ induced by all the branches in $\mathcal{S}_b^{\beta-}$. In effect, $\mathcal{H}$ is $\mathcal{S}_b^{\beta-}$ with all its bordering $(\beta+)$-nodes stripped away. In fact, $\mathcal{H}$ is an α-graph. Since $\mathcal{G}^\nu$ satisfies Conditions 4.2-2, $\mathcal{H}$ will too — except that ν is replaced by α. Consequently, we have

Corollary 4.2-6. *Let $\mathcal{G}^\nu$ be the ν-graph of Theorem 4.2-4. For any rank β with $1 \leq \beta \leq \nu$, let $\mathcal{S}_b^{\beta-}$ be any $(\beta-)$-subsection of $\mathcal{G}^\nu$. Let α be the internal rank of $\mathcal{S}_b^{\beta-}$, and assume that $\mathcal{S}_b^{\beta-}$ has an infinity of internal α-nodes. Then, given any internal δ-node n^δ ($\delta \leq \alpha$) in $\mathcal{S}_b^{\beta-}$, there is at least one one-ended α-path in $\mathcal{S}_b^{\beta-}$ starting at n^δ.*

The case where $\alpha = 0$ in Corollary 4.2-6 is simply Konig's lemma.

There is another generalization of Konig's lemma, which we will also need in Sections 4.5 and 6.3.

Theorem 4.2-7. *Assume that the $\vec{\omega}$-graph $\mathcal{G}^{\vec{\omega}}$ satisfies Conditions 4.2-2 with $\nu = \vec{\omega}$. Assume furthermore that $\mathcal{G}^{\vec{\omega}}$ has no $\vec{\omega}$-node. Then, $\mathcal{G}^{\vec{\omega}}$ contains an $\vec{\omega}$-path.*

Proof. The rank of every node in $\mathcal{G}^{\vec{\omega}}$ is a natural number. Moreover, there is no natural number that uniformly bounds all the ranks of all the nodes of $\mathcal{G}^{\vec{\omega}}$. Choose any maximal node $n_0^{\mu_0}$ in $\mathcal{G}^{\vec{\omega}}$ of any rank μ_0. We can choose a natural number μ_1 and a (μ_1-)-subsection $\mathcal{S}_b^{\mu_1-}$ such that $\mu_1 > \mu_0$ and $n_0^{\mu_0}$ is an internal node of $\mathcal{S}_b^{\mu_1-}$. Proceeding inductively, for every positive natural number $k = 1, 2, 3, \ldots$, we can choose a natural number μ_k and a (μ_k-)-subsection $\mathcal{S}_b^{\mu_k-}$ such that the bordering nodes of $\mathcal{S}_b^{\mu_k-}$ are internal nodes of $\mathcal{S}_b^{\mu_{k+1}-}$. This can be done because every $\mathcal{S}_b^{\mu_k-}$ has only finitely many bordering nodes, by virtue of Condition 4.2-2(d). It follows that $\mu_k < \mu_{k+1}$, that $\mathcal{S}_b^{\mu_k-} \subset \mathcal{S}_b^{\mu_{k+1}-}$, and that μ_{k+1} is larger than the maximum rank among all the bordering nodes (indeed, among all the nodes) of $\mathcal{S}_b^{\mu_k -}$.

Now, consider $\mathcal{G}^{\vec{\omega}} \backslash \mathcal{S}_b^{\mu_1 -}$, the subgraph induced by all the branches that are not in $\mathcal{S}_b^{\mu_1 -}$. This may be $\vec{\omega}$-connected, but in general it will consist of no more than finitely many components because $\mathcal{S}_b^{\mu_1 -}$ has only finitely many bordering nodes. To save words, we shall say that "a component has unbounded ranks" if there is no natural number larger than all the ranks of all the nodes of that component. Thus, there is at least one component $\mathcal{C}_1$ of $\mathcal{G}^{\vec{\omega}} \backslash \mathcal{S}_b^{\mu_1 -}$ having unbounded ranks. $\mathcal{C}_1$ will also contain at least one maximal bordering node $n_1^{\mu_1 +}$ of $\mathcal{S}_b^{\mu_k -}$. Let $P_0^{\mu_1 -}$ be a (μ_1-)-path in $\mathcal{S}_b^{\mu_1 -}$ starting at $n_0^{\mu_0}$, reaching $n_1^{\mu_1 +}$, but not reaching any other (μ_1+)-node (see the second conclusion of Lemma 4.2-3).

Next, consider $\mathcal{G}^{\vec{\omega}} \backslash \mathcal{S}_b^{\mu_2 -}$. This too will have only finitely many components (perhaps just one). At least one of them $\mathcal{C}_2$ will be a subgraph of $\mathcal{C}_1$, will have unbounded ranks, and will have a maximal bordering node $n_2^{\mu_2 +}$ of $\mathcal{S}_b^{\mu_2 -}$. Moreover, $\mathcal{C}_1 \cap \mathcal{S}_b^{\mu_2 -}$ will be a (μ_2-)-subsection of $\mathcal{G}^{\vec{\omega}} \backslash \mathcal{S}_b^{\mu_1 -}$. Therefore, we can choose in $\mathcal{C}_1 \cap \mathcal{S}_b^{\mu_2 -}$ a (μ_2-)-path $P_1^{\mu_2 -}$ starting at $n_1^{\mu_1 +}$, reaching $n_2^{\mu_2 +}$, but not reaching any other (μ_2+)-node (Lemma 4.2-3). In fact, $P_1^{\mu_2 -}$ will lie in $\mathcal{S}_b^{\mu_2 -} \backslash \mathcal{S}_b^{\mu_1 -}$. Thus, $P_0^{\mu_1 -}$ and $P_1^{\mu_2 -}$ both reach $n_1^{\mu_1 +}$ but are otherwise totally disjoint from each other; that is, they along with $n_1^{\mu_1 +}$ comprise a path.

This process can be continued recursively for all k. Just replace 2 by $k+1$ and 1 by k in the preceding paragraph. This yields a component $\mathcal{C}_{k+1}$ of $\mathcal{G}^{\vec{\omega}} \backslash \mathcal{S}_b^{\mu_{k+1} -}$, which is a subgraph of $\mathcal{C}_k$ and which has unbounded ranks and a maximal bordering node $n_{k+1}^{\mu_{k+1} +}$ of $\mathcal{S}_b^{\mu_{k+1} -}$. Moreover, $\mathcal{C}_k \cap \mathcal{S}_b^{\mu_{k+1} -}$ will be a $(\mu_{k+1}-)$-subsection of $\mathcal{G}^{\vec{\omega}} \backslash \mathcal{S}_b^{\mu_k -}$, and therefore, by Lemma 4.2-3 again, there will be a $(\mu_{k+1}-)$-path $P_k^{\mu_{k+1} -}$ in $\mathcal{C}_k \cap \mathcal{S}_b^{\mu_{k+1} -}$ reaching $n_k^{\mu_k +}$ and $n_{k+1}^{\mu_{k+1} +}$. In fact, $P_k^{\mu_{k+1} -}$ will lie in $\mathcal{S}_b^{\mu_{k+1} -} \backslash \mathcal{S}_b^{\mu_k -}$. All the $P_l^{\mu_{l+1} -}$ $(l = 1, \ldots, k)$ along with the intervening nodes $n_l^{\mu_l +}$ $(l = 0, \ldots, k+1)$ will comprise a path.

With k increasing indefinitely, we will generate in this way a one-ended path

$$\{n_0^{\mu_0}, P_0^{\mu_1 -}, n_1^{\mu_1 +}, P_1^{\mu_2 -}, \ldots\}$$

that sequentially meets infinitely many nodes whose natural-number ranks increase beyond every natural number. It now follows from Lemma 4.1-5(ii) that the rank of this one-ended path must be larger than every natural number. Since it lies in $\mathcal{G}^{\vec{\omega}}$, its rank must be $\vec{\omega}$. ♣

4.3 Isolating Sets and Cuts

In forthcoming chapters we will be applying Kirchhoff's current law to nodes of ranks higher than 0. However, for any given β-node n^β ($\beta \geq 1$), this cannot be done to n^β alone because the non-elementary tips embraced by n^β do not meet it with incident branches. That law will instead be applied to a set of 0-nodes that "isolate" n^β from all the other $(\beta+)$-nodes. To do so, we equate to 0 the algebraic sum of the currents in the branches incident to the isolating set of 0-nodes but separated from n^β by those 0-nodes. That set of branches will be called a "cut" for n^β and is in fact a generalization of a "cutset" in a conventional graph that isolates a single node. Furthermore, we will be using cuts for other purposes as well and therefore will define them narrowly enough to fulfill all our needs. These require various conditions on our given ν-graph, which we specify now and impose throughout the rest of this chapter.

Conditions 4.3-1. *$\mathcal{G}^\nu$ is a ν-graph with $1 \leq \nu \leq \omega$, and $\mathcal{G}^\nu$ satisfies Conditions 4.2-2. In addition, $\mathcal{G}^\nu$ has no self-loops and no parallel branches. Every branch of $\mathcal{G}^\nu$ is incident to at least one maximal 0-node. Finally, every 0-node (whether or not maximal) has only finitely many incident branches, and, for each $\beta = 1, \ldots, \nu$, every $(\beta+)$-node (whether or not maximal) is incident to only finitely many $(\beta-)$-subsections.*

All but one of these restrictions beyond those of Conditions 4.2-2 are inconsequential and are being imposed just for convenience in our discussions of random walks and their related electrical theory. For instance, self-loops do not contribute anything essential to the behavior of an electrical network; they can in fact be removed without changing the voltages and currents in the other branches. Similarly, a parallel connection of branches can be combined into a single branch without disturbing the voltage-current regime elsewhere; moreover, they would serve no purpose in our theory of transfinite random walks among the nodes of $\mathcal{G}^\nu$. Furthermore, any branch that is not incident to a maximal 0-node can be split into two series-connected branches by inserting a maximal 0-node; this too will not alter the voltages and currents elsewhere. The only essential restriction beyond those of Conditions 4.2-2 is given by the last sentence of Conditions 4.3-1; it strengthens our generalization of local-finiteness for transfinite graphs because without that restriction a $(\beta+)$-node may be incident to an infinity of $(\beta-)$-subsections having no other incident $(\beta+)$-nodes. Finally, let us note again that Condition 4.2-2(d) implies that every $(\beta-)$-subsection has only finitely many bordering maximal $(\beta+)$-nodes.

We now prepare for our definition of an "isolating set." Let $\mathcal{G}_s$ be any subgraph of $\mathcal{G}^\nu$ $(\nu \leq \omega)$, and let $\mathcal{N}_1$, $\mathcal{N}_2$, and $\mathcal{N}_3$ be three sets of nodes in $\mathcal{G}_s$. The nodes of these sets need not be maximal, and their ranks need not be the same — even within a single set. $\mathcal{N}_3$ is said to *separate* $\mathcal{N}_1$ *and* $\mathcal{N}_2$ *in* $\mathcal{G}_s$ if every path in $\mathcal{G}_s$ that meets $\mathcal{N}_1$ and $\mathcal{N}_2$ also meets a node of $\mathcal{N}_3$. This definition allows nodes of $\mathcal{N}_3$ to embrace nodes of $\mathcal{N}_1$ and/or $\mathcal{N}_2$, and conversely. For instance, if n_a^α is embraced by n_b^β, then n_a^α separates n_b^β from all other nodes, and n_b^β does the same for n_a^α. Similarly, two subgraphs of $\mathcal{G}_s$ are said to be *separated by* $\mathcal{N}_3$ *in* $\mathcal{G}_s$ if $\mathcal{N}_3$ separates their node sets. Finally, $\mathcal{N}_3$ will *separate* a subgraph of $\mathcal{G}_s$ from a node n_0 of $\mathcal{G}_s$ if it separates the node set of the subgraph from $\{n_0\}$.

If the $(\beta-)$-subsection $\mathcal{S}_b^{\beta-}$ is incident to the $(\beta+)$-node $n^{\beta+}$ through one or more non-elementary tips (that is, if the core of $\mathcal{S}_b^{\beta-}$ is incident to $n^{\beta+}$), we let $\mathcal{V}$ denote a nonvoid finite set of maximal 0-nodes in the core of $\mathcal{S}_b^{\beta-}$; none of these 0-nodes will be bordering nodes of $\mathcal{S}_b^{\beta-}$. If, in addition, $\mathcal{S}_b^{\beta-}$ is incident to $n^{\beta+}$ also through one or more elementary tips (that is, if a branch of $\mathcal{S}_b^{\beta-}$ is incident to $n^{\beta+}$), then $n^{\beta+}$ will embrace a 0-node n^0, and we set $\mathcal{W} = \mathcal{V} \cup \{n^0\}$; otherwise, we set $\mathcal{W} = \mathcal{V}$. On the other hand, if $\mathcal{S}_b^{\beta-}$ is incident to $n^{\beta+}$ only through elementary tips, we set $\mathcal{W} = \{n^0\}$ and make $\mathcal{V}$ void. In every case, $\mathcal{W}$ is not void.

Furthermore, when the core of $\mathcal{S}_b^{\beta-}$ is incident to $n^{\beta+}$, that is, when $\mathcal{V}$ is not void, we let $\mathcal{A}$ be the subgraph of $\mathcal{S}_b^{\beta-}$ induced by all branches in the core of $\mathcal{S}_b^{\beta-}$ satisfying the following: Either the branch is incident to two nodes of $\mathcal{V}$ or the branch is connected to $n^{\beta+}$ by a path in the core of $\mathcal{S}_b^{\beta-}$ that does not meet $\mathcal{V}$. On the other hand, if $\mathcal{V}$ is void, we take $\mathcal{A}$ to be void as well.

If $\mathcal{V}$ is not void, neither is the branch set of $\mathcal{A}$. Indeed, $n^{\beta+}$ will be incident to $\mathcal{S}_b^{\beta-}$ through a tip having a representative lying in the core of $\mathcal{S}_b^{\beta-}$, and that representative can meet $\mathcal{V}$ only finitely often; hence, $\mathcal{A}$ will have an infinity of branches. The *complement* $\tilde{\mathcal{A}} = \mathcal{S}_b^{\beta-} \backslash \mathcal{A}$ *of* $\mathcal{A}$ *in* $\mathcal{S}_b^{\beta-}$ is the subgraph of $\mathcal{S}_b^{\beta-}$ induced by all the branches of $\mathcal{S}_b^{\beta-}$ not in $\mathcal{A}$. Thus, within the core of $\mathcal{S}_b^{\beta-}$, $\mathcal{V}$ separates $\tilde{\mathcal{A}}$ from $\mathcal{A}$ and also separates $\tilde{\mathcal{A}}$ from $n^{\beta+}$. (Presently, $\tilde{\mathcal{A}}$ may be void, but $\tilde{\mathcal{A}}$ will be nonvoid under some more conditions imposed below.)

Now, let $\mathcal{D}$ denote the set of all nodes of $\mathcal{S}_b^{\beta-}$ that are 0-adjacent to $n^{\beta+}$.

$\mathcal{D}$ will be void if $\mathcal{W} = \mathcal{V}$ and will be nonvoid if $\mathcal{W} = \mathcal{V} \cup \{n^0\}$. In the latter case, $\mathcal{D}$ will be a finite set of maximal 0-nodes because all 0-nodes are of finite degree and because every branch is incident to at least one maximal 0-node. Let $\mathcal{X} = \mathcal{V} \cup \mathcal{D}$. Thus, $\mathcal{X} = \mathcal{V}$ if $\mathcal{W} = \mathcal{V}$, and $\mathcal{X} = \mathcal{D}$ if $\mathcal{W} = \{n^0\}$. Moreover, every branch incident to $n^{\beta+}$ and to a node of $\mathcal{D}$ is a member of $\tilde{\mathcal{A}}$.

Definition of an isolating set in a subsection: $\mathcal{W}$ is called an *isolating set for $n^{\beta+}$ in $\mathcal{S}_b^{\beta-}$* if the following three conditions are satisfied whenever the finite set $\mathcal{V}$ of maximal 0-nodes is nonvoid (so that $\mathcal{S}_b^{\beta-}$ is incident to $n^{\beta+}$ through at least one nonelementary tip — and possibly through elementary tips as well).

Conditions 4.3-2.

(a) Within the core of $\mathcal{S}_b^{\beta-}$, $\mathcal{V}$ separates $\mathcal{A}$ from all the $(\beta+)$-nodes incident to that core and totally disjoint from $n^{\beta+}$ (if there are any) — as well as from all the nodes of the core that are 0-adjacent to $(\beta+)$-nodes including $n^{\beta+}$ (again if there are any). Moreover, no node of $\mathcal{V}$ is 0-adjacent to any $(\beta+)$-node (thus, $\mathcal{V} \cap \mathcal{D} = \emptyset$).

(b) For every node n^0 of $\mathcal{V}$ there is a $(\beta-)$-path in $\mathcal{A}$ that meets n^0 and reaches $n^{\beta+}$ but does not meet $\mathcal{V} \backslash \{n^0\}$.

(c) Every node of $\mathcal{V}$ is incident to a branch in $\tilde{\mathcal{A}}$.

On the other hand, if $\mathcal{V}$ is void (so that $\mathcal{W} = \{n^0\}$, $n^{\beta+}$ is incident to $\mathcal{S}_b^{\beta-}$ only through one or more elementary tips embraced by n^0, and $\mathcal{A}$ is void), we call $\mathcal{W}$ the *trivial isolating set for $n^{\beta+}$ in $\mathcal{S}_b^{\beta-}$*.

Under these conditions (whether or not $\mathcal{V}$ is void), $\mathcal{X}$ will be called the *conjoining set for $n^{\beta+}$ in $\mathcal{S}_b^{\beta-}$ corresponding to $\mathcal{W}$*, the subgraph $\mathcal{A}$ will be called the *arm in $\mathcal{S}_b^{\beta-}$ for $\mathcal{V}$* or *for $\mathcal{W}$*, and $\mathcal{V}$ will be called the *base of $\mathcal{A}$* or the *base of $\mathcal{W}$*. Note that $\mathcal{V}$ and $\mathcal{X}$ are both finite sets of maximal 0-nodes; so too is $\mathcal{W}$ except for the 0-node embraced by $n^{\beta+}$ if such exists — that one will not be maximal. It follows from these definitions that, within $\mathcal{S}_b^{\beta-}$, the isolating set $\mathcal{W}$ separates $n^{\beta+}$ from every other $(\beta+)$-node that is incident to $\mathcal{S}_b^{\beta-}$, and so too does $\mathcal{X}$. Moreover, if $n^{\beta+}$ is incident to the core of $\mathcal{S}_b^{\beta-}$ (i.e., if $\mathcal{V}$ is not void), then within that core $\mathcal{V}$ separates $n^{\beta+}$ from $\mathcal{D}$ as well as from every other $(\beta+)$-node that is incident to that core. It also follows now that $\tilde{\mathcal{A}}$ is not void.

Example 4.3-3. Figure 4.3 (on page 99) illustrates these ideas. Part (a) shows a 1-graph consisting of a two-way-infinite ladder along with an extra branch b_0. All the 0-tips on the extreme left (or extreme right) are shorted through the 1-node n_1^1 (resp. n_2^1). Furthermore, b_0 is incident to n_1^1 through the embraced 0-node n_1^0; it is also incident to the maximal 0-node n_4^0. The entire graph is a $(1-)$-subsection. Its core is the ladder without b_0. $\mathcal{W} = \{n_1^0, n_2^0, n_3^0\}$ is an isolating set for n_1^1, and $\mathcal{X} = \{n_2^0, n_3^0, n_4^0\}$ is the corresponding conjoining set. The corresponding base is $\mathcal{V} = \{n_2^0, n_3^0\}$, and the arm $\mathcal{A}$ for $\mathcal{W}$ is induced by all ladder branches to the left of and between n_2^0 and n_3^0. On the other hand, $\{n_2^0, n_3^0\}$ is not an isolating set for the 1-node n_2^1 because of the presence of b_0. Were b_0 incident to n_6^0 and n_1^0 instead of n_4^0 and n_1^0, $\{n_2^0, n_3^0\}$ would be an isolating set for n_2^1, but then $\{n_1^0, n_2^0, n_3^0\}$ would not be an isolating set for n_1^1.

Part (b) of Figure 4.3 is a 2-graph consisting of a ladder of ladders, an extra ladder L, and three additional branches b_0, b_1, and b_2. The entire graph is a $(2-)$-subsection with two maximal bordering 2-nodes n_1^2 and n_2^2. The internal rank of the $(2-)$-subsection is 1. In each of three ladders, we have selected a pair of maximal 0-nodes as in part (a). This yields the isolating set

$$\mathcal{W} = \{n_1^0, n_2^0, n_3^0, n_4^0, n_5^0, n_6^0, n_7^0\}$$

for n_1^2. In this case, $\mathcal{V} = \{n_2^0, \ldots, n_7^0\}$; moreover, $\mathcal{A}$ is induced by all branches to the left and between the nodes of $\mathcal{V}$ (but not b_1 and b_2). The conjoining set $\mathcal{X}$ corresponding to $\mathcal{W}$ is $\{n_2^0, \ldots, n_7^0, n_{10}^0\}$. The core of the $(2-)$-subsection is induced by all the branches except b_1. This 2-graph has an infinity of $(1-)$-subsections. b_1 and b_2 together induce one of them. Each of the other $(1-)$-subsections consists of a single ladder, except for the two ladders connected by b_0; those two ladders along with b_0 comprise a single $(1-)$-subsection. The internal ranks of the $(1-)$-subsections are all 0. Note that every $(1-)$-subsection has only finitely many bordering nodes. Were branches like b_0 connected to all adjacent ladders, the ladder of ladders would become a single $(1-)$-subsection, but then that $(1-)$-subsection would have an infinity of bordering 1-nodes — in violation of Condition 4.2-2(d). It may also be worth noting that the conjoining set $\mathcal{X} = \{n_2^0, \ldots, n_7^0, n_{10}^0\}$ for n_1^2 happens to be an isolating set for the other 2-node n_2^2. ♣

Example 4.3-4. Two examples of isolating sets for an $\vec{\omega}$-node $n^{\vec{\omega}}$ in an $\vec{\omega}$-graph satisfying Conditions 4.3-1 are illustrated in Figure 4.4 (on page

102), which is a modification of Figure 2.9(b). It is understood that the graph extends downward through a mirror-image extension of the part that is shown. The sets $\{m_0\}$ and $\{m_1^0, m_2^0, m_3^0\}$ of maximal 0-nodes are two of the many isolating sets for $n^{\vec{\omega}}$. They separate $n^{\vec{\omega}}$ from another $\vec{\omega}$-node in the mirror-image extension.

To be sure, this $\vec{\omega}$-graph also has two $\vec{\omega}$-tips; one of them has a representative along the diagonal $\vec{\omega}$-path $P^{\vec{\omega}}$. But, this does not alter the fact that $n^{\vec{\omega}}$ is separated from all $(\vec{\omega}+)$-nodes by the said isolating sets because there are no nodes of ranks higher than $\vec{\omega}$. Were we to convert this $\vec{\omega}$-graph into an ω-graph by appending two singleton ω-nodes that contain the two $\vec{\omega}$-tips, then $n^{\vec{\omega}}$ would not be separated from one of them by any finite set of maximal 0-nodes and therefore would not have an isolating set. ♣.

Lemma 4.3-5. *Let $\mathcal{V}$ be a nonvoid base of an isolating set in a $(\beta-)$-subsection, and let $\mathcal{A}$ be its arm. Then, every branch b in $\mathcal{A}$ is $(\beta-)$-connected within $\mathcal{A}$ to a node of $\mathcal{V}$.*

Proof. $\mathcal{A}$ is not void because $\mathcal{V}$ is not void. b and $\mathcal{V}$ both reside in the core of $\mathcal{S}_b^{\beta-}$. By Lemma 4.1-3(iv) and the fact that $\mathcal{V}$ consists of 0-nodes, there is a two-ended $(\beta-)$-path $P^{\beta-}$ in that core that connects a node of b to a node of $\mathcal{V}$. Moreover, within that core, $\mathcal{V}$ separates $\mathcal{A}$ from its complement $\tilde{\mathcal{A}}$. Hence, a tracing of $P^{\beta-}$ from a node of b to the first node of $\mathcal{V}$ that $P^{\beta-}$ meets yields a two-ended $(\beta-)$-path within $\mathcal{A}$. ♣

So far we have focused on a $(\beta-)$-subsection and considered a bordering node $n^{\beta+}$ in an ancillary way. We now shift our attention to a maximal β-node n^{β} and treat its incident $(\beta-)$-subsections in an ancillary fashion. Such a node must be incident to the core of at least one $(\beta-)$-subsection, with that core having an infinity of branches and an infinity of internal nodes; this is because that node will have at least one $(\beta-1)$-tip $t^{\beta-1}$ and because Condition 4.2-2(c) insures that a representative of $t^{\beta-1}$ will lie in a single $(\beta-)$-subsection. If each of the $(\beta-)$-subsections incident to n^{β} has an isolating set for n^{β}, we can consider n^{β} as being "isolated" by the union of those isolating sets. For subsequent purposes, we want that union to be finite; this is the reason for the last sentence in Conditions 4.3-1.

Definition of an isolating set for a node: Given a maximal β-node n^{β}, a set $\mathcal{W}$ of 0-nodes is called an *isolating set for* n^{β} if $\mathcal{W} = \bigcup_{k=1}^{K} \mathcal{W}_k$, where, for each k, $\mathcal{W}_k$ is an isolating set for n^{β} in a $(\beta-)$-subsection $\mathcal{S}_{b,k}^{\beta-}$ incident to n^β and where the $\mathcal{S}_{b,k}^{\beta-}$ $(k = 1, \ldots, K)$ are all the $(\beta-)$-subsections incident to n^β.

Thus, every isolating set for a maximal β-node n^β separates n^β from all the other maximal $(\beta+)$-nodes in $\mathcal{G}^\nu$.

With $\mathcal{V}_k$ denoting the base for $\mathcal{W}_k$ and with $\mathcal{A}_k$ denoting the corresponding arm, we call $\mathcal{V} = \bigcup_{k=1}^K \mathcal{V}_k$ the *base of* $\mathcal{W}$ and call $\mathcal{A} = \bigcup_{k=1}^K \mathcal{A}_k$ the *arm for* $\mathcal{W}$ or *the arm for* $\mathcal{V}$. We also refer to $\mathcal{A}$ as an *arm for* n^β. Finally, with $\mathcal{X}_k$ denoting the conjoining set in $\mathcal{S}_{b,k}^{\beta-}$ corresponding to $\mathcal{W}_k$, we call $\mathcal{X} = \bigcup_{k=1}^K \mathcal{X}_k$ the *conjoining set for* n^β *corresponding to* $\mathcal{W}$.

Note that the $\mathcal{V}_k$ are disjoint from each other because each $\mathcal{V}_k$ is a set of internal maximal 0-nodes of $\mathcal{S}_{b,k}^{\beta-}$ and because the $\mathcal{S}_{b,k}^{\beta-}$ meet only at bordering nodes. $\mathcal{V} = \bigcup_{k=1}^K \mathcal{V}_k$ is a finite set of maximal 0-nodes within the cores of the $\mathcal{S}_{b,k}^{\beta-}$. However, $\mathcal{V}$ cannot be void — in contrast to any single $\mathcal{V}_k$ — because the core of at least one of the $\mathcal{S}_{b,k}^{\beta-}$ will be incident to n^β, as was noted above. Furthermore, $\mathcal{W} = \mathcal{V} \cup \{n^0\}$ if n^β embraces a 0-node n^0; otherwise $\mathcal{W} = \mathcal{V}$. Finally, with $\mathcal{D}$ denoting the set of all the 0-nodes that are 0-adjacent to n^β, we have $\mathcal{X} = \mathcal{V} \cup \mathcal{D}$; $\mathcal{D}$ may be void, but, if it is not void, all its 0-nodes are maximal.

Our isolating sets have been so defined that maximal nodes are "isolated" not only by the nodes of those isolating sets but also by certain branches incident to those latter nodes. These branches may play a role analogous to that played by the branches incident to an maximal 0-node. For instance, Kirchhoff's current law might be applicable to them, as we shall see when we discuss electrical networks.

Definition of a cut in a subsection: Under the notations and conditions for an isolating set in a subsection, let $\mathcal{C}$ be the set of all branches in $\tilde{\mathcal{A}}$ that are incident to $\mathcal{W}$. Then, $\mathcal{C}$ is called a *cut for* $n^{\beta+}$ *at* $\mathcal{W}$ *in* $\mathcal{S}_b^{\beta-}$.

$\mathcal{C}$ is a finite set because $\mathcal{W}$ is finite and every 0-node is of finite degree.

Example 4.3-6. In Figure 4.3(a), $\mathcal{C} = \{b_0, b_1, b_2\}$ is a cut for n_1^1 at $\mathcal{W} = \{n_1^0, n_2^0, n_3^0\}$. That same set $\mathcal{C}$ is also a cut for n_2^1 at $\mathcal{W}'$, where $\mathcal{W}' = \mathcal{V}' = \{n_4^0, n_5^0\}$. With regard to Figure 4.3(b), let $\mathcal{C}$ be b_1 along with the six branches incident to n_2^0 through n_7^0 and lying to the right of those nodes. Then, $\mathcal{C}$ is a cut for the 2-node n_1^2 at $\mathcal{W} = \{n_1^0, n_2^0, \ldots, n_7^0\}$. On the other hand, b_1 along with the branches incident to n_2^0 through n_7^0 and lying to the left of those nodes comprise a cut for the other 2-node n_2^2 at the isolating set $\{n_2^0, \ldots, n_7^0, n_{10}^0\}$ for n_2^2. ♣

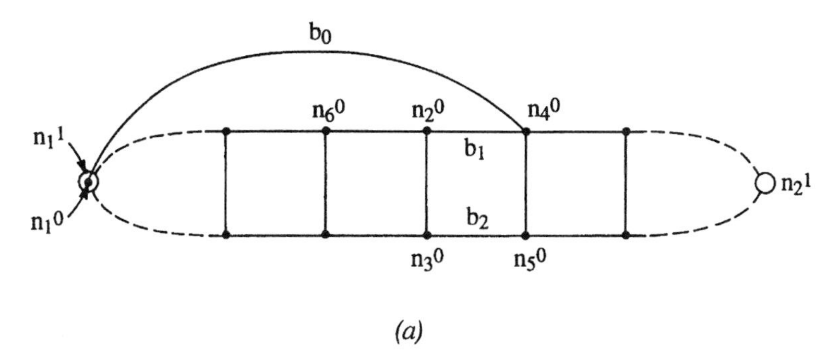

(a)

Figure 4.3. **(a)** A 1-graph consisting of a two-way infinite ladder shorted at its extremities by two 1-nodes and with an appended branch b_0. The heavy dots denote 0-nodes, the two small circles denote 1-nodes, and the line segments are branches.

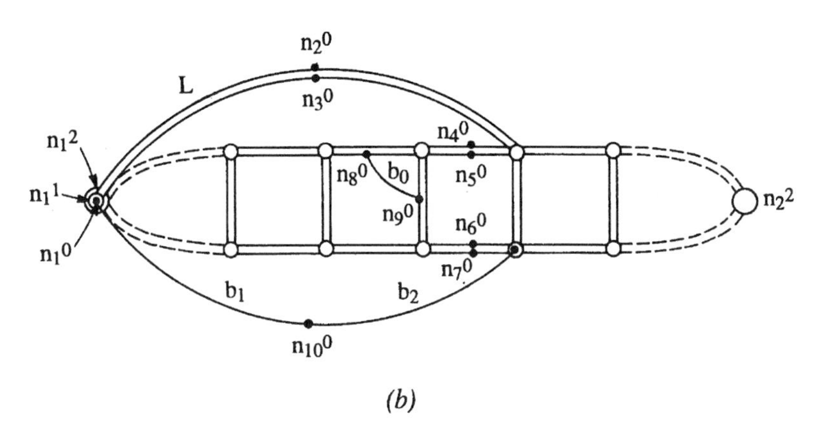

(b)

Figure 4.3. **(b)** A 2-graph consisting of a ladder of ladders, to which are appended three extra branches b_0, b_1, and b_2 and also another ladder L. Each bar of two closely spaced lines denotes a ladder of branches like that of part (a). The two larger circles denote 2-nodes at the two extremities of this 2-graph. The dots and small circles signify 0-nodes and 1-nodes again.

Definition of a cut for a node: Under the notations and conditions for an isolating set for a node, let $\mathcal{C}_k$ be the cut for the β-node n^β at $\mathcal{W}_k$ in $\mathcal{S}_{b,k}^{\beta-}$ for each $k = 1, \ldots, K$. Then, $\mathcal{C} = \bigcup_{k=1}^{K} \mathcal{C}_k$ is called a *cut for* n^β *at* $\mathcal{W}$. We also say that $\mathcal{C}$ *isolates* n^β *from all other* $(\beta+)$*-nodes.*

Note that $\mathcal{C}$ resides in $\tilde{\mathcal{A}} = \bigcup_{k=1}^{K} \tilde{\mathcal{A}}_k = \bigcup_{k=1}^{K} (\mathcal{S}_{b,k}^{\beta-} \backslash \mathcal{A}_k)$.

4.4 Contractions

Remember that the ν-graph $\mathcal{G}^\nu$ satisfies Conditions 4.3-1. For the moment, let us consider once again just a single $(\beta-)$-subsection $\mathcal{S}_b^{\beta-}$ in $\mathcal{G}^\nu$. Let $n^{\beta+}$ be a maximal bordering $(\beta+)$-node for $\mathcal{S}_b^{\beta-}$.

Definition of a contraction in a subsection.

Case 1: First, assume that $n^{\beta+}$ is incident to $\mathcal{S}_b^{\beta-}$ through one or more non-elementary tips (and possibly through elementary tips as well). A *contraction to $n^{\beta+}$ in $\mathcal{S}_b^{\beta-}$* is a sequence $\{\mathcal{W}_p\}_{p=1}^{\infty}$ of isolating sets $\mathcal{W}_p$ for $n^{\beta+}$ in $\mathcal{S}_b^{\beta-}$ satisfying the following two conditions, wherein $\mathcal{A}_p$ is the arm for $\mathcal{W}_p$ and $\mathcal{V}_p$ is its base.

Conditions 4.4-1.

(a) *Given any branch b, there is a p such that b is not in $\mathcal{A}_q$ for all $q \geq p$. Moreover, for $q > p$, $\mathcal{A}_q \subset \mathcal{A}_p$ and $\mathcal{V}_q \cap \mathcal{V}_p = \emptyset$.*

(b) *There is a finite set $\{P_k^{\beta-}\}_{k=1}^{m}$ of one-ended $(\beta-)$-paths in the core of $\mathcal{S}_b^{\beta-}$ such that $P_k^{\beta-}$ reaches $n^{\beta+}$ and also meets exactly one node of $\mathcal{V}_p$ for every p. Moreover, every node of $\mathcal{V}_p$ is met by at least one of the $P_k^{\beta-}$.*

The $P_k^{\beta-}$ are called the *contraction paths for $\{\mathcal{W}_p\}_{p=1}^{\infty}$*.

Case 2: On the other hand, if $n^{\beta+}$ is incident to $\mathcal{S}_b^{\beta-}$ only through elementary tips, set $\mathcal{W}_p = \{n^0\}$ for every p, where n^0 is the 0-node that contains those elementary tips. Then, $\{\mathcal{W}_p\}_{p=1}^{\infty}$ is called the *trivial contraction to $n^{\beta+}$ in $\mathcal{S}_b^{\beta-}$*.

In Case 1, the ranks of the $P_k^{\beta-}$ need not all be the same. Furthermore, with $\overline{\overline{\mathcal{Y}}}$ denoting the cardinality of a set $\mathcal{Y}$, we have $0 < \overline{\overline{\mathcal{V}_p}} \leq m$ and $0 < \overline{\overline{\mathcal{W}_p}} \leq m+1$ for every p. Of course, $\overline{\overline{\mathcal{V}_p}} \leq \overline{\overline{\mathcal{W}_p}}$. Thus, in saying that there is a contraction to $n^{\beta+}$ in $\mathcal{S}_b^{\beta-}$, we are in fact imposing more structure upon $\mathcal{S}_b^{\beta-}$. Note also that the part of a contraction path between $\mathcal{V}_p$ and $n^{\beta+}$ will lie entirely within $\mathcal{A}_p$, for otherwise $\mathcal{V}_p$ would not separate $n^{\beta+}$ from $\tilde{\mathcal{A}}_p = \mathcal{S}_b^{\beta-} \backslash \mathcal{A}_p$.

In Case 2, $\mathcal{V}_p = \emptyset$ for all p, and there are no contraction paths and no arms corresponding to the trivial contraction $\{\mathcal{W}_p\}_{p=1}^{\infty}$.

Lemma 4.4-2. *In Case 1 of the definition of a contraction in a subsection, every node n_0 that is totally disjoint from $n^{\beta+}$ will not be incident to $\mathcal{A}_q$ for all sufficiently large q.*

Proof. The conclusion is obviously true if n_0 is not incident to $\mathcal{S}_b^{\beta-}$; so, assume it is incident to $\mathcal{S}_b^{\beta-}$. If n_0 is incident to $\mathcal{S}_b^{\beta-}$ through an elementary tip, it is incident to a branch of $\mathcal{S}_b^{\beta-}$ that will be excluded from $\mathcal{A}_q$ for all sufficiently large q. Hence, n_0 will be, too.

So, suppose n_0 is incident to $\mathcal{A}_{q_i}$ only through one or more non-elementary tips for infinitely many q_i $(i = 0, 1, 2, \ldots)$ with $q_0 < q_1 < q_2 < \cdots$. Since $\mathcal{A}_{q_i} \subset \mathcal{A}_q$ whenever $q_i > q$, n_0 is incident to $\mathcal{A}_q$ for all $q \geq 1$. Therefore, any path in $\mathcal{A}_1$ that reaches n_0 must meet $\mathcal{V}_q$ for all sufficiently large q. Consequently, within $\mathcal{A}_1$ there is a one-ended δ-path P^δ which meets every $\mathcal{V}_q$ for all sufficiently large q and whose δ-tip t^δ is shorted to n_0. Since the cardinalities of all the $\mathcal{V}_q$ are uniformly bounded by the natural number m, P^δ meets a contraction path in $\mathcal{A}_1$ infinitely often; that is, t^δ is nondisconnectable from the tip t_0 of that contraction path. By Condition 4.2-2(b), t^δ and t_0 are shorted together. Since t_0 is embraced by $n^{\beta+}$, n_0 and $n^{\beta+}$ cannot be totally disjoint. This contradicts our hypothesis. So, our supposition is false. ♣

Definition of a contraction to a node: Let n^β be a maximal β-node and let $\mathcal{S}_{b,k}^{\beta-}$ $(k = 1, \ldots, K)$ be its incident $(\beta-)$-subsections. Assume that there is a contraction $\{\mathcal{W}_{k,p}\}_{p=1}^\infty$ to n^β in each $\mathcal{S}_{b,k}^{\beta-}$. Set $\mathcal{W}_p = \bigcup_{k=1}^K \mathcal{W}_{k,p}$ for each p. Under these conditions, $\{\mathcal{W}_p\}_{p=1}^\infty$ is called a *contraction to* n^β. A *contraction path for* $\{\mathcal{W}_p\}_{p=1}^\infty$ is simply a contraction path for $\{\mathcal{W}_{k,p}\}_{p=1}^\infty$ in one of the $\mathcal{S}_{b,k}^{\beta-}$.

Here again, for all p, $\overline{\overline{\mathcal{V}_p}}$ and $\overline{\overline{\mathcal{W}_p}}$ are uniformly bounded by a natural number, and $\overline{\overline{\mathcal{V}_p}} \leq \overline{\overline{\mathcal{W}_p}}$. Also, for $q > p$, we have again $\mathcal{A}_q \subset \mathcal{A}_p$ and $\mathcal{V}_q \cap \mathcal{V}_p = \emptyset$.

Example 4.4-3. We can construct a contraction to the 1-node n_1^1 in Figure 4.3(a) by choosing an infinite sequence of vertically adjacent 0-node pairs, such as the base $\{n_2^0, n_3^0\}$, that shift progressively leftwards. Each such base along with the embraced 0-node n_1^0 comprise one of the isolating sets in the contraction to n_1^1. The corresponding contraction paths lie along the upper and lower horizontal parts of the ladder.

An example of a contraction to the 2-node n_1^2 in Figure 4.3(b) can be obtained by shifting the base $\{n_2^0, \ldots, n_7^0\}$ progressively leftwards and appending n_1^0 to each such base to get a sequence of isolating sets for n_1^2. Now, we have six contraction paths lying along the three upper parts and three lower parts of the horizontal ladders and the ladder L.

Actually, a node of any rank other than the rank $\vec{\omega}$ may (but need not) have a contraction in a ν-graph $\mathcal{G}^\nu$ satisfying Conditions 4.3-1. Examples are easily constructed. For instance, if $\mathcal{G}^\omega$ is simply a one-ended $\vec{\omega}$-path incident to an ω-node n^ω, a contraction can be chosen for n^ω by referring to Figure 2.9(c) and selecting a maximal 0-node from each μ-path P^μ ($\mu = 0, 1, 2, \ldots$). ♣

However, there is a peculiarity concerning the $\vec{\omega}$-nodes. We have noted by means of Figure 4.4 that in an $\vec{\omega}$-graph satisfying Conditions 4.3-1 an $\vec{\omega}$-node $n^{\vec{\omega}}$ can have an isolating set. But, there is no contraction to $n^{\vec{\omega}}$ in that figure because it is impossible to choose the $\mathcal{V}_p$ ($p = 1, 2, 3, \ldots$) with uniformly finitely bounded cardinalities. This is a particular case of the following general result.

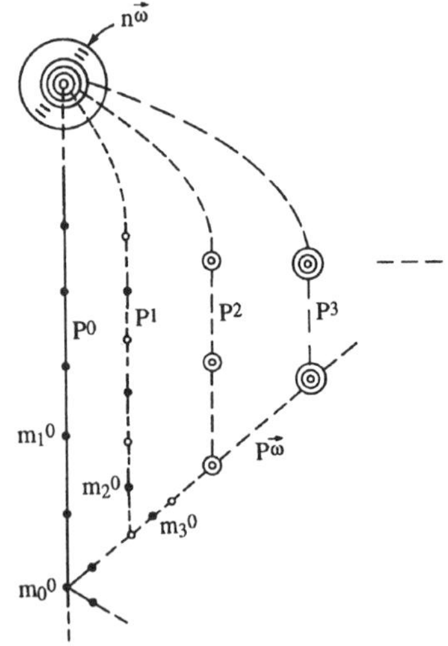

Figure 4.4. The $\vec{\omega}$-graph discussed in Example 4.3-4. As always, heavy dots are 0-nodes, single circles are 1-nodes, double circles are 2-nodes, and so on. The sets $\{m_0^0\}$ and $\{m_1^0, m_2^0, m_3^0\}$ are isolating sets for the $\vec{\omega}$-node $n^{\vec{\omega}}$. The P^μ($\mu = 0, 1, 2, \ldots$) are one-ended μ-paths whose μ-tips are embraced by $n^{\vec{\omega}}$. The diagonal path $P^{\vec{\omega}}$ is an $\vec{\omega}$-path whose $\vec{\omega}$-tip is not embraced. It is understood that this graph extends downward through the mirror image of the part shown.

Theorem 4.4-4. *If the ν-graph $\mathcal{G}^\nu$ ($\nu = \vec{\omega}$ or ω) satisfies Conditions 4.3-1 and if the node n embraces an $\vec{\omega}$-node $n^{\vec{\omega}}$ (i.e., either n is $n^{\vec{\omega}}$ or n contains $n^{\vec{\omega}}$), then there does not exist a contraction to n.*

Proof. Suppose there is a contraction to n. Since $n^{\vec{\omega}}$ embraces infinitely many tips with infinitely many ranks, we can select an infinite sequence of tips of strictly increasing ranks and then a single representative for each tip in the sequence. In fact, we can and do choose those representatives to satisfy Conditions 4.2-2(c); that is, each such representative of any rank μ does not reach any node of rank greater than μ other than those embraced by n. This yields an infinite set $\mathcal{R}$ of representatives with strictly increasing natural number ranks. For each representative in $\mathcal{R}$ there will be a natural number $\hat{p}$ depending upon the choice of the representative such that the representative meets each base $\mathcal{V}_p$ of $\mathcal{W}_p$ for all $p \geq \hat{p}$. Note that a representative may meet a $\mathcal{V}_p$ several times but not more than $\overline{\overline{\mathcal{V}_p}}$ times. Since for all p the $\overline{\overline{\mathcal{V}_p}}$ are uniformly bounded by the natural number m according to our supposition, there will be a representative P^β that meets infinitely many representatives in $\mathcal{R}$ with ranks larger than β and moreover meets each of them infinitely many times along some or all of the bases $\mathcal{V}_p$. Indeed, if this were not the case, the $\overline{\overline{\mathcal{V}_p}}$ could not be uniformly bounded by m for all p.

For any such representative P^λ ($\lambda > \beta$), we can trace along P^β starting at some node n_0 until we reach P^λ at a 0-node of some $\mathcal{V}_p$ and then can trace along P^λ until we first meet a maximal node of rank larger than β. Moreover, we can do this for infinitely many choices of λ larger than β, where for each choice we choose another base sufficiently closer to n in order to find still another maximal node of rank larger than β. This implies that there is an infinite set of maximal $((\beta+1)+)$-nodes that are reached by $((\beta+1)-)$-paths that start at n_0. Consequently, there is a $((\beta+1)-)$-subsection having infinitely many bordering nodes — in violation of Condition 4.2-2(d). ♣

Corollary 4.4-5. *If an ω-graph $\mathcal{G}^\omega$ satisfies Conditions 4.3-1 and if an ω-node in $\mathcal{G}^\omega$ has a contraction, then n^ω does not contain an $\vec{\omega}$-node (that is, n^ω embraces only finitely many nodes).*

This last corollary will be significant when we discuss transfinite random walks because Conditions 4.3-1 will then be in force and all ω-nodes will have contractions. None of the graphs to be considered at that time will have $\vec{\omega}$-nodes.

4.5 Finitely Structured ν-Graphs

Here now is the special structure of a transfinite graph upon which our theory of transfinite random walks will be based.

Definition of a finitely structured ν-graph: A ν-graph $\mathcal{G}^\nu$ will be called *locally finitely structured* if $\mathcal{G}^\nu$ satisfies Conditions 4.3-1 (and thereby Conditions 4.2-2 as well), if there is a contraction to n^β for every maximal β-node n^β of every rank β with $1 \leq \beta \leq \nu$, and if every tip is embraced by a node. $\mathcal{G}^\nu$ will be called *finitely structured* if it is locally finitely structured and has at most finitely many ν-nodes. (If ν is a natural number or ω, the finiteness of the number of ν-nodes insures that every tip is embraced because there cannot be any tip of rank ν or higher.)

Lemma 4.5-1.

(i) *A locally finitely structured graph cannot have any $\vec{\omega}$-node.*

(ii) *No $\vec{\omega}$-graph can be finitely structured.*

(iii) *A locally finitely structured ν-graph $\mathcal{G}^\nu$ $(\nu \neq \vec{\omega})$ is perforce finitely structured.*

Proof. (i) This is an immediate consequence of Theorem 4.4-4 and the requirement in the above definition concerning the existence of contractions.

(ii) Suppose the $\vec{\omega}$-graph $\mathcal{G}^{\vec{\omega}}$ is locally finitely structured. By (i), $\mathcal{G}^{\vec{\omega}}$ cannot have any $\vec{\omega}$-node. So, by Theorem 4.2-7, $\mathcal{G}^{\vec{\omega}}$ has an $\vec{\omega}$-tip. But then, by definition again, there will be an ω-node embracing that $\vec{\omega}$-tip. This contradicts the fact that $\mathcal{G}^{\vec{\omega}}$ is an $\vec{\omega}$-graph (not an ω-graph).

(iii) Suppose $\mathcal{G}^\nu$ has infinitely many ν-nodes. Then, by Theorem 4.2-4, it has a ν-tip. That tip must be embraced by a $(\nu+1)$-node. This contradicts the fact that $\mathcal{G}^\nu$ is a ν-graph, not a graph of higher rank. ♣

Lemma 4.5-2. *Assume $\mathcal{G}^\nu$ is finitely structured. Then, $\mathcal{G}^\nu$ has only finitely many $(\nu-)$-subsections.*

Proof. $\mathcal{G}^\nu$ has only finitely many ν-nodes, all of which are perforce maximal. Moreover, every ν-node is incident to only finitely many $(\nu-)$-subsections (Conditions 4.3-1). Furthermore, every $(\nu-)$-subsection is incident to at least one ν-node since $\mathcal{G}^\nu$ is ν-connected . Our conclusion follows. ♣

Example 4.5-3. Figure 4.5 simulates several $(\beta-)$-subsections comprising a part of a finitely structured ν-graph. When $\beta = 1$, each crosshatched area denotes a $(1-)$-subsection , which is in fact a conventional finite or infinite graph of branches and 0-nodes. Each $(1-)$-subsection is incident to only finitely many bordering $(1+)$-nodes. Bordering 1-nodes are represented by small circles, but there may also be nodes of higher ranks bordering the $(1-)$-subsections. The $(2+)$-nodes are indicated by double circles. This pattern of $(1-)$-subsections may extend infinitely to approach $(2+)$-nodes bordering a $(2-)$-subsection . The latter is denoted by the outer ring of dashed lines. However, it may happen that a particular $(2-)$-subsection consists of only finitely many $(1-)$-subsections, in which case each of its bordering $(2+)$-nodes must be incident to at least one of those $(1-)$-subsections; such a $(2+)$-node is indicated at the bottom of the figure. In any case, every $(2+)$-node must be approachable through an infinity of $(1-)$-subsections within at least one of its incident $(2-)$-subsections.

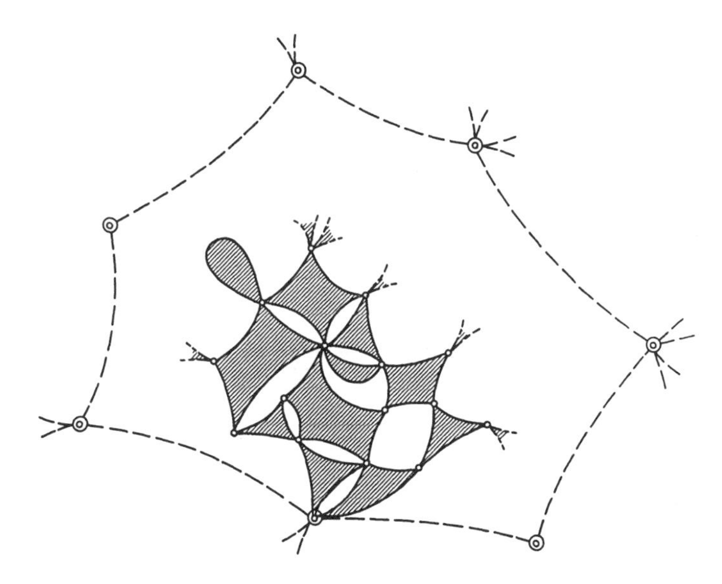

Figure 4.5. An impression of a portion of a finitely structured ν-graph. It is described in Example 4.5-3.

This pattern also holds for the higher ranks. For example, let us shift our perspective one rank higher. The crosshatched areas now represent $(2-)$-subsections, the small circles represent 2-nodes, and the double circles represent $(3+)$-nodes.

And so it goes for the higher ranks. If for any natural number μ there are only finitely many μ-nodes, then there can be no nodes of higher ranks, and we will have a finitely structured μ-graph. But, if there are infinitely many μ-nodes for every natural number μ, then the ν-graph will be a finitely structured graph of rank ω (or perhaps higher were we to allow such higher ranks). This is — so to speak — a "reverse fractal structure"; instead of the pattern repeating itself as our view focuses more and more microscopically, it repeats itself as our view extends more and more telescopically. However, this reversed analogy should not be interpreted too strictly; the patterns repeat themselves only in a general way but not in any precise fashion.

Of course, a finitely structured graph has still more structure than that illustrated in Figure 4.5. Indeed, every maximal node of every rank has a contraction, and Conditions 4.2-2 and 4.3-1 are fulfilled as well. ♣

Given an arm $\mathcal{A}$ for a maximal β-node, we say that a one-ended α-path P^α $(\alpha < \beta)$ *eventually lies in* $\mathcal{A}$ if in P^α there is a one-ended α-path that lies entirely in $\mathcal{A}$.

Lemma 4.5-4. *Let $\mathcal{G}^\nu$ be finitely structured. For every rank β $(0 < \beta \leq \nu)$ and for every maximal β-node in $\mathcal{G}^\nu$, choose an arm for that β-node. Then, every one-ended α-path P^α $(0 \leq \alpha < \nu)$ will eventually lie in the arm $\mathcal{A}$ for the maximal node n^β $(\alpha < \beta)$ that embraces the α-tip of P^α.*

Proof. Let $\mathcal{V}$ be the base of $\mathcal{A}$. P^α cannot pass infinitely often into and out of $\mathcal{A}$ because each such passage must be through a different node of $\mathcal{V}$ and $\mathcal{V}$ has only finitely many nodes. Since n^β embraces the α-tip of P^α, P^α must eventually lie in $\mathcal{A}$. ♣

Recall that for two subgraphs $\mathcal{H}$ and $\mathcal{M}$ of $\mathcal{G}^\nu$, $\mathcal{H}\backslash\mathcal{M}$ denotes the subgraph induced by all the branches in $\mathcal{H}$ that are not in $\mathcal{M}$.

Lemma 4.5-5. *Let $\mathcal{G}^\nu$ be finitely structured. Let $\{\mathcal{W}_p\}_{p=1}^\infty$ be a contraction to the maximal β-node n^β $(\beta > 0)$, and let $\{\mathcal{A}_p\}_{p=1}^\infty$ be the corresponding sequence of arms. Then, for any $q > p$ the subgraph $\mathcal{A}_p\backslash\mathcal{A}_q$ has at most finitely many nodes of rank $\beta - 1$ and no node of higher rank.*

Proof. First note that, if $\beta = \omega$, $\beta - 1$ denotes $\vec{\omega}$. By Lemma 4.5-1(i), there are no $\vec{\omega}$-nodes in $\mathcal{G}^\nu$. Furthermore, $\mathcal{A}_p\backslash\mathcal{A}_q$ will not have any ω-node.

So, consider the case where β is a natural number. Let $\mathcal{V}_p$ be the base of $\mathcal{A}_p$, and let $\mathcal{S}_c$ denote the union of the cores of all the $(\beta-)$-subsections to which n^β is incident. Now, all the nodes of $\mathcal{A}_p \backslash \mathcal{A}_q$ are interior nodes of $\mathcal{S}_c$ and hence have ranks no larger than $\beta - 1$. Also, $\mathcal{V}_p \cup \mathcal{V}_q$ separates $\mathcal{A}_p \backslash \mathcal{A}_q$ from all $(\beta+)$-nodes, and all the nodes of $\mathcal{V}_p \cup \mathcal{V}_q$ are maximal 0-nodes. It follows that no $(\beta+)$-node is incident to $\mathcal{A}_p \backslash \mathcal{A}_q$.

Now, suppose that $\mathcal{A}_p \backslash \mathcal{A}_q$ has an infinity of $(\beta - 1)$-nodes. Each one has to be maximal. $\mathcal{A}_p \backslash \mathcal{A}_q$ can have only finitely many components because each branch of $\mathcal{A}_p \backslash \mathcal{A}_q$ is $(\beta-)$-connected to a node of $\mathcal{V}_p \cup \mathcal{V}_q$ according to Lemma 4.3-5 and because $\mathcal{V}_p \cup \mathcal{V}_q$ is a finite set. Thus, one of these components, say, $\mathcal{M}$, has an infinity of $(\beta - 1)$-nodes. Moreover, $\mathcal{M}$ is $(\beta - 1)$-connected since it is $(\beta-)$-connected . Since $\mathcal{G}^\nu$ is finitely structured, by Condition 4.2-2(d), every $(\beta - 1)$-node in $\mathcal{M}$ is $(\beta - 1)$-adjacent to only finitely many $(\beta - 1)$-nodes. In fact, all of the hypothesis of Theorem 4.2-4 is fulfilled when $\mathcal{G}^\nu$ is replaced by $\mathcal{A}_p \backslash \mathcal{A}_q$ and ν is replaced by $\beta - 1$. Consequently, $\mathcal{M}$ contains at least one one-ended $(\beta - 1)$-path. Hence, $\mathcal{M}$ and thereby $\mathcal{A}_p \backslash \mathcal{A}_q$ has a $(\beta - 1)$-tip. Since $\mathcal{A}_p \backslash \mathcal{A}_q$ is a subgraph of a ν-graph where $\nu \geq \beta$, $\mathcal{A}_p \backslash \mathcal{A}_q$ must have a β-node. This contradicts our prior conclusion that no $(\beta+)$-node is incident to $\mathcal{A}_p \backslash \mathcal{A}_q$. Thus, our supposition is wrong, and $\mathcal{A}_p \backslash \mathcal{A}_q$ has at most finitely many $(\beta - 1)$-nodes and no node of higher rank. ♣

Another pertinent result is that a finitely structured graph is countable (i.e., has a countable branch set). Moreover, this countability derives from Conditions 4.2-2 and 4.3-1 alone, since the other attributes of a locally finitely structured graph are superfluous in this respect. The proof of this is much like that of Theorem 3.6-2 except that now we argue in terms of subsections rather than sections. We now need to define the "adjacency" of a subsection. Two $(\beta-)$-subsections will be called *β-adjacent* if they share a common bordering node, that is, if they meet the same $(\beta+)$-node. Moreover, the *β-adjacency* of a $(\beta-)$-subsection $\mathcal{S}_b^{\beta-}$ is the set of all other $(\beta-)$-subsections that are β-adjacent to $\mathcal{S}_b^{\beta-}$. The special case of $\beta = 0$ yields the same adjacency idea as for a $\vec{0}$-section. That is, a $(0-)$-subsection is a single branch (as is a $\vec{0}$-section), and its 0-adjacency is the set of branches that are 0-adjacent to it.

Lemma 4.5-6. *A finitely structured ν-graph $\mathcal{G}^\nu$ is countable.*

Proof. Consider any $(1-)$-subsection $\mathcal{S}_b^{1-}$ and choose any branch b_0 in it. Set $\mathcal{H}_0 = \{b_0\}$. Let $\mathcal{H}_1$ be the set of all branches in $\mathcal{S}_b^{1-}$ that are 0-

adjacent to b_0. $\mathcal{H}_1$ is a finite set since all 0-nodes are of finite degree. Proceeding inductively, let us assume that $\mathcal{H}_0, \mathcal{H}_1, \ldots, \mathcal{H}_{k-1}$ have been chosen as finite sets of branches in $\mathcal{S}_b^{1-}$. Let $\mathcal{H}_k$ be the set of branches in $\mathcal{S}_b^{1-}$ that are 0-adjacent to branches of $\mathcal{H}_{k-1}$ and are not in $\bigcup_{l=0}^{k-1} \mathcal{H}_l$. Since all 0-nodes are of finite degree and since $\mathcal{H}_{k-1}$ is a finite set, $\mathcal{H}_k$ is a finite set too. Moreover, every branch in $\mathcal{S}_b^{1-}$ will lie in some $\mathcal{H}_k$ because it is 0-connected to b_0 through a two-ended 0-path that does not meet any $(1+)$-node. It now follows that $\mathcal{S}_b^{1-}$ is countable.

Next, let us assume that, for some natural number μ, every $(\mu-)$-subsection is countable. Consider any $((\mu+1)-)$-subsection $\mathcal{S}_b^{(\mu+1)-}$. It is partitioned by a set of $(\mu-)$-subsections according to Lemma 4.1-1. Observe that, for each $(\mu-)$-subsection $\mathcal{S}_b^{\mu-}$ in $\mathcal{S}_b^{(\mu+1)-}$, there are at most finitely many $(\mu-)$-subsections in $\mathcal{S}_b^{(\mu+1)-}$ that are μ-adjacent to $\mathcal{S}_b^{\mu-}$; this is a consequence of Condition 4.2-2(d) and the last sentence of Conditions 4.3-1. Now, let $\mathcal{H}_0$ be any $(\mu-)$-subsection in $\mathcal{S}_b^{(\mu+1)-}$. Let $\mathcal{H}_1$ be the union of all $(\mu-)$-subsections in $\mathcal{S}_b^{(\mu+1)-}$ that are μ-adjacent to $\mathcal{H}_0$. Recursively, having chosen $\mathcal{H}_0, \mathcal{H}_1, \ldots, \mathcal{H}_{k-1}$, we let $\mathcal{H}_k$ be the union of all the $(\mu-)$-subsections in $\mathcal{S}_b^{(\mu+1)-}$ that are μ-adjacent to $(\mu-)$-subsections in $\mathcal{H}_{k-1}$ but are not in $\bigcup_{l=0}^{k-1} \mathcal{H}_l$. Since all branches of $\mathcal{S}_b^{(\mu+1)-}$ are pairwise μ-connected by two-ended μ-paths that do not meet $((\mu+1)+)$-nodes, $\bigcup_{k=0}^{\infty} \mathcal{H}_k$ will be $\mathcal{S}_b^{(\mu+1)-}$. (It may happen that the $\mathcal{H}_k$ are void for all sufficiently large k.) By our above observation, there are only finitely many $(\mu-)$-subsections in each $\mathcal{H}_k$. Hence, there are only countably many $(\mu-)$-subsections in each $\mathcal{S}_b^{(\mu+1)-}$, each of which is countable by our inductive assumption. Consequently, $\mathcal{S}_b^{(\mu+1)-}$ is countable too. Now, if $\mathcal{G}^\nu$ is a μ-graph (i.e., $\nu = \mu$), $\mathcal{G}^\nu$ is a $((\mu+1)-)$-subsection by itself and therefore is countable.

We do not need to consider any $(\vec{\omega}-)$-subsection because by Lemma 4.5-1 a finitely structured graph does not have any $\vec{\omega}$-nodes.

So consider next an $(\omega-)$-subsection $\mathcal{S}_b^{\omega-}$ in $\mathcal{G}^\nu$, where now $\nu = \omega$. Any two branches in $\mathcal{S}_b^{\omega-}$ are connected by a two-ended μ-path that meets them but does not meet any ω-node. Consequently, upon choosing any $(1-)$-subsection $\mathcal{S}_b^{1-}$ in $\mathcal{S}_b^{\omega-}$ and letting $\mathcal{S}_b^{\mu-}$ be the unique $(\mu-)$-subsection in which $\mathcal{S}_b^{1-}$ lies (Lemma 4.1-1), we obtain

$$\mathcal{S}_b^{\omega-} = \bigcup_{1 \le \mu < \vec{\omega}} \mathcal{S}_b^{\mu-}.$$

Since each $\mathcal{S}_b^{\mu-}$ is countable, so to is $\mathcal{S}_b^{\omega-}$.

Finally, $\mathcal{G}^\omega$ can have only countably many $(\omega-)$-subsections because of Condition 4.2-2(d), Conditions 4.3-1, and the argument given in the second paragraph of this proof. Thus, $\mathcal{G}^\nu$ is countable in this case too. ♣

It is worth mentioning again that all of our results can be extended to graphs of ranks higher than ω. For instance, we can repeat the arguments of this chapter to get finitely structured graphs of ordinal ranks $\omega+1, \omega+2, \ldots, \omega \cdot 2$. Further repetitions will reach still higher ranks. However, for the arrow ranks, there will be no nodes with contractions and no finitely structured graphs.

4.6 Spanning Trees

As with a conventional graph, a *spanning tree* in a ν-connected ν-graph $\mathcal{G}^\nu$ is a ν-connected subgraph that meets every node and has no loops. It is a fact that every connected 0-graph (whether finite or infinite) has a spanning tree [29, Theorem 9.1]. Does every connected transfinite graph have one? The answer is "no." Example 4.6-1 below shows this. Nonetheless, we will prove that, if the transfinite graph is finitely structured, it will have a spanning tree. This too is established inductively by working from lower ranks for subsections to higher ranks and by building a spanning tree for each subsection by expanding a tree through subsections of lower ranks.

Example 4.6-1. Perhaps the simplest transfinite graph having no spanning tree is the graph of Figure 3.1. The difficulty arises again from the fact that the graph has nondisconnectable tips that are not shorted together. In particular, if a tree were to meet any 1-node in this graph, it would have to traverse a representative for the single 0-tip in that 1-node. For any two such 1-nodes, those representatives would differ for some pair of branches a_k and b_k for some k — that is, the tree would contain a loop, an impossibility. This problem would not disappear were we to require only that the tree be "essentially spanning" in the sense that it be required to meet only the nonsingleton 1-nodes; in this case it would still have to meet the 1-nodes n_a^1 and n_b^1 and would again have to contain a loop with branches a_k and b_k for some k. ♣

The *distance* between two 0-nodes of a connected 0-graph is the number of branches in the shortest path between them.

Theorem 4.6-2. *Every finitely structured 1-graph $\mathcal{G}^1$ has a spanning tree.*

Proof. Choose any 0-subsection of $\mathcal{G}^1$. For a subsequent purpose, we shall denote it by $\mathcal{S}^0_{b,0,1}$. $\mathcal{S}^0_{b,0,1}$ is either a 0-graph or a 1-graph, but its internal rank is 0. So, let us consider at first the 0-graph of $\mathcal{S}^0_{b,0,1}$, that is, the graph consisting of all the branches and 0-nodes of $\mathcal{S}^0_{b,0,1}$. (The 1-nodes of $\mathcal{S}^0_{b,0,1}$ will all be bordering nodes and are only finitely many in number.) In that 0-graph we can choose a spanning tree T^0_0 as follows. Choose any internal 0-node n^0 in $\mathcal{S}^0_{b,0,1}$ along with its finitely many incident branches. This is a star graph. We now proceed inductively as follows. Assume that a spanning tree has been chosen for the subgraph induced by all branches of $\mathcal{S}^0_{b,0,1}$ whose 0-nodes are all at distances from n^0 less than m ($m \geq 2$). Let $n^0_{m,k}$ ($k = 1, \ldots, K_m$) be the finitely many 0-nodes that are at a distance m from n^0. Set $\mathcal{N}^0_m = \{n_{m,k} : k = 1, \ldots, K_m\}$, and similarly for $\mathcal{N}^0_{m-1}$. Choose any branch connecting $n^0_{m,1}$ to a 0-node in $\mathcal{N}^0_{m-1}$; there will be such a branch. Proceed along the nodes of $\mathcal{N}^0_m$ as follows. Having chosen one such branch for each $n^0_{m,l}$ ($1 \leq l < k \leq K_m$), choose one branch connecting $n^0_{m,k}$ to a node in $\mathcal{N}^0_{m-1}$. At each k there will be at least one such branch, and — whatever be that choice — no loop will be formed with the previously chosen branches. Proceeding in this fashion along the nodes of each $\mathcal{N}^0_m$ and then successively for $m = 2, 3, \ldots$, we obtain a spanning tree T^0_0 for the 0-graph of $\mathcal{S}^0_{b,0,1}$.

Now, consider any bordering node n^1 of $\mathcal{S}^0_{b,0,1}$. If n^1 contains a 0-node n^0, then the tree T^0_0 will meet n^1 through a branch incident to n_0. Otherwise, the tree will reach n^1 through one or more 0-tips. However, the arms $\mathcal{A}_p$ ($p = 1, 2, \ldots$) for any contraction to n^1 within $\mathcal{S}^0_{b,0,1}$ will have bases $\mathcal{V}_p$ whose cardinalities are all bounded by some natural number q. It follows that the tree T^0_0 can reach n^1 through no more than q 0-tips because the representatives of those 0-tips must be totally disjoint if T^0_0 is to have no loops. Since $\mathcal{S}^0_{b,0,1}$ has only finitely many bordering nodes, the tree T^0_0 will have only finitely many 0-tips.

Next, consider the 1-loops and 0-loops created by the shorts imposed by the bordering nodes of $\mathcal{S}^0_{b,0,1}$ upon the 0-tips and finitely many of the elementary tips of T^0_0. There can be only finitely many such 1-loops and

0-loops. We can break all of them one-by-one by opening finitely many branches without disconnecting the resulting graphs. This yields a spanning tree T for $\mathcal{S}^0_{b,0,1}$. (T may have 1-nodes.)

Repeat this procedure to get such a spanning tree for every 0-subsection of $\mathcal{G}^1$. Let us denote the 0-subsections by $\mathcal{S}^0_{b,m,l}$ ($m = 0, 1, 2, \ldots; l = 1, \ldots, L_m; L_0 = 1$), where, for each $m \geq 1$ and each l, $\mathcal{S}^0_{b,m,l}$ is 1-adjacent to at least one subsection $\mathcal{S}^0_{b,m-1,l}$ but not 1-adjacent to any $\mathcal{S}^0_{b,p,l}$, where $p < m - 1$. The union of any finite number of spanning trees for the $\mathcal{S}^0_{b,m,l}$ will contain no more than finitely many 0-loops or 1-loops. These too can be broken one-by-one by opening branches without disconnecting the resulting graphs. We do so by proceeding sequentially starting with a spanning tree for $\mathcal{S}^0_{b,0,1}$, then appending all the spanning trees for the $\mathcal{S}^0_{b,1,l}$ and breaking loops, then appending all the spanning trees for the $\mathcal{S}^0_{b,2,l}$ and breaking loops, and so on. Since $\mathcal{G}^1$ is finitely structured, it has only finitely many 0-subsections. Thus, the procedure ends, and the result is a spanning tree for $\mathcal{G}^1$. ♣

Corollary 4.6-3. *Let $\mathcal{S}^1_b$ be a 1-subsection in a finitely structured ν-graph ($\nu \geq 2$). Then, $\mathcal{S}^1_b$ has a spanning tree.*

Note. Here, the superscript 1 indicates that the internal rank of the subsection is 1. Thus, $\mathcal{S}^1_b$ is a (2−)-subsection.

Proof. By working with the 0-subsections of $\mathcal{S}^1_b$ as in the preceding proof, we can construct a spanning tree T for the 1-graph of $\mathcal{S}^1_b$, but in this case the procedure may continue indefinitely for $m = 1, 2, \ldots$ since $\mathcal{S}^1_b$ may have infinitely many 0-subsections — but no more than countably many of them (Lemma 4.5-6). Nonetheless, the said spanning tree will be achieved.

Every bordering node of $\mathcal{S}^1_b$ will impose shorts among some of the tips (i.e., elementary tips, 0-tips, and 1-tips) of T. But, here again $\mathcal{S}^1_b$ has only finitely many bordering nodes, and T will reach them through no more than finitely many 0-tips and 1-tips because of the finite structuring. Therefore, only finitely many loops will be created by the shorts imposed by the bordering nodes of $\mathcal{S}^1_b$. So, here again we can break them one-by-one by opening branches without disconnecting the resulting graphs. In the end we will have a spanning tree for $\mathcal{S}^1_b$. ♣

This procedure for constructing a spanning tree can be extended inductively and without much alteration to graphs with higher ranks. The induction is on the internal ranks of subsections. We start with the natural-number ranks.

Theorem 4.6-4. *Every finitely structured μ-graph $\mathcal{G}^\mu$, where μ is a natural number, has a spanning tree.*

Proof. Consider any α-subsection $\mathcal{S}_b^\alpha$ of $\mathcal{G}^\mu$, where α is the internal rank of $\mathcal{S}_b^\alpha$ $(2 \le \alpha \le \mu)$. $\mathcal{S}_b^\alpha$ is partitioned by its $(\alpha-)$-subsections. We can arrange the latter by choosing any one of them, say, $\mathcal{S}_{b,0,1}^{\alpha-}$, then letting $\mathcal{S}_{b,1,l}^{\alpha-}$ $(l = 1, \ldots, L_1)$ be the $(\alpha-)$-subsections that are α-adjacent to $\mathcal{S}_{b,0,1}^{\alpha-}$, and similarly, for each $m = 2, 3, \ldots$, letting $\mathcal{S}_{b,m,l}^{\alpha-}$ $(l = 1, \ldots, L_m)$ be the $(\alpha-)$-subsections that are α-adjacent to at least one of the $\mathcal{S}_{b,m-1,l}^{\alpha-}$ but not α-adjacent to any $\mathcal{S}_{b,p,l}^{\alpha-}$, where $p < m - 1$.

Let us assume that a spanning tree $T_{m,l}$ has been constructed for each $(\alpha-)$-subsection $\mathcal{S}_{b,m,l}^{\alpha-}$ $(m = 0, 1, 2, \ldots; l = 1, \ldots, L_m; L_0 = 1)$. Thus, $T_{0,1}$ is a spanning tree for $\mathcal{S}_{b,0,1}^{\alpha-}$. Then, $T_{0,1} \cup \left(\bigcup_{l=1}^{L_1} T_{1,l}\right)$ will have only finitely many loops. These can be broken by opening no more than finitely many branches without disconnecting the resulting graph. We get a spanning tree T_1 for $\mathcal{S}_{b,0,1}^{\alpha-} \cup \left(\bigcup_{l=1}^{L_1} \mathcal{S}_{b,1,l}^{\alpha-}\right)$.

We proceed inductively for $m = 2, 3, \ldots$. Let T_{m-1} be the spanning tree obtained after appending all the spanning trees $T_{p,l}$ up to and including the index $p = m - 1$ and breaking loops. Again $T_{m-1} \cup \left(\bigcup_{l=1}^{L_m} T_{m,l}\right)$ will have only finitely many loops, and these too can be broken by opening finitely many branches to get a spanning tree T_m for $\bigcup_{p=0}^{m} \bigcup_{l=1}^{L_p} \mathcal{S}_{b,p,l}^{\alpha-}$. Continuing this procedure through all m, we obtain a spanning tree for the α-graph of $\mathcal{S}_b^\alpha$, where again α is the internal rank of $\mathcal{S}_b^\alpha$. As the next step, we break no more than finitely many loops — those created by the bordering nodes of $\mathcal{S}_b^\alpha$ — to get a spanning tree for $\mathcal{S}_b^\alpha$.

In view of Corollary 4.6-3, we can now conclude that, for each $\alpha = 1, \ldots, \mu$, every α-subsection of $\mathcal{G}^\mu$ has a spanning tree, and in particular $\mathcal{G}^\mu$ itself has a spanning tree. ♣

Since no $\vec{\omega}$−graph can be finitely structured (Lemma 4.5-1(ii)), let us proceed to an ω-graph.

Theorem 4.6-5. *Every finitely structured ω-graph $\mathcal{G}^\omega$ has a spanning tree.*

Proof. Every $(\omega-)$-subsection $\mathcal{S}_b^{\omega-}$ of $\mathcal{G}^\omega$ has a spanning tree. We can conclude this by induction on α as in the last paragraph of the preceding proof, where now α progresses through all the natural numbers. (Remember that every loop in $\mathcal{S}_b^{\omega-}$ must have a natural number rank, and therefore every loop created in our procedure of appending trees will be broken at some natural number for α.) Next, upon considering the shortings of tips imposed by the finitely many ω-nodes of $\mathcal{G}^\omega$, we can break finitely many loops again to obtain a spanning tree for $\mathcal{G}^\omega$. ♣

We have thus constructed a spanning tree for any finitely structured graph of any rank up to and including ω. Clearly, our procedure can be continued to still higher ranks.

Chapter 5

Transfinite Electrical Networks

An electrical network is a graph whose branches are assigned certain analytical structures, which in turn are described by certain physically motivated laws. Since graphs have now been extended transfinitely, the transfinite extension of electrical networks beckons. Toward this we proceed, but, as with graphs, there are surprises in store for us. The laws that govern finite resistive networks, namely Kirchhoff's voltage and current laws, break when stretched transfinitely. A more fundamental principle will be needed to bring transfinite networks to heel. Another peculiarity is that a pure voltage source or a pure current source cannot in general be applied to any two nodes of a purely resistive network, in contrast to finite networks with positive resistances in all branches. We will find ourselves searching for special kinds of transfinite networks that allow arbitrary applications of sources. Still another difficulty is that node voltages may fail to exist throughout a transfinite electrical network, and, even when they do exist, they may not be uniquely determined. This is an obstacle to the development of a theory of random walks on transfinite networks, and we will once again have to restrict our transfinite networks in order to avoid it.

Such is the agenda for this chapter.

5.1 Kirchhoff's Laws, Tellegen's Equation, and Ohm's Law

Our objective now is to convert any finite or transfinite graph into an "electrical network." As the first step, we assign two real numbers to each branch, one of which is designated as the *branch voltage* v and the other as the *branch current* i. The branch is assigned an orientation, in accordance with which v is interpreted as the "drop in electrical potential" and i is interpreted as the "current in the branch." The conventional symbolisms for these are shown in Figure 5.1. The drop in potential is from the plus sign $+$ to the minus sign $-$, and the direction of current is indicated by an arrow. The arrow is chosen to point from $+$ to $-$, and is in fact the *orientation* of the branch. If v (or i) is negative, the drop in potential (resp. the direction of current) is understood to be opposite to that shown. The units for v and i are respectively *volts*, symbolized by V, and *amperes*, symbolized by A.

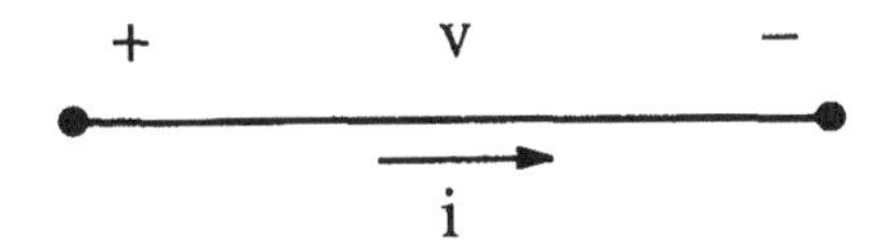

Figure 5.1. The conventions for measuring a branch voltage v and a branch current i in accordance with an orientation of the branch from left to right.

As with graphs, the branches are indexed by the symbol j drawn from a branch-index set J. J and therefore the set of branches may be finite, denumerable, or uncountable. All entities associated with a branch will carry the same index as the branch; thus, v_j and i_j are the branch voltage and branch current of the jth branch b_j.

Let n be any maximal 0-node of finite degree. *Kirchhoff's current law* asserts that the currents in the finitely many branches incident to n satisfy

$$\sum_{(n)} \pm i_j = 0 \tag{5.1}$$

where the summation is over the indices of the branches incident to n and where $+$ (or $-$) is used if the jth branch is oriented away from (resp. toward) n. With this assignment of signs, we call the left-hand side of (5.1) the *algebraic sum of the branch currents incident away from* n. (Were n of infinite degree, (5.1) would have infinitely many terms and the summability of (5.1) would become an issue. We shall see that Kirchhoff's current law cannot always be extended to infinite maximal 0-nodes.)

Now, let L denote any 0-loop. Thus, L has only finitely many branches. Assign an orientation to L. *Kirchhoff's voltage law* asserts that the voltages for those branches satisfy

$$\sum_{(L)} \pm v_j \;=\; 0 \tag{5.2}$$

where the summation is over the indices for the branches in L and where $+$ (or $-$) is used if the jth branch's orientation agrees (resp. disagrees) with the orientation of L. We call the left-hand side of (5.2) the *algebraic sum of the branch voltages for* L. (Were L a transfinite loop, (5.2) would have infinitely many terms and might not be summable. In this case too, we shall show that Kirchhoff's voltage law cannot always be extended to transfinite loops.)

To insure that (5.1) and (5.2) remain finite sums, let us now restrict our attention to a finite 0-connected 0-graph $\mathcal{G}^0$ having q branches and $p+1$ nodes. Let us disallow self-loops just to avoid trivial cases. Furthermore, let us choose some node n_g, call it the *ground node* or simply *ground*, and then number all the other nodes from 1 to p. The *incidence matrix* A for $\mathcal{G}^0$ with the chosen ground node is the $p \times q$ matrix $[a_{kj}]$, where $k = 1, \ldots, p$ and $j = 1, \ldots, q$, such that a_{kj} is $+1$ (or -1) if branch b_j is incident to node n_k and oriented away from (resp. toward) n_k and a_{kj} is 0 if b_j is not incident to n_k.

With the superscript T denoting matrix transpose, we let the column vectors

$$\mathbf{v} \;=\; [v_1, \ldots, v_q]^T \tag{5.3}$$

and

$$\mathbf{i} \;=\; [i_1, \ldots, i_q]^T \tag{5.4}$$

represent the vector of branch voltages and the vector of branch currents, respectively. It follows immediately that the p scalar equations obtained by applying Kirchhoff's current law (5.1) to all the nodes other than ground can be expressed as the matrix equation

$$A\mathbf{i} \;=\; 0. \tag{5.5}$$

Furthermore, that law will also hold at the ground node so long as it holds at all the other nodes. This can be seen by adding the equations for all the other nodes.

Kirchhoff's voltage law can also be expressed by means of A but in a more complicated way through *node voltages*, real numbers assigned to the nodes of $\mathcal{G}^0$. We will always assign 0 V to the ground node n_g. For any other node n_k, let us choose its node voltage u_k arbitrarily, at least for the moment. Then, the branch voltage for a branch incident to, say, the nodes n_k and n_l and oriented from n_k to n_l will be taken to be $u_k - u_l$, where now either n_k or n_l may be ground. Let $\mathbf{u} = [u_1, \ldots, u_p]^T$ be the vector of node voltages at all the nodes other than ground. Then, the branch-voltage vector $\mathbf{v}$ is determined from the chosen node-voltage vector $\mathbf{u}$ through the matrix equation

$$\mathbf{v} \;=\; A^T\mathbf{u}. \tag{5.6}$$

Lemma 5.1-1. *Kirchhoff's voltage law holds around every 0-loop in the finite graph $\mathcal{G}^0$ if and only if the branch-voltage vector $\mathbf{v}$ is determined from some node-voltage vector $\mathbf{u}$ through (5.6).*

Proof. *If:* Choose $\mathbf{u}$ arbitrarily and let $\mathbf{v}$ be given by (5.6). Let L be any 0-loop in $\mathcal{G}^0$. We can assign an orientation to L by choosing a 1 A current that flows along L and nowhere else. Thus, the jth component i_j of the branch current vector $\mathbf{i}$ is $+1$ (or -1) if branch b_j lies in L with an orientation that agrees (resp. disagrees) with the 1 A flow around L and i_j is 0 if b_j is not in L. Now, with $\langle \cdot, \cdot \rangle_m$ denoting the inner product for m-dimensional real Euclidean space R^m, we can prove that Kirchhoff's voltage law holds around L by showing that $\langle \mathbf{v}, \mathbf{i} \rangle_q = 0$. By (5.6), we can write

$$\langle \mathbf{v}, \mathbf{i} \rangle_q \;=\; \langle A^T\mathbf{u}, \mathbf{i} \rangle_q \;=\; \langle \mathbf{u}, A\mathbf{i} \rangle_p \;=\; \langle \mathbf{u}, \mathbf{0} \rangle_p \;=\; 0 \tag{5.7}$$

because the 1 A current around L fulfills Kirchhoff's current law $A\mathbf{i} = \mathbf{0}$.

Only if: Now, let $\mathbf{v}$ be a branch-voltage vector satisfying Kirchhoff's voltage law (5.2). We wish to find a node-voltage vector $\mathbf{u}$ satisfying (5.6). As always, we assign 0 V to the ground node n_g. To obtain the node voltage u_k at any other node n_k, choose a path P from n_k to n_g oriented that way. Let u_k be the algebraic sum

$$u_k \;=\; \sum_{(P)} \pm v_j$$

where the summation is for the branch voltages v_j along P and $+$ (or $-$) is used if the branch's orientation agrees (resp. disagrees) with the orientation of P. This value for u_k does not depend upon the choice of P from n_k to n_g because the union of two different such paths consists of at most finitely many 0-loops, around each of which Kirchhoff's voltage law holds. Now, let b_j be a branch incident away from node n_k and toward node n_l. (Either n_k or n_l may be the ground node.) The branch voltage v_j is given by $v_j = u_k - u_l$; this, too, can be seen by taking the union of a path from n_k to n_g, a path from n_l to n_g, and the branch b_j and then applying Kirchhoff's voltage law to the finitely many 0-loops comprising that union. This result is none other than (5.6). ♣

Actually, the manipulations (5.7) continue to hold for any branch-current vector $\mathbf{i}$ satisfying (5.5) and any branch-voltage vector $\mathbf{v}$ given by (5.6). This yields through Lemma 5.1-1 the following fundamental result.

Theorem 5.1-2. *(Tellegen's equation [22], [28]): For the finite graph $\mathcal{G}^0$, any branch-current vector $\mathbf{i}$ that satisfies Kirchhoff's current law and any branch-voltage vector $\mathbf{v}$ that satisfies Kirchhoff's voltage law are orthogonal, that is,*

$$\langle \mathbf{v}, \mathbf{i} \rangle_q = 0. \tag{5.8}$$

Conversely, Kirchhoff's laws are derivable from Tellegen's equation in the following ways.

Theorem 5.1-3. *For the finite graph $\mathcal{G}^0$, if a branch-voltage vector $\mathbf{v}$ is orthogonal to every branch-current vector that satisfies Kirchhoff's current law, then $\mathbf{v}$ satisfies Kirchhoff's voltage law.*

Proof. Given any oriented 0-loop L, choose a current vector $\mathbf{i}$ that represents a 1 A current flowing around L. $\mathbf{i}$ satisfies Kirchhoff's current law. By hypothesis, $\langle \mathbf{v}, \mathbf{i} \rangle_q = 0$. But, $\langle \mathbf{v}, \mathbf{i} \rangle_q$ is the algebraic sum of the branch voltages around L. So, (5.2) is satisfied. ♣

Theorem 5.1-4. *For the finite graph $\mathcal{G}^0$, if a branch-current vector $\mathbf{i}$ is orthogonal to every branch-voltage vector that satisfies Kirchhoff's voltage law, then $\mathbf{i}$ satisfies Kirchhoff's current law.*

Proof. Choose any node n. Assign to it the node voltage 1 V, and assign to all the other nodes the node voltage 0 V. Let $\mathbf{v}$ be the corresponding branch-voltage vector, as given by (5.6). Then, $\langle \mathbf{v}, \mathbf{i} \rangle_q$ is the algebraic sum of all the components of $\mathbf{i}$ for the branches incident at n, with signs attached as in the left-hand side of (5.1). By Lemma 5.1-1 and our hypothesis, $\langle \mathbf{v}, \mathbf{i} \rangle_q = 0$. ♣

As was mentioned above and will be shown below, Kirchhoff's laws do not always hold for transfinite networks. Nevertheless, Tellegen's equation can and will be used as the fundamental relationship establishing voltage-current regimes in transfinite electrical networks. Moreover, it will yield Kirchhoff's laws under certain circumstances.

Let us turn now to another class of relationships governing the voltages and currents in electrical networks. They are the possible rules coupling the voltage v and the current i in a given branch.

One of them is Ohm's law: $v = ri$. Here, r is a positive (nonzero) real number and is called the *resistance* or the *resistor* for the branch. The units for resistance is *ohms*, symbolized by Ω. The circuit symbol and the graph of a resistor are shown in Figure 5.2(a). The reciprocal $g = 1/r$ of r is called the *conductance* of the branch, and its units are *mhos*, symbolized by ℧. Ohm's law is represented by a straight line through the origin of the voltage-current plane having a slope g with respect to the voltage axis. The product vi is called the *power dissipated in* r; it has the units of *watts*, symbolized by W.

Throughout this book, the graph for a voltage-current relationship for any branch will always be a straight line. However, we have not yet exhausted all possibilities. Another one is a straight line with a negative slope; this we disallow. There will be no "negative resistances" in this book.

Another possibility is a line parallel to the current axis and passing through the voltage axis at the value $-e$. This is the graph of a *pure voltage source*. Its circuit symbol is shown in Figure 5.2(b). It fixes v at $v = -e$ but does not restrict i. We will always follow the convention that the $+$ to $-$ direction within a voltage source is the reverse of that for the branch voltage v. The branch for a pure voltage source is not allowed to be a self-loop. Moreover, when $e = 0$, we have the special case of a *short circuit* or simply a *short*; in this case, the branch is discarded and the two nodes incident to the short are coalesced into a single node by taking the union of their tips and combining their exceptional elements appropriately. This conforms with our previous use of the word "short." Finally, the product ei is called the *power generated by* e; it too has the units of *watts*, symbolized by W.

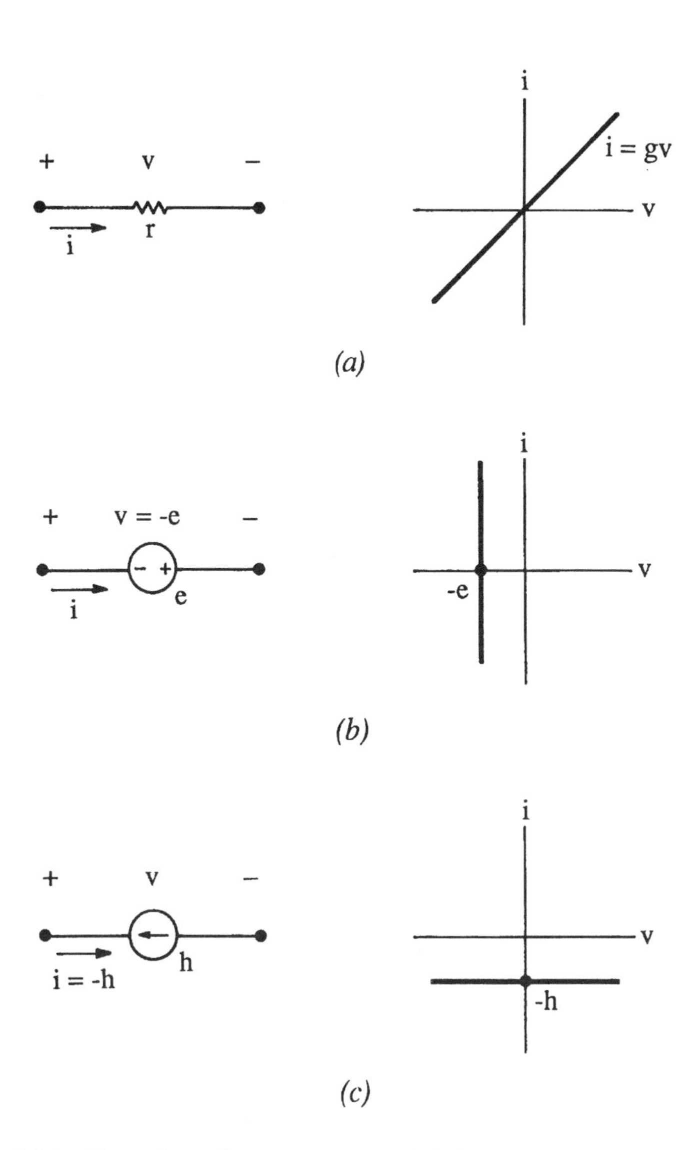

Figure 5.2. Three branch parameters and their graphs. **(a)** A resistor r. **(b)** A pure voltage source e. **(c)** A pure current source h.

Still another possibility is a line parallel to the voltage axis and passing through the current axis at the value $-h$. This is a *pure current source*, and its circuit symbol is shown in Figure 5.2(c). It fixes i at $i = -h$ but does not restrict v. The arrow within a current source and the orientation of the branch in which the current source resides (that is, the direction of the branch current i) will always point in opposite directions. When $h \neq 0$, we will require that there be a path connecting the nodes of the current source through the rest of the graph. When $h = 0$, we have the special case of an *open circuit* or simply an *open*; this terminology does not conflict with our prior use of the word "open." Also, the product hv is called the *power generated by* h and has *watts* (W) as its units.

The last possibility for a straight-line relationship between v and i is a straight line of positive slope passing through the voltage axis at $e = -hr$ and through the current axis at $h = -ge$. (Remember that a negative slope is never allowed.) This case can be obtained by combining a resistor r with a source. There are two ways of doing this. One is shown in Figure 5.3(a); it is called *Thevenin's equivalent branch* for the Ohm's law relationship, $v + e = ri$, or simply the *Thevenin form* for the branch. The other is shown in Figure 5.3(b); it is called *Norton's equivalent branch* or the *Norton form* for the same relationship, which can be rewritten as $i + h = gv$ so long as $g = 1/r$ and $h = -ge$. In these forms, r and g are positive numbers, but e and h are any real numbers, possibly 0. In short, both forms are entirely equivalent to each other so far as the relationship between v and i is concerned.

We are now ready to state what we mean by a (generally transfinite) "electrical network" or synonymously by a "ν-network," where ν denotes the rank of the graph from which the ν-network is constructed. We lose nothing by assuming that the ν-graph has only one component; this we do throughout the rest of this book by assuming that the ν-graph is ν-connected.

Definition of a ν-network: A *ν-network* $\mathbf{N}^\nu$ is a ν-connected ν-graph $\mathcal{G}^\nu$ whose every branch is assigned electrical parameters including a branch voltage v and a branch current i, oriented as in Figure 5.1 and related in accordance with one of the graphs of Figures 5.2 and 5.3. The set of all branch voltages and branch currents is called the *voltage-current regime* or simply the *regime* for $\mathbf{N}^\nu$.

All the graph-theoretic terminology applicable to $\mathcal{G}^\nu$ will be applied to $\mathbf{N}^\nu$ as well. For example, we will call ν the *rank* of $\mathbf{N}^\nu$ and may speak of the nodes, paths, loops, sections, and subsections in $\mathbf{N}^\nu$ and possibly of its isolating sets, contractions, and finite-structuring. Networks and subnetworks will be denoted by boldface capital letters — in contrast to the calligraphic letters used for graphs. Thus, if $\mathcal{A}$ is an arm in a graph, then $\mathbf{A}$ will denote $\mathcal{A}$ when the branches of $\mathcal{A}$ are assigned electrical parameters.

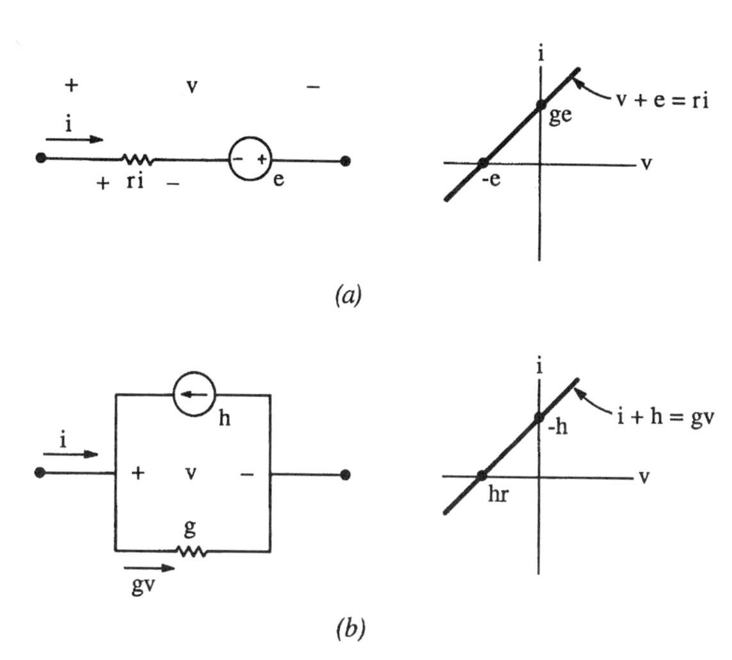

Figure 5.3. **(a)** Thevenin's form of a branch. r is positive, but e may be any real number, possibly 0. **(b)** Norton's form for a branch. g is positive, but h may be any real number, possibly 0.

Kirchhoff's laws, Tellegen's equation, and Ohm's law express various restrictions upon the voltage-current regime. The question facing us is whether these restrictions determine a unique voltage-current regime for $\mathbf{N}^\nu$.

Example 5.1-5. The answer for a finite network is "yes." In fact, just Kirchhoff's laws and Ohm's law will do.

Let $\mathbf{N}^0$ be a 0-connected 0-network with finitely many branches b_j, $j = 1, \ldots, q$. Assume that every branch is in the Norton form; thus, $g_j > 0$ for all j. Let $\mathbf{v}$ and $\mathbf{i}$ be the branch-voltage vector (5.3) and the branch-current vector (5.4), respectively, and let $\mathbf{h} = [h_1, \ldots, h_q]^T$ be the branch-current-source vector. Choose any node as ground n_g with 0 V as its node voltage, and let n_k, where $k = 1, \ldots, p$, be the remaining nodes. Let $\mathbf{u} = [u_1, \ldots, u_p]^T$ be the node-voltage vector. Finally, let G be the $q \times q$ diagonal matrix $G = \text{diag}\,[g_1, \ldots, g_q]$, whose main-diagonal entries are the branch conductances g_j.

By Ohm's law and the Norton form of Figure 5.3(b), $\mathbf{i} = G\mathbf{v} - \mathbf{h}$. So, by Kirchhoff's current law as given by (5.5), we have $A(G\mathbf{v} - \mathbf{h}) = 0$. On the other hand, by Kirchhoff's voltage law and Lemma 5.1-1, there is a $\mathbf{u}$ such that (5.6) holds. Therefore,

$$AGA^T\mathbf{u} \;=\; A\mathbf{h}. \qquad (5.9)$$

The $p \times p$ matrix AGA^T is positive-definite. Indeed, upon letting $\langle \cdot, \cdot \rangle_m$ be the inner product for m-dimensional real Euclidean space R^m, we can write, for any $\mathbf{x} \in R^p$,

$$\langle AGA^T\mathbf{x}, \mathbf{x} \rangle_p \;=\; \langle GA^T\mathbf{x}, A^T\mathbf{x} \rangle_q \;>\; 0$$

whenever $A^T\mathbf{x} \neq \mathbf{0}$ because the diagonal matrix G has strictly positive main-diagonal entries. Furthermore, $A^T\mathbf{x} \neq \mathbf{0}$ whenever $\mathbf{x} \neq \mathbf{0}$ because the ground node is at 0 V, and therefore there must be at least one branch incident to two nodes with differing node voltages under the nonzero node-voltage vector $\mathbf{x}$. Our assertion follows.

Thus, AGA^T is an invertible $p \times p$ matrix. By (5.6) and (5.9), the branch-voltage vector $\mathbf{v}$ is uniquely given by

$$\mathbf{v} \;=\; A^T(AGA^T)^{-1}A\mathbf{h}. \qquad (5.10)$$

This in turn uniquely determines the branch-current vector $\mathbf{i}$ through $\mathbf{i} = G\mathbf{v} - \mathbf{h}$. We have hereby shown that Kirchhoff's laws and Ohm's law determine a unique voltage-current regime in any finite 0-connected 0-network with a positive resistor in every branch. (Nothing so simple will work for transfinite networks.) ♣

Example 5.1-6. We now show by example that Kirchhoff's current law may collapse at an infinite maximal 0-node when a certain finite-power condition is imposed. Consider the infinite 0-network of Figure 5.4. Let us require that any allowable voltage-current regime must produce no more than a finite total amount of power dissipation in all the resistors. This implies that $v = 0$, for otherwise an infinite amount of power would be dissipated in the infinite parallel circuit of 1 Ω resistors. Consequently, i is 1 A in the source branch, whereas the current in each of the other 1 Ω resistors is 0 A. This violates Kirchhoff's current law at both of the infinite nodes n_1 and n_2 because an infinite sum of 0's is 0. ♣

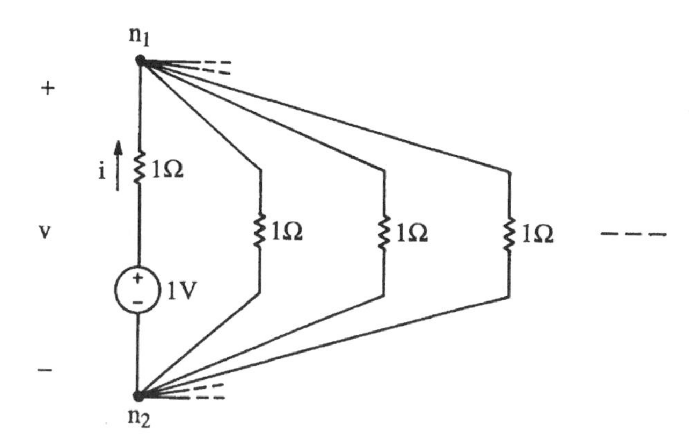

Figure 5.4. A parallel circuit of denumerably many 1 Ω resistors on a Thevenin-form source branch consisting of a 1 V voltage source and another 1 Ω resistor.

Example 5.1-7. The "dual" (in the circuit-theoretic sense) of the 0-network of Figure 5.4 is the 1-network of Figure 5.5. Let us again require that the total power dissipated in all the resistors be finite. Hence, $i = 0$ and $v = 1$. Therefore, the algebraic sum of the voltages around the 1-loop equals 1 instead of 0. Thus, Kirchhoff's voltage law cannot hold around this 1-loop under our finite-power condition. ♣

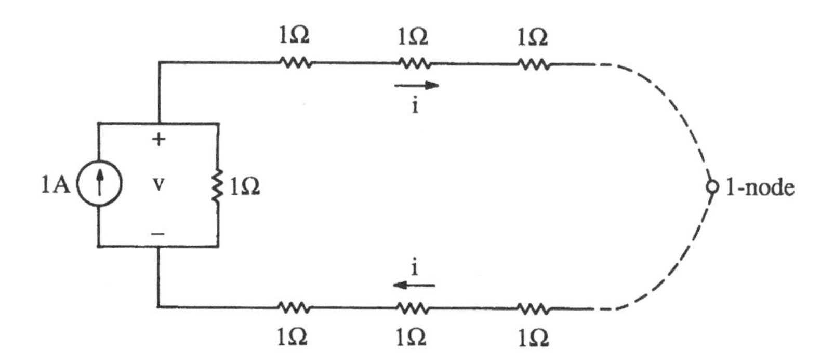

Figure 5.5. A 1-network, which is a 1-loop with denumerably many 1 Ω resistors in series with a Norton-form source branch consisting of a 1 A current source and another 1 Ω resistor.

5.2 The Voltage-Current Regime

In this section we shall establish some broad conditions under which a ν-network will have a unique voltage-current regime. This will be accomplished through a generalization of Tellegen's equation along with the requirement that the total power available from all the sources be finite. Our analysis is an extension to transfinite networks of some ideas introduced by Flanders [9] for the analysis of conventional infinite networks.

We let $\mathbf{N}^\nu$ denote a ν-network as defined in the preceding section. We assume furthermore that every branch is in the Thevenin form, that is, the form shown in Figure 5.3(a) with $r_j > 0$ and $e_j \in R^1$. The branch set of $\mathbf{N}^\nu$ can have any cardinality; thus, it can be finite, denumerable, or uncountable, but the finite case can occur only when $\nu = 0$.

With a current i_j assigned to each branch b_j, we let $\mathbf{i} = \{i_j\}_{j \in J}$ be the set of all branch currents, and we call $\mathbf{i}$ the *branch-current vector.* As always, J is the branch-index set. Unless something else is indicated, the symbol $\sum$ will denote a summation over J. $\mathcal{I}$ will denote the set of all branch-current vectors $\mathbf{i}$ such that

$$\sum i_j^2 r_j \;<\; \infty. \tag{5.11}$$

The left-hand side of (1.11) is called the *total dissipated power for* $\mathbf{i}$ and is a summable series (see Appendix B). Consequently, for any $\mathbf{i} \in \mathcal{I}$, only countably many of the branches b_j can have nonzero i_j (see Appendix B9). However, for different choices of $\mathbf{i} \in \mathcal{I}$, the countable branch sets for which this is so will be different in general. The *support of* $\mathbf{i}$, denoted supp $\mathbf{i}$, is the set of branches b_j for which $i_j \neq 0$. Moreover, $\mathcal{I}$ is a linear space with the linear operations defined componentwise on the current vectors in $\mathcal{I}$.

We convert $\mathcal{I}$ into an inner-product space by assigning to it the *inner product*

$$(\mathbf{i}, \mathbf{s}) \;=\; \sum r_j i_j s_j; \quad \mathbf{i}, \mathbf{s} \in \mathcal{I}. \tag{5.12}$$

A standard proof (see for instance [35, pages 63-64]) shows that $\mathcal{I}$ is complete and therefore a Hilbert space. The *norm* of any $\mathbf{i} \in \mathcal{I}$ is then

$$\|\mathbf{i}\| \;=\; \left(\sum i_j^2 r_j\right)^{1/2}. \tag{5.13}$$

Lemma 5.2-1. *Convergence in $\mathcal{I}$ implies componentwise convergence.*

Proof. Let $\mathbf{i}_k \to \mathbf{i}$ in $\mathcal{I}$ as $k \to \infty$. Thus, with i_{kj} being the jth component of $\mathbf{i}_k$ and i_j the jth component of $\mathbf{i}$, we have

$$\|\mathbf{i}_k - \mathbf{i}\|^2 \;=\; \sum (i_{kj} - i_j)^2 r_j \;\to\; 0.$$

Since $r_j > 0$ for all j, we must have $i_{kj} \to i_j$ for each j. ♣.

$\mathcal{I}$ is not a suitable space in which to search for a current vector $\mathbf{i}$ for our network $\mathbf{N}^\nu$ because $\mathcal{I}$ is too large. It allows any current vector producing finite power dissipation. We would rather have a finite-power current vector that satisfies Kirchhoff's current law wherever possible, certainly at maximal 0-nodes of finite degree. For this reason, we shall now construct a subspace $\mathcal{K}$ of $\mathcal{I}$ which contains only the latter kind of current vectors and yet is big enough to be complete under the norm of $\mathcal{I}$.

Let γ be any rank no larger than ν and not equal to $\vec{\omega}$. A "γ-loop current" is a current vector that represents a current flowing around a γ-loop and no place else. (As was pointed out in Section 2.3, there is no such thing as an $\vec{\omega}$-loop, hence our barring of $\vec{\omega}$.) Let us be more precise. Let L be a γ-loop and let us assign to L an *orientation*, that is, one of the two possible directions of tracing L. Also, let a be a real number. A current vector $\mathbf{i} = \{i_j\}_{j \in J}$ is called a *γ-loop current of value a* if $i_j = +a$ (or $-a$) whenever the branch b_j is in L with the orientations of b_j and L in agreement (resp. disagreement) and if $i_j = 0$ whenever b_j is not in L.

If $a \neq 0$, then the γ-loop current $\mathbf{i}$ of value a is a member of $\mathcal{I}$ if and only if $\sum_{(L)} r_j < \infty$, where $\sum_{(L)}$ is a summation over the indices of the branches in L. Indeed, the total power dissipated by $\mathbf{i}$ is $a^2 \sum_{(L)} r_j$, which must be finite if $\mathbf{i}$ is to be in $\mathcal{I}$. We shall call a γ-loop L *permissive* whenever $\sum_{(L)} r_j < \infty$. Similarly, a path is called *permissive* whenever the sum of its resistances is finite. Finally, a network is called *permissive* if between every two of its nodes there is a permissive path. (In prior works, we said "perceptible" instead of "permissive." Permissive and perceptible are synonyms so far as transfinite networks are concerned.)

A set of branches will be said to have a *finite γ-diameter* if there is a natural number p such that every two branches in the set are connected by a path of rank no larger than γ having no more than p γ-nodes.

Definition of a basic current: Let γ be a rank no larger than ν and different from $\vec{\omega}$. A *γ-basic current in* $\mathbf{N}^\nu$ is a branch-current vector $\mathbf{i}$ such that $\mathbf{i} = \sum_{m \in M} \mathbf{i}_m$ where

(a) $\{\mathbf{i}_m\}_{m\in M}$ is a set of γ-loop currents $\mathbf{i}_m$,

(b) any branch or any maximal 0-node is embraced by no more than finitely many of the γ-loops of the $\mathbf{i}_m$, and

(c) the support of $\mathbf{i}$ has a finite γ-diameter.

Finally, a *basic current in* $\mathbf{N}^\nu$ is a γ-basic current of any rank γ with $0 \leq \gamma \leq \nu$ and $\gamma \neq \vec{\omega}$.

Condition (b) insures two things: the sum $\sum_{m\in M} \mathbf{i}_m$ truly defines a current for every branch, and every basic current satisfies Kirchhoff's current law at every maximal 0-node.

The purpose of condition (c) is to prevent a γ-basic current from becoming in effect a basic current of higher rank. How this can happen can be seen from the following example.

Example 5.2-2. Consider the ladder of Figure 5.6, which we shall treat as a 0-network by not specifying any 1-nodes. For this example, we let the resistance values be arbitrary. For each $m = 1, 2, 3, \ldots$, let $\mathbf{i}_m$ be the 0-loop current of value 1 A passing through the resistances r_{2m-1}, r_{2m+1}, and the two resistors of value $r_{2m}/2$, as in Figure 5.6(a). Then, the sum $\mathbf{i} = \sum_{m=1}^{\infty} \mathbf{i}_m$ of all the $\mathbf{i}_m$ does not have a support of finite 0-diameter; indeed, $\mathbf{i}$ is the current indicated in Figure 5.6(b), which is identical to a 1-loop current of value 1 A passing through the two 0-tips of the upper and lower horizontal paths. Thus, $\mathbf{i}$ satisfies conditions (a) and (b) but not (c). Condition (c), were it satisfied for $\gamma = 0$, would prevent this unintended creation of what is effectively a basic 1-current from a sum of 0-basic currents. ♣

It is possible for a basic current $\mathbf{i} = \sum_{m\in M} \mathbf{i}_m$ to be of finite total dissipated power even though every $\mathbf{i}_m$ is not like that; that is, we can have $\mathbf{i} \in \mathcal{I}$ even though $\mathbf{i}_m \notin \mathcal{I}$ for every m. Such is the case in the next example.

Example 5.2-3. Consider the 1-network of Figure 5.7(a) consisting of an infinite binary tree with the apex 0-node n^0 and another branch b incident to n^0 and to a 1-node (not shown) that embraces all the 0-tips of the binary tree. Let us assume that every branch has a 1 Ω resistor. Figure 5.7(b) shows a 1-basic current $\mathbf{i} = \sum_{m\in M} \mathbf{i}_\mathrm{m}$ where each $\mathbf{i}_m$ is a 1-loop current represented by one of the dotted lines. There is an infinity of such 1-loop currents, only the first few of which are shown. No 1-loop current is a member of $\mathcal{I}$ because it passes through an infinity of 1 Ω resistors, that is, its 1-loop is not permissive. However, the branchwise sum of the $\mathbf{i}_m$ yields the branch currents indicated in Figure 5.7(a). Clearly, the vector $\mathbf{i}$ of these branch currents is a member of $\mathcal{I}$. (As another example, [35, pages 75-77] presents a 1-basic current $\mathbf{i} = \sum \mathbf{i}_m$ in a two-dimensional grid for which $\mathbf{i} \in \mathcal{I}$ but $\mathbf{i}_m \notin \mathcal{I}$ for every m.) ♣

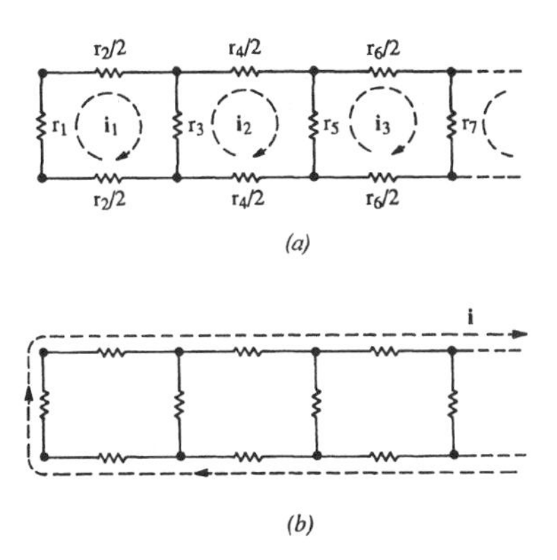

Figure 5.6. **(a)** A ladder network with $r_j = 2^{-j}$ and with an infinity of 0-loop currents $\mathbf{i}_m$, each of a value 1 A. The support of the sum of the $\mathbf{i}_m$ does not have a finite 0-diameter. **(b)** The sum of $\mathbf{i}$ of all the $\mathbf{i}_m$ in part (a). $\mathbf{i}$ is in effect a 1-loop current of value 1 A passing through an apparent 1-node at infinity, even though such a 1-node has not been stipulated.

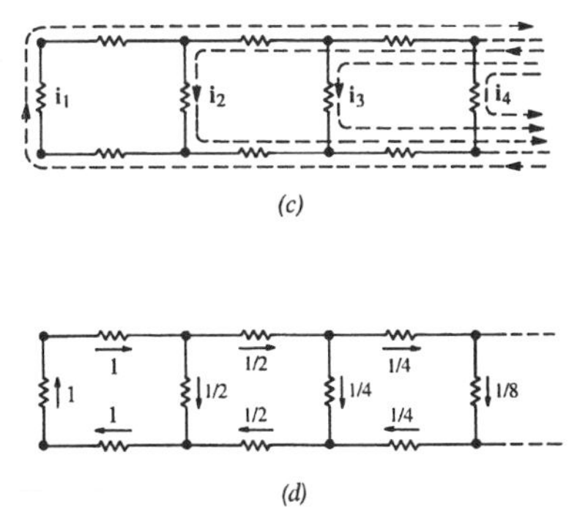

Figure 5.6. **(c)** Now, every resistance is 1 Ω. Each oriented dotted line indicates a 1-loop current $\mathbf{i}_m$ of value 2^{-m+1} for $m = 1, 2, 3, \ldots$, and $\mathbf{i} = \sum_{m=0}^{\infty} \mathbf{i}_m$ is a 1-basic current in $\mathcal{K}^0$. **(d)** The branch of currents for the 1-basic current of part (c).

Let $\mathcal{K}^o$ be the span of all basic currents that reside in $\mathcal{I}$. Thus, $\mathcal{K}^o \subset \mathcal{I}$. Moreover, let $\mathcal{K}$ be the closure of $\mathcal{K}^o$ in $\mathcal{I}$. Thus, $\mathcal{K}^o \subset \mathcal{K} \subset \mathcal{I}$. We assign to $\mathcal{K}$ the same inner product as that of $\mathcal{I}$. This makes $\mathcal{K}$ a Hilbert space by itself. $\mathcal{K}$ is the space of current vectors within which we shall search for the unique current vector dictated by a generalization of Tellegen's equation.

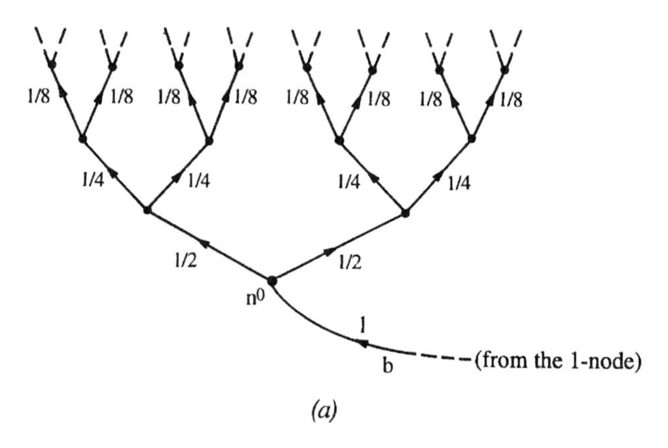

Figure 5.7. **(a)** A 1-network consisting of an infinite binary tree, whose apex node is n^0, along with an additional branch b incident to n^0 and to a 1-node (not shown) that embraces all the 0-tips of the tree. Every branch has a 1 Ω resistor. The numbers are the branch currents for a current vector **i** in $\mathcal{I}$.

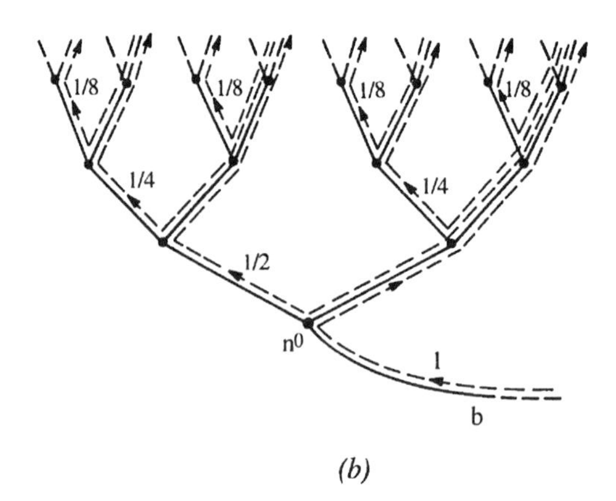

Figure 5.7. **(b)** The decomposition of **i** into a 1-basic current $\mathbf{i} = \sum_{m \in M} \mathbf{i}_m$ of 1-loop currents $\mathbf{i}_m$. No $\mathbf{i}_m$ is a member of $\mathcal{I}$, but **i** is in $\mathcal{I}$.

Example 5.2-4. Consider the ladder of Figure 5.6(a) again. We will maintain this as a network $\mathbf{N}^0$ of rank 0 by not specifying any 1-nodes. Let $r_j = 2^{-j}$ for every $j = 1, 2, 3, \ldots$. Furthermore, for each $m = 1, 2, 3, \ldots$, let $\mathbf{i}_m$ be the 0-loop current of value 1 A indicated in Figure 5.6(a) and specified in Example 5.2-2. Also, for each $k = 1, 2, 3, \ldots$, let $\mathbf{i}_k = \sum_{m=1}^{k} \mathbf{i}_m$; $\mathbf{i}_k$ is a 0-loop current that passes through r_1, r_{2k+1}, and all the $r_l/2$ for which l is even and $1 < l < 2k+1$. Thus, $\mathbf{i}_k \in \mathcal{K}^o$. Finally, let $\mathbf{i}$ be the current vector indicated in Figure 5.6(b); the support of $\mathbf{i}$ is the set of all horizontal branches along with the branch for r_1. Because of our choice of resistance values, $\mathbf{i} \in \mathcal{I}$. As $k \to \infty$, the $\mathbf{i}_k$ converge in $\mathcal{I}$ to $\mathbf{i}$; indeed,

$$\|\mathbf{i} - \mathbf{i}_k\|^2 \;=\; r_{2k+1} + r_{2k+2} + r_{2k+4} + r_{2k+6} + \cdots \;\to\; 0.$$

Since $\mathcal{K}$ is the closure of $\mathcal{K}^o$ in $\mathcal{I}$, we can conclude that $\mathbf{i} \in \mathcal{K}$. In effect, a 1-node has been introduced at infinity just by taking the closure of $\mathcal{K}^o$ in $\mathcal{I}$. Intuitively, we might say that the structure and resistance values of the ladder are such that they simulate a 1-node at infinity. In this case, nothing will change if we specify that 1-node by shorting all the 0-tips of the ladder.

By way of contrast, let us now set $r_j = 2^{-j}$ for all even j as before but set $r_j = 1$ for all odd j. In this case, $\|\mathbf{i} - \mathbf{i}_k\|$ tends to 1 as $k \to \infty$; in fact, $\mathbf{i}$ cannot be in $\mathcal{K}$ because there is no sequence of basic currents in $\mathcal{K}^o$ that converges to $\mathbf{i}$. (Remember that there is no 1-node at infinity, and therefore every basic current is a sum of 0-loop currents, whose support has finite 0-diameter.)

Finally, let us alter both the graph and the resistance values as follows. We stipulate that all the 0-tips of the ladder are shorted together by a 1-node. This changes the network's rank from 0 to 1. Moreover, let us set all resistances equal to 1 Ω. In this case, every 0-basic current, in particular, every 0-loop current is a member of $\mathcal{K}^o$, but the current vector $\mathbf{i}$ of Figure 5.6(b) cannot be in $\mathcal{K}$ because its total dissipated power is infinite. In fact, there is no nonzero 1-loop current in $\mathcal{K}$. However, a 1-basic current $\mathbf{i} = \sum \mathbf{i}_m$ does exist in $\mathcal{K}$; some of the $\mathbf{i}_m$ are indicated in Figure 5.6(c), and some of the branch currents for $\mathbf{i}$ are shown in Figure 5.6(d). Actually, the current regime of Figure 5.6(d) can also be obtained from an infinite superposition of 0-loop currents $\mathbf{i}_m$ $(m = 1, 2, 3, \ldots)$ of values 2^{-m+1}, with the 0-loops as in Figure 5.6(a). Thus, the 1-basic current of Figure 5.6(c), which requires a 1-node, yields a current regime, which does not need a 1-node. ♣

We can conclude from this last example that there is a fundamental difference between transfinite graphs and transfinite networks with regard to their connectivities. Connectivity in the graphical sense is completely determined by the specification of the nodes of various ranks. However, that specification is not enough to determine the possible ways current may flow. The resistance values in the network also affect which current vectors are in $\mathcal{K}$, the space of permissible current vectors; that is, with regard to nodes of ranks larger than 0, the resistances may on the one hand effectively expand nodes or introduce new nodes and may on the other hand effectively contract nodes or even eliminate them — so far as the flow of current is concerned.

Let us now define a coupling between any voltage vector $\mathbf{w} = \{\mathbf{w}_j\}_{j\in J}$ and any current vector $\mathbf{s} = \{\mathbf{s}\}_{j\in J}$ by

$$\langle \mathbf{w}, \mathbf{s}\rangle \;=\; \sum w_j s_j. \tag{5.14}$$

Here, w_j may be any voltage (possibly e_j or $r_j i_j$) assigned to the jth branch, and s_j may be any current in that branch. Therefore, the right-hand side of (5.14) may not be summable, but, when it is, it will be electrical power. Given any $\mathbf{w}$, (5.14) defines $\langle \mathbf{w}, \cdot\rangle$ as a functional on the set of all $\mathbf{s}$ for which $\sum w_j s_j$ is summable.

We need two more definitions before we can state our fundamental theorem for the network $\mathbf{N}^\nu$, whose every branch, remember, is in the Thevenin form. First, the *resistance operator* R of $\mathbf{N}^\nu$ is a mapping that assigns to each $\mathbf{i} \in \mathcal{I}$ the vector $R\mathbf{i}$ of voltages $r_j i_j$ across each resistor r_j in $\mathbf{N}^\nu$; thus, $R\mathbf{i} = \{r_j i_j\}_{j\in J}$. $R\mathbf{i}$ expresses Ohm's law for $\mathbf{N}^\nu$.

Second, the *branch-voltage-source vector* $\mathbf{e} \;=\; \{e_j\}_{j\in J}$ is the vector of source voltages e_j measured in accordance with the polarity shown in Figure 5.3(a). Given any $\mathbf{e}$, the sum $\sum e_j^2 g_j$ will be called the *total isolated source power*; it is the total power generated by all the sources when a short is placed across every branch of $\mathbf{N}^\nu$, thereby making every branch a self-loop. We shall show below (Corollary 5.2-9) that $\sum e_j^2 g_j$ is an upper bound on the total power that $\mathbf{N}^\nu$ extracts from all its sources.

Lemma 5.2-5. *If*

$$\sum e_j^2 g_j < \infty, \tag{5.15}$$

then $\langle \mathbf{e}, \cdot\rangle$ *is a continuous linear functional on* $\mathcal{I}$ *and therefore on* $\mathcal{K}$*, too.*

Proof. We first show that $\sum e_j i_j$ is summable whenever $\mathbf{i} \in \mathcal{I}$. By Appendix B6, $\sum e_j i_j$ is summable if and only if $\sum |e_j i_j|$ is summable. Moreover, we have that $\sum e_j^2 g_j$ and $\sum i_j^2 r_j$ are both summable. So, by the

Schwarz inequality (Appendix B12),

$$\sum |e_j i_j| = \sum |e_j| g_j^{1/2} |i_j| r_j^{1/2} \leq \left[\sum e_j^2 g_j \sum i_j^2 r_j \right]^{1/2} < \infty. \tag{5.16}$$

By Appendix B7, $\sum |e_j i_j|$ is summable.

Now, let $\mathbf{i}, \mathbf{s} \in \mathcal{I}$ and let $a, b \in R^1$. Then,

$$\langle \mathbf{e}, a\mathbf{i} + b\mathbf{s} \rangle = \sum (a e_j i_j + b e_j s_j). \tag{5.17}$$

Since $\sum e_j i_j$ and $\sum e_j s_j$ are both summable, we have from Appendix B5 that the right-hand side of (5.17) is equal to

$$a \sum e_j i_j + b \sum e_j s_j = a\langle \mathbf{e}, \mathbf{i} \rangle + b\langle \mathbf{e}, \mathbf{s} \rangle.$$

Thus, $\langle \mathbf{e}, \cdot \rangle$ is a linear functional on $\mathcal{I}$.

Finally, we have from (5.16) that, for every $\mathbf{i} \in \mathcal{I}$,

$$|\langle \mathbf{e}, \mathbf{i} \rangle| \leq \sum |e_j i_j| \leq \left[\sum e_j^2 g_j \right]^{1/2} \|\mathbf{i}\|.$$

Consequently, the linear functional $\langle \mathbf{e}, \cdot \rangle$ is also continuous on $\mathcal{I}$.

Since $\mathcal{K}$ is a closed subspace of $\mathcal{I}$ with the same inner product, our conclusion holds for $\mathcal{K}$ as well. ♣

Our fundamental theorem is based upon the following standard result.

Theorem 5.2-6. *The Riesz-Representation Theorem: If f is a continuous linear functional on a Hilbert space $\mathcal{X}$ with the inner product $(\cdot, \cdot)$, then there exists a unique $y \in \mathcal{X}$ such that $f(x) = (x, y)$ for every $x \in \mathcal{X}$.*

Here at last is our fundamental theorem. For future reference, we state its hypothesis separately.

Conditions 5.2-7. $\mathbf{N}^\nu$ *is a ν-network (possibly, $\nu = \vec{\omega}$). Every branch b_j of $\mathbf{N}^\nu$ is in the Thevenin form with $0 < r_j < \infty$ and $e_j \in R^1$ for every j. Furthermore, its total isolated source power is finite: $\sum e_j^2 g_j < \infty$.*

Theorem 5.2-8. *Under Conditions 5.2-7, there exists a unique branch-current vector $\mathbf{i} \in \mathcal{K}$ and thereby a unique branch-voltage vector $\mathbf{v} = R\mathbf{i} - \mathbf{e}$ such that*

$$\langle \mathbf{e}, \mathbf{s} \rangle = \langle R\mathbf{i}, \mathbf{s} \rangle \tag{5.18}$$

for every $\mathbf{s} \in \mathcal{K}$. This equation insures the uniqueness of $\mathbf{i}$ and $\mathbf{v}$ even when $\mathbf{s}$ is restricted to $\mathcal{K}^o$.

Proof. By Lemma 5.2-5 $\langle \mathbf{e}, \cdot \rangle$ is a continuous linear functional on $\mathcal{K}$, and by Theorem 5.2-6 there is a unique $\mathbf{i} \in \mathcal{K}$ such that $\langle \mathbf{e}, \mathbf{s} \rangle = (\mathbf{s}, \mathbf{i})$ for every $\mathbf{s} \in \mathcal{K}$. But, $(\mathbf{s}, \mathbf{i}) = \sum r_j s_j i_j = \langle R\mathbf{i}, \mathbf{s} \rangle$. Thus, (5.18) insures the existence and uniqueness of $\mathbf{i}$ and therefore of $\mathbf{v} = R\mathbf{i} - \mathbf{e}$. Finally, since $\mathcal{K}^o$ is dense in $\mathcal{K}$, a knowledge of $\langle \mathbf{e}, \cdot \rangle$ on $\mathcal{K}^o$ determines $\langle \mathbf{e}, \cdot \rangle$ on all of $\mathcal{K}$ — whence the second conclusion. ♣.

We may rewrite (5.18) as $\langle \mathbf{v}, \mathbf{s} \rangle = 0$, which can be viewed as a certain extension of Tellegen's equation to transfinite networks. Indeed, $\mathbf{s}$ can be any vector in $\mathcal{K}$. Moreover, we are free to vary $\mathbf{e}$ — so long as its total isolated power remains finite — in order to obtain a variety of voltage vectors $\mathbf{v} = R\mathbf{i} - \mathbf{e}$ for which $\langle \mathbf{v}, \mathbf{s} \rangle = 0$.

Corollary 5.2-9. *Under Conditions 5.2-7 and with* $\mathbf{i}$ *being the current vector dictated by Theorem 5.2-8, the total power dissipation* $\sum i_j^2 r_j$ *is equal to the total power* $\sum e_j i_j$ *generated by all the sources and is no larger than the total isolated power* $\sum e_j^2 g_j$ *of those sources.*

Proof. Upon setting $\mathbf{s} = \mathbf{i}$ in (5.18), we get $\sum i_j^2 r_j = \sum e_j i_j$. Then, by manipulating as in (5.16) and canceling the factor $\left[\sum i_j^2 r_j\right]^{1/2}$, we obtain $\sum i_j^2 r_j \leq \sum e_j^2 g_j$. ♣

Because of (5.11) and (5.15), we shall say that the voltage-current regime dictated by Theorem 5.2-8 is a *finite-power regime.*

Corollary 5.2-10. *Under Conditions 5.2-7, the superposition principle holds for the sources; that is, if* $\mathbf{e}_1$ *and* $\mathbf{e}_2$ *are two voltage-source vectors of finite total isolated power, if* $\mathbf{i}_1$ *and* $\mathbf{i}_2$ *are the corresponding current vectors dictated by Theorem 5.2-8, and if* a *and* b *are any real numbers, then* $a\mathbf{i}_1 + b\mathbf{i}_2$ *is the current vector corresponding to* $a\mathbf{e}_1 + b\mathbf{e}_2$.

Proof. This follows immediately from the fact that the summations inherent in (5.18) are summable and can therefore be rearranged; see Appendix B5. (Or, in other words, superposition is simply a reflection of the linearity of the operations inherent in (5.18).) ♣

Note that Kirchhoff's laws play no role in Theorem 5.2-8. In fact, both laws may be violated in particular cases, as was indicated in Examples 5.1-6 and 5.1-7. Let us now verify that the voltage-current regimes of those examples are truly those dictated by Theorem 5.2-8. This is easy.

Example 5.2-11. Consider the 0-network of Figure 5.4. We concluded previously that the current i in the source branch is 1 A and the currents in all the other branches are 0 A. The space $\mathcal{K}^0$ is simply the span of all 0-loop currents in this 0-network. So, we need merely show that (5.18) holds for each 0-loop current s of value 1 A. If the 0-loop current s does not pass through the source branch, both sides of (5.18) are equal to 0. When the 0-loop current s passes through the source branch, say, in the upward direction, both sides of (5.18) are equal to 1. ♣

Example 5.2-12. Now, consider the 1-network of Figure 5.5. Our conclusion in this case was the $i = 0$ and $v = 1$. Upon converting the Norton-form of the source branch into the Thevenin form, we get a 1 V voltage source (positive polarity sign on top) in series with a 1 Ω resistor. Now, for this network, $\mathcal{K}^o$ is the set having just one vector s, all of whose components are 0, because there is only a single loop — a nonpermissive 1-loop. Thus, both sides of (5.18) are equal to 0 for the only s in $\mathcal{K}^o$. ♣

So far, we have established the voltage-current regime of $\mathbf{N}^\nu$ by searching for a branch-current vector $\mathbf{i} = \{i_j\}_{j \in J}$ that satisfies a generalized form of Tellegen's equation (5.18) and then have obtained the corresponding voltage vector $\mathbf{v} = \{v_j\}_{j \in J}$ through Ohm's law $\mathbf{v} = R\mathbf{i} - \mathbf{e}$. In this approach every branch is taken to be in its Thevenin form (Figure 5.3(a)).

An entirely equivalent analysis, which we shall call the *dual analysis*, can be performed by converting every branch into its Norton form (Figure 5.3(b)) and then by searching for a branch-voltage vector $\mathbf{v} = \{v_j\}_{j \in J}$ that satisfies a similarly generalized Tellegen's equation. The corresponding branch-current vector is then given by Ohm's law again as $\mathbf{i} = G\mathbf{v} - \mathbf{h}$. Here, G is the *conductance operator*: $G\{v_j\}_{j \in J} = \{g_j v_j\}_{j \in J} = \{i_j + h_j\}_{j \in J}$. Also, $\mathbf{h} = \{h_j\}_{j \in J}$ is the vector of branch-current sources and, by virtue of the Thevenin-to-Norton transformation, is given by $\mathbf{h} = -G\mathbf{e}$. Note that $\sum e_j^2 g_j = \sum h_j^2 r_j$, where $\sum h_j^2 r_j$ is the power dissipated in all the conductances g_j when every branch is isolated by open circuits. It follows that $\mathbf{e}$ satisfies its finite total isolated source power condition if and only if $\mathbf{h}$ satisfies the equivalent condition: $\sum h_j^2 r_j < \infty$. In fact, the following is entirely equivalent to Conditions 5.2-7.

Conditions 5.2-13. $\mathbf{N}^\nu$ *is a* ν*-network (possibly,* $\nu = \vec{\omega}$*). Every branch* b_j *of* $\mathbf{N}^\nu$ *is in the Norton form with* $0 < g_j < \infty$ *and* $h_j \in R^1$ *for every* j. *Furthermore, its total isolated source power is finite:* $\sum h_j^2 r_j < \infty$.

In order to pursue this dual approach, we set up a space $\mathcal{V}$ of all branch-voltage vectors $\mathbf{w} = \{w_j\}_{j \in J}$ satisfying the following two conditions:

(i) $\mathbf{w}$ is of finite power dissipation: $\sum w_j^2 g_j < \infty$.

(ii) $\mathbf{w}$ satisfies Tellegen's equation:

$$\langle \mathbf{w}, \mathbf{s} \rangle \;=\; 0 \tag{5.19}$$

for every $\mathbf{s} \in \mathcal{K}$.

We can make $\mathcal{V}$ into a Hilbert space by defining the linear operations componentwise on the members of $\mathcal{V}$ and by assigning to $\mathcal{V}$ the inner product: $(\mathbf{w}, \mathbf{w}') = \sum g_j w_j w_j'$. A standard proof shows that $\mathcal{V}$ is complete.

In this dual approach, Tellegen's equation (5.19) holds for all $\mathbf{w} \in \mathcal{V}$ and for all $\mathbf{s} \in \mathcal{K}$ simply through the definition of $\mathcal{V}$. Moreover, a proof virtually the same as that of Lemma 5.2-5 shows that, if $\sum h_j^2 r_j < \infty$, then $\langle \cdot, \mathbf{h} \rangle$ is a continuous linear functional on $\mathcal{V}$. Since $\mathcal{K}^o$ is dense in $\mathcal{K}$, (5.19) holds for all $\mathbf{s} \in \mathcal{K}$ if and only if it holds for all $\mathbf{s} \in \mathcal{K}^o$.

Theorem 5.2-14. *Under Conditions 5.2-13, there exists a unique branch-voltage vector* $\mathbf{v} \in \mathcal{V}$ *such that*

$$\langle \mathbf{w}, \mathbf{h} \rangle \;=\; \langle \mathbf{w}, G\mathbf{v} \rangle \tag{5.20}$$

for every $\mathbf{w} \in \mathcal{V}$. *The corresponding current vector* $\mathbf{i} = G\mathbf{v} - \mathbf{h}$ *is precisely the branch-current vector dictated by Theorem 5.2-8.*

Proof. Let $\mathbf{i} \in \mathcal{K}$ be the branch-current vector dictated by Theorem 5.2-8 and consider the corresponding branch-voltage vector $\mathbf{v} = R\mathbf{i} - \mathbf{e}$ given by Ohm's law. We may write

$$\left(\sum v_j^2 g_j\right)^{1/2} \;=\; \left[\sum (r_j i_j - e_j)^2 g_j\right]^{1/2} \;=\; \left[\sum (i_j \sqrt{r_j} - e_j \sqrt{g_j})^2\right]^{1/2}$$

$$\leq\; \left(\sum i_j^2 r_j\right)^{1/2} + \left(\sum e_j^2 g_j\right)^{1/2}$$

where the inequality is Minkowski's (Appendix B12). Since $\mathbf{i}$ is a member of $\mathcal{K}$ and $\mathbf{e}$ is of finite total isolated power, the right-hand side is finite. Thus, $\sum v_j^2 g_j < \infty$. Furthermore, by (5.18), $\mathbf{v} = R\mathbf{i} - \mathbf{e}$ satisfies $\langle \mathbf{v}, \mathbf{s} \rangle = 0$ for all $\mathbf{s} \in \mathcal{K}$. Thus, $\mathbf{v} \in \mathcal{V}$.

By the Thevenin-to-Norton transformation, $\mathbf{i} = G(\mathbf{v} + \mathbf{e}) = G\mathbf{v} - \mathbf{h}$. Upon setting $\mathbf{s} = \mathbf{i}$ in (5.19), we obtain (5.20) for every $\mathbf{w} \in \mathcal{V}$. Altogether, we have shown so far that there is at least one branch-voltage vector $\mathbf{v} \in \mathcal{V}$ satisfying (5.20). The corresponding $\mathbf{i} = G\mathbf{v} - \mathbf{h}$ is the branch-current vector designated by Theorem 5.2-8.

To demonstrate uniqueness, let $\mathbf{v}$ and $\mathbf{v}'$ be two members of $\mathcal{V}$ satisfying (5.20) for the given $\mathbf{h}$. Then, $\langle \mathbf{w}, G\mathbf{v} \rangle = \langle \mathbf{w}, \mathbf{h} \rangle = \langle \mathbf{w}, G\mathbf{v}' \rangle$. Therefore,

$$0 \;=\; \langle \mathbf{w}, G(\mathbf{v} - \mathbf{v}') \rangle \;=\; \sum w_j g_j (v_j - v_j') \;=\; (\mathbf{w}, \mathbf{v} - \mathbf{v}')$$

for every $\mathbf{w} \in \mathcal{V}$. Consequently, $\mathbf{v} = \mathbf{v}'$. ♣

In much the same way as Corollaries 5.2-9 and 5.2-10 were proven, we can also prove

Corollary 5.2-15. *Under Conditions 5.2-13 and with* $\mathbf{v}$ *being given by Theorem 5.2-14, the total power dissipation* $\sum v_j^2 g_j$ *is equal to the total generated power* $\sum v_j h_j$ *and is no larger than the total isolated source power* $\sum h_j^2 r_j$. *Furthermore, the superposition principle holds with regard to different branch-current-source vectors.*

There is one more consequence of Theorem 5.2-14 we shall be needing — a version of the *reciprocity principle*:

Corollary 5.2-16. *Under Conditions 5.2-13, let* $\mathbf{h} = (h_1, h_2, \ldots)$ *and* $\mathbf{k} = (k_1, k_2, \ldots)$ *be two current-source vectors, let* $\mathbf{v} = (v_1, v_2, \ldots)$ *be the voltage vector produced by* $\mathbf{h}$, *and let* $\mathbf{w} = (w_1, w_2, \ldots)$ *be the voltage vector produced by* $\mathbf{k}$. *Then,*

$$\sum v_j k_j \;=\; \sum w_j h_j. \tag{5.21}$$

Proof. Along with (5.20) we also have $\langle \mathbf{v}, \mathbf{k} \rangle = \langle \mathbf{v}, G\mathbf{w} \rangle$. Hence,

$$\langle \mathbf{w}, \mathbf{h} \rangle \;=\; \langle \mathbf{w}, G\mathbf{v} \rangle \;=\; \sum w_j g_j v_j \;=\; \langle \mathbf{v}, G\mathbf{w} \rangle \;=\; \langle \mathbf{v}, \mathbf{k} \rangle.$$

This is (5.21). ♣

The reason (5.21) is called "reciprocity" is because of the following special case. If a current source h in branch b_1 produces a voltage v in branch b_2, then a current source of the same value h in branch b_2 will produce the same voltage v in branch b_1.

5.3 Conditions Legitimizing Kirchhoff's Laws

Although Kirchhoff's laws do not always hold for transfinite networks, they will do so in special cases. Let us first consider Kirchhoff's current law:

$$\sum_{(n)} \pm i_j = 0 \tag{5.22}$$

where the symbols have the same meanings as they do in (5.1) except that now the degree of the node n may have any cardinality. In particular, with J_n denoting the branch-index set for the branches incident to n, $\sum_{(n)}$ denotes a summation over J_n.

Throughout this section, we continue to assume Conditions 5.2-7; thus, every branch b_j has a positive finite conductance g_j. We shall say that n is *restraining* if $\sum_{(n)} g_j < \infty$. Thus, n will be restraining if it is of finite degree. On the other hand, n cannot be restraining if its degree is uncountable because the sum of uncountably many positive conductances incident to n will be infinite according to Appendix B9.

Theorem 5.3-1. *If n is a restraining maximal 0-node in $\mathbf{N}^\nu$, then, under the regime dictated by Theorem 5.2-8, Kirchhoff's current law (5.22) is satisfied. Moreover, the left-hand side of (5.22) is summable whenever the degree of n is infinite (perforce, denumerable).*

Proof. Since n is restraining and the current vector $\mathbf{i}$ for the regime is in $\mathcal{I}$,

$$\sum_{(n)} |i_j| = \sum_{(n)} g_j^{1/2} |i_j| r^{1/2} \le \left(\sum_{(n)} g_j \sum_{(n)} i_j^2 r_j \right)^{1/2} < \infty.$$

So, by Appendix B6, we have the asserted summability.

Since $\mathcal{K}^o$ is dense in $\mathcal{K}$, we can choose a sequence $\{\mathbf{i}_m\}_{m=0}^{\infty}$ in $\mathcal{K}^o$ that converges in $\mathcal{K}$ to $\mathbf{i}$. Let i_{mj} denote the jth component of $\mathbf{i}_m$. Since every basic current and therefore every member of $\mathcal{K}^o$ satisfies Kirchhoff's current law at every maximal 0-node, $\sum_{(n)} \pm i_{mj} = 0$. Consequently,

$$\left| \sum_{(n)} \pm i_j \right| = \left| \sum_{(n)} \pm i_j - \sum_{(n)} \pm i_{mj} \right| \le \sum_{(n)} |i_j - i_{mj}|$$

$$\leq \left(\sum_{(n)} g_j \sum_{(n)} (i_j - i_{mj})^2 r_j \right)^{1/2} \leq \left(\sum_{(n)} g_j \right)^{1/2} \|\mathbf{i} - \mathbf{i}_m\| \to 0$$

as $m \to \infty$. ♣

The hypothesis of Theorem 5.3-1 is merely sufficient but not in general necessary. Kirchhoff's current law may still be satisfied at a nonrestraining node as a result of the configuration and conductance values of the branches not incident to the node. For similar reasons, the next two theorems also assert only sufficient conditions.

Our next objective is to extend Kirchhoff's current law to a certain kind of "cut" in $\mathbf{N}^\nu$. (Another extension to a different kind of "cut" will be given in Section (6.2).) Choose any set $\mathcal{N}_s^0$ of maximal 0-nodes in $\mathbf{N}^\nu$. Let $\mathcal{B}_s$ be the set of branches in $\mathbf{N}^\nu$, each of which is incident only to nodes of $\mathcal{N}_s^0$. Finally, let $\mathbf{N}_s^\gamma$ be the subnetwork induced by $\mathcal{B}_s$. We shall call $\mathbf{N}_s^\gamma$ the *subnetwork induced by* $\mathcal{N}_s^0$ or a *0-node-induced subnetwork*, the maximality of the 0-nodes being understood. $\mathbf{N}_s^\gamma$ may have nodes of ranks higher than 0, as will be illustrated in the next example.

Next, let $\mathcal{B}_c$ be the set of all branches, each of which is incident to a 0-node of $\mathcal{N}_s^0$ and also to a 0-node (possibly nonmaximal) not in $\mathcal{N}_s^0$. Let us call $\mathcal{B}_c$ the *0-cut for* $\mathbf{N}_s^\gamma$ or a *0-cut for a 0-node-induced subnetwork*. In general, it is possible for a path to lie outside of $\mathbf{N}_s^\gamma$ and nonetheless reach $\mathbf{N}_s^\gamma$ through a tip of rank 0 or larger without embracing any branch of $\mathcal{B}_c$. This too is illustrated in the next example.

Example 5.3-2. Make the $\vec{\omega}$-graph of Figure 4.4 into an $\vec{\omega}$-network $\mathbf{N}^{\vec{\omega}}$ by assigning electrical parameters to its branches; thus, every branch of $\mathbf{N}^{\vec{\omega}}$ now has an orientation. Let $\mathcal{N}_s^0$ be the 0-nodes embraced by the $\vec{\omega}$-path $P^{\vec{\omega}}$ and lying between m_0^0 and m_3^0 including those two nodes. Such 0-nodes are all maximal. Then, $\mathbf{N}_s^\gamma$ is the 1-path between and terminating at m_0^0 and m_3^0. $\mathbf{N}_s^\gamma$ has exactly one 1-node. Moreover, the 0-cut $\mathcal{B}_c$ for $\mathbf{N}_s^\gamma$ consists of the four branches that are incident to m_0^0 and m_3^0 but are not in $\mathbf{N}_s^\gamma$. Note that there is a one-ended 0-path starting at the 0-node m_2^0 and reaching the 1-node of $\mathbf{N}_s^\gamma$; that 0-path lies outside $\mathbf{N}_s^\gamma$ and does not embrace any branch of $\mathcal{B}_c$. ♣

Returning to the general case wherein $\mathcal{B}_c$ may be an infinite set, let us orient the 0-cut $\mathcal{B}_c$ away from $\mathbf{N}_s^\gamma$; thus, each branch of $\mathcal{B}_c$ has a cut-orientation in addition to its own orientation. An assertion of Kirchhoff's current law for $\mathcal{B}_c$ is

$$\sum_{(c)} \pm i_j = 0 \tag{5.23}$$

where the summation is over the branch-index set J_c for $\mathcal{B}_c$ and $+$ (or $-$) is used if the cut orientation and the branch orientation agree (resp. disagree). This law (5.23) need not hold for $\mathcal{B}_c$, but it will hold under the next theorem's hypothesis.

In the following, J_s will denote the branch-index set for $\mathbf{N}_s^\gamma$, and $\sum_{(sc)}$ will denote a summation over $J_s \cup J_c$.

Theorem 5.3-3. *Let* $\mathbf{N}_s^\gamma$ *be a 0-node-induced subnetwork of the* ν*-network* $\mathbf{N}^\nu$*. Assume that the set* $\{g_j\}_{j\in J_s\cup J_c}$ *of all the conductances in* $\mathbf{N}_s^\gamma$ *and in* $\mathcal{B}_c$ *is summable. Then, under the regime dictated by Theorem 5.2-8, Kirchhoff's current law (5.23) holds for the cut* $\mathcal{B}_c$*.*

Note. By virtue of the summability of $\{g_j\}_{j\in J_s\cup J_c}$, both $\mathcal{B}_c$ and the branch set of $\mathbf{N}_s^\gamma$ are now countable because no more than countably many positive conductances can be summable. Remember that every branch must have a positive conductance.

Proof. We first argue that the set $\{i_j\}_{j\in J_s\cup J_c}$ of all the branch currents in $\mathbf{N}_s^\gamma$ and in $\mathcal{B}_c$ is summable. By Schwarz's inequality,

$$\sum_{(sc)} |i_j| = \sum_{(sc)} g_j^{1/2}|i_j|r_j^{1/2} \le \left(\sum_{(sc)} g_j \sum_{(sc)} i_j^2 r_j\right)^{1/2}.$$

The right-hand side is finite because of the hypothesis on the g_j and the fact that the current vector $\mathbf{i}$ for $\mathbf{N}^\nu$ is a member of $\mathcal{I}$. So truly, $\{i_j\}_{j\in J_s\cup J_c}$ is summable (Appendix B6).

Furthermore, every maximal 0-node in $\mathbf{N}_s^\gamma$ is restraining by virtue of our assumption on the g_j. Hence, Kirchhoff's current law is satisfied at each maximal 0-node of $\mathbf{N}_s^\gamma$. Let us add together all the expressions (5.22) of that law at all the maximal 0-nodes of $\mathbf{N}_s^\gamma$. Since each expression equals 0, the result equals 0, too. Moreover, every term of the result is a current in a branch of $\mathbf{N}_s^\gamma$ or of $\mathcal{B}_c$. By virtue of our signs convention for (5.22), every such branch current appears twice with opposite signs if the branch is in $\mathbf{N}_s^\gamma$ and appears once with the appropriate sign for (5.23) if the branch is in $\mathcal{B}_c$. Since $\{i_j\}_{j\in J_s\cup J_c}$ is summable, so too is the set of terms obtained by adding the said Kirchoff's current laws (Appendix B6). Hence, we may rearrange and add those terms in any fashion without altering the result (Appendix B1). It now follows that the current in every branch of $\mathbf{N}_s^\gamma$ cancels out of the result, whereas the current in every branch of $\mathcal{B}_c$ appears as it does in (5.23). We have established (5.23). ♣

Consider now Kirchhoff's voltage law:

$$\sum_{(L)} \pm v_j \; = \; 0 \tag{5.24}$$

where the symbols have the same meanings as they do in (5.2) except that now the oriented loop L is allowed to be of any rank. As before, L will be called *permissive* (synonymously with "perceptible") if $\sum_{(L)} r_j < \infty$. Every 0-loop in $\mathbf{N}^\nu$ is permissive because it has only finitely many branches. On the other hand, as a consequence of how transfinite loops are defined, no loop can have uncountably many branches; this stands in contrast to a transfinite node, which can have an uncountable degree (in which case that node cannot be restraining).

Theorem 5.3-4. *If L is a permissive loop in* $\mathbf{N}^\nu$*, then, under the regime dictated by Theorem 5.2-8, Kirchhoff's voltage law (5.24) is satisfied. Moreover, the left-hand side of (5.24) is summable whenever L is a permissive transfinite loop.*

Proof. Since every branch of $\mathbf{N}^\nu$ is in the Thevenin form, $v_j = r_j i_j - e_j$ for every branch b_j. We can check the summability of the said series by considering $\sum_{(L)} r_j |i_j|$ and $\sum_{(L)} |e_j|$ separately and then invoking Appendix B6. By the Schwarz inequality,

$$\sum_{(L)} r_j |i_j| \;\leq\; \left(\sum_{(L)} r_j \sum_{(L)} i_j^2 r_j \right)^{1/2} \;<\; \infty$$

because $\mathbf{i} \in \mathcal{I}$. Similarly,

$$\sum_{(L)} |e_j| \;=\; \sum_{(L)} r_j^{1/2} |e_j| g_j^{1/2} \;\leq\; \left(\sum_{(L)} r_j \sum_{(L)} e_j^2 g_j \right)^{1/2} \;<\; \infty$$

because $\mathbf{e}$ is of finite total isolated power. Whence the asserted summability.

Finally, let s be a loop current of value 1 A flowing around L. Upon substituting this s into (5.18), we obtain (5.24). ♣

Here too, Kirchhoff's voltage law may be satisfied around a nonpermissive loop if the rest of the network is structured properly.

Let us now take note of a certain peculiarity exhibited by uncountable networks.

Theorem 5.3-5. *If* $\mathbf{N}^\nu$ *has uncountably many branches and satisfies Conditions 5.2-7, then, except for at most countably many of its branches, every one of its other uncountably many branches will have both a zero branch voltage and a zero branch current under the regime dictated by Theorem 5.2-8.*

Note. In this way, an uncountable network mimics a countable one.

Proof. According to Appendix B9, only countably many of the terms in a summable series can be nonzero. Since both $\sum i_j^2 r_j$ and $\sum e_j^2 g_j$ are finite and have nonnegative terms, these two series are summable (Appendix B7). Therefore, only countable many of the i_j and e_j are nonzero. Thus, the same is true for $v_j = i_j r_j - e_j$. ♣

For our next observation, we need some more definitions: *To open a branch* b_j in a given ν-network $\mathbf{N}^\nu$ will mean that b_j is deleted from the branch set and that its two $\vec{0}$-tips are deleted from the set of all $\vec{0}$-tips. *To short a self-loop* b_j will mean that b_j is opened. *To short a branch* b_j that is not a self loop will refer to the following two-step procedure: First, the two maximal nodes n_1^γ and n_2^δ to which b_j is incident are combined into a single node m^λ such that $\lambda = \max(\gamma, \delta)$ and, for each $\alpha = \vec{0}, 0, \ldots, \lambda - 1$, the set of α-tips embraced by m^λ consists of all the α-tips embraced by either n_1^γ or n_2^δ; second, the resulting self-loop that b_j becomes is opened. (These definitions of opening and shorting a branch do not conflict with our prior definitions of "shorted tips" and "open tips.") Furthermore, to *remove a branch* b_j will mean that the branch is either opened or shorted. Finally, we shall say that b_j is a *null branch* if its branch voltage v_j and its branch current i_j are both equal to 0.

One might be tempted to surmise that all the null branches can be removed from $\mathbf{N}^\nu$ through any combinations of removals (such as opening all of them or alternatively shorting all of them) without affecting the voltage-current regime in the rest of $\mathbf{N}^\nu$. After all, this is so for a finite network

$\mathbf{M}^0$ because $\mathbf{M}^0$ is governed by Kirchhoff's laws and the sum of currents at a finite node or the sum of voltages around a finite loop is not affected by the null branches. However, this conjecture is generally false for transfinite networks, as the next example shows. Here is another case where our habits of thought conditioned by finite networks can be misleading.

Example 5.3-6. Let $\mathbf{N}^\nu$ contain a two-ended γ-path P^γ that has no sources and reaches $\mathbf{N}^\nu\backslash P^\gamma$ only through the terminal nodes of P^γ. Assume that the resistances in P^γ sum to infinity. Then, no nonzero loop current can pass through P^γ, for otherwise infinite power would be dissipated. Thus, every branch of P^γ is a null branch. We may open all of those branches without affecting $\mathcal{K}^o$ and thereby $\mathcal{K}$ and the voltage -current regime of $\mathbf{N}^\nu\backslash P^\gamma$. On the other hand, let us interpret the *shorting of* P^γ as the shorting of all the branches in P^γ and the coalescing of the two terminal nodes of P^γ into a single node as above. Then, that shorting of P^γ may affect the regime because new nonzero loop currents may thereby be introduced into $\mathcal{K}^o$, allowing the regime to change. Note, however, that before this shorting, the contribution of P^γ to Kirchhoff's voltage law around any loop is zero.

A similar situation holds for an infinite parallel circuit Q^0 in $\mathbf{N}^\nu$. Assume that Q^0 has no sources and that the conductances in Q^0 sum to infinity. Then, the voltage across Q^0 must be zero if the power dissipated in it is to be finite. Consequently, all of the branches of Q^0 are null branches. We may short all of those branches without affecting the regime in $\mathbf{N}^\nu\backslash Q^0$ because such a shorting will not affect the restrictions of the members of $\mathcal{K}^o$ to $\mathbf{N}^\nu\backslash Q^0$. However, the opening of all of the branches may affect the regime in $\mathbf{N}^\nu\backslash Q^0$ because this might disrupt a nonzero loop current passing through Q^0. (For instance, if we open all the purely resistive branches in the parallel circuit of Figure 5.4, the current in the source branch will change from 1 A to 0 A.) Nonetheless, before this opening, the contribution of Q^0 to Kirchhoff's current law at any maximal 0-node is zero.

These two circuits show in another way that Kirchhoff's laws are inadequate for a theory of unique regimes in transfinite networks. ♣

5.4 The Regime Induced by a Pure Current Source

Up to now, the ν-network $\mathbf{N}^\nu$ we have been considering has not had any pure sources. For certain developments that we shall discuss later on, we need to consider the case where $\mathbf{N}^\nu$ is excited only by a single pure current source. We can obtain a voltage-current regime in this case by using as a definition a technique valid for finite networks. This involves the transferring of the pure current source onto the branches of a path connecting the two nodes of the source [21, pages 131-132], and it will work in the transfinite case if the path is permissive.

The technique we shall discuss herein is a special case of one used for defining the regime corresponding to an excitation of $\mathbf{N}^\nu$ by many pure current sources and pure voltage sources. The latter has been presented in [35, Sections 3.6 and 3.7], wherein it is shown that a unique regime exists under certain finite-power assumptions. This more general situation requires a rather complicated analysis, especially when both pure current sources and pure voltages sources occur simultaneously. A considerably simpler analysis can be given when there is only one source, a pure current source. Such is the objective of this section.

As always, we restrict ν to ranks no larger than ω. Furthermore, we will call a tip *permissive* if it has a permissive representative; in this case, all the representatives embraced by that permissive one will also be permissive.

Condition 5.4-1. *If two tips of ranks less than ν (possibly differing ranks) are permissive and nondisconnectable, then those tips are shorted together.*

Note that, although Conditions 3.5-1 and 5.4-1 are similar, neither implies the other. However, we will use Condition 5.4-1 in such circumstances that we will be able to invoke the consequences of Condition 3.5-1.

To proceed: Let all the branches of $\mathbf{N}^\nu$ be purely resistive (no sources) and append to $\mathbf{N}^\nu$ a pure current source h, whose incident nodes are n_1 and n_2. Assume furthermore that there is a permissive two-ended path P^γ in $\mathbf{N}^\nu$ terminating at n_1 and n_2. We shall show that h induces a unique voltage-current regime in $\mathbf{N}^\nu$ by transferring h onto the branches of P^γ, converting those branches from the obtained Norton forms into their equivalent Thevenin forms, invoking Theorem 5.2-8 to obtain a voltage-current regime, and finally converting that regime into the corresponding one for the original network $\mathbf{N}^\nu$. Actually, there may be more than one permissive path in $\mathbf{N}^\nu$ terminating at n_1 and n_2; in that case we have to show that the final regime for $\mathbf{N}^\nu$ is independent of the choice of the permissive path.

The transference of h onto the branches of P^γ is illustrated in Figure 5.8, parts (a) and (b). The transference removes h from its original position in part (a) and converts the purely resistive branches of P^γ into Norton-form branches as shown in part (b). This does not violate Kirchhoff's current law at any maximal 0-node or at any cut. The next step is to convert those Norton-form branches into the Thevenin-form branches shown in part (c) of Figure 5.8. We now have many voltage sources, but their total isolated power is finite. Indeed, with Π being the index set for the branches of the permissive path P^{γ}, we have $e_j g_j = -h$ for $j \in \Pi$, and therefore

$$\sum_{j\in\Pi} e_j^2 g_j \;=\; \sum_{j\in\Pi} h_j^2 r_j \;=\; h^2 \sum_{j\in\Pi} r_j \;<\; \infty.$$

Consequently, we can invoke Theorem 5.2-8 to obtain a unique voltage-current regime for the excitation indicated in Figure 5.8(c).

We now wish to convert the voltage-current regime obtained for part (c) into a voltage-current regime for part (a). To be more specific at this point, let us consider Figure 5.9, which shows the jth branch b_j of $\mathbf{N}^{\nu}$; that branch may or may not be in P^{γ}. The superscript "(1)" indicates that the transfer of h is with regard to the choice of a particular permissive path P_1^{γ}. In part (a), we show b_j before h is transferred; thus, $i_{oj}^{(1)}$ and $v_{oj}^{(1)}$ are the current and voltage for b_j in its original form. By Ohm's law, $v_{oj}^{(1)} \;=\; r_j i_{oj}^{(1)}$. In part (b), h has been transferred, and the branch appears in the Norton form; $h_j^{(1)}$ is the resulting current source. If b_j is in P_1^{γ}, $h_j^{(1)}$ is either $+h$ or $-h$ depending upon the orientation of b_j; on the other hand, if b_j is not in P_1^{γ}, $h_j^{(1)}$ is 0. Furthermore, we have that $i_{oj}^{(1)} \;=\; i_j^{(1)} + h_j^{(1)}$, where now $i_j^{(1)}$ is the branch current for this Norton form. Part (c) of Figure 5.9 shows the corresponding Thevenin form for b_j; here, $i_j^{(1)}$ and $v_j^{(1)}$ are the current and voltage dictated by Theorem 5.2-8. Now, $v_j^{(1)} \;=\; (i_j^{(1)} + h_j^{(1)}) r_j = v_{oj}^{(1)}$, as is required by the equivalence between the Norton and Thevenin forms. Thus, the specification of $i_j^{(1)}$ and $v_j^{(1)}$ by Theorem 5.2-8 uniquely determines the current $i_{oj}^{(1)}$ and voltage $v_{oj}^{(1)}$ in the original branch of part (a). This is how we specify the voltage-current regime when $\mathbf{N}^{\nu}$ is excited only by a single appended pure current source.

However, in the event that there are several permissive paths connecting n_1 and n_2, we must now show that this regime is independent of the choice of permissive path onto which h is transferred. This will be accomplished when we prove that $i_{oj}^{(1)}$ is independent of the choice of path, because then $v_{oj}^{(1)} \;=\; i_{oj}^{(1)} r_j$ will be similarly independent as well.

For this purpose, note first of all that the space $\mathcal{K}$ pertaining to $\mathbf{N}^{\nu}$ does not change as the pure current source is transferred and converted. This is because $\mathcal{K}$ depends only upon the graph and the resistance values of $\mathbf{N}^\nu$ — not on the sources.

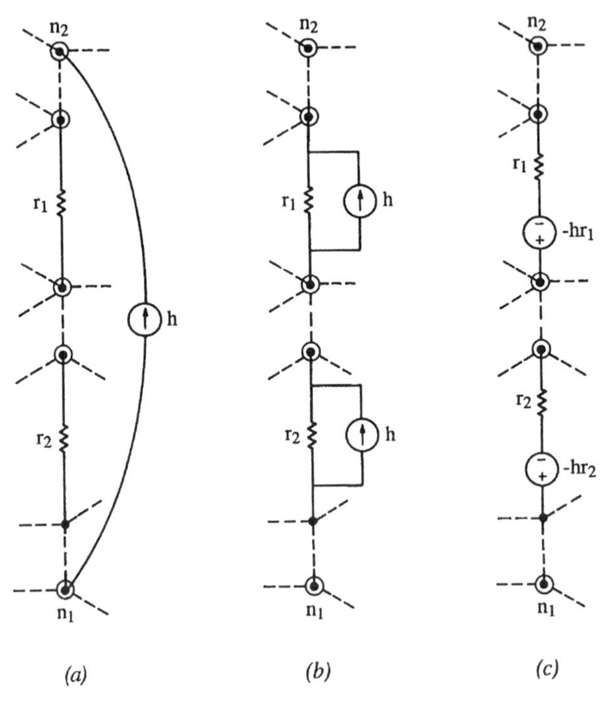

Figure 5.8. The transference and conversion of a pure current source. **(a)** A pure current source h incident to the terminal nodes n_1 and n_2 of an entirely resistive, permissive path P^γ. **(b)** The result of transferring h onto the branches of P^γ to obtain Norton forms for those branches. **(c)** The result of converting the Norton forms into Thevenin forms.

Now assume that there are several permissive two-ended paths connecting the nodes n_1 and n_2. Let P_1^γ denote one of them. Also, let $\mathbf{h}^{(1)}$ denote the branch-current-source vector resulting from a transference of the single pure current source h onto P_1^γ. Thus, the component of $\mathbf{h}^{(1)}$ corresponding to a branch of P_1^γ is $+h$ or $-h$ depending upon the orientation of the branch and is 0 for a branch not in P_1^γ. The branch-voltage-source vector resulting from the Norton-to-Thevenin conversion is $\mathbf{e}^{(1)} = -R\mathbf{h}^{(1)}$. Let $\mathbf{i}^{(1)}$ be the corresponding branch-current-vector dictated by Theorem 5.2-8. Thus, $\mathbf{i}^{(1)} \in \mathcal{K}$, and

$$\langle R(\mathbf{h}^{(1)} + \mathbf{i}^{(1)}), \mathbf{s}\rangle = 0 \tag{5.25}$$

for all $\mathbf{s} \in \mathcal{K}$.

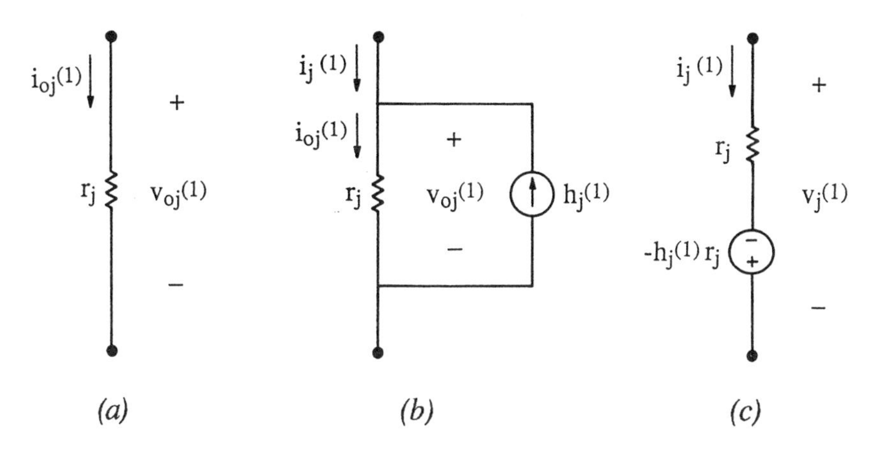

Figure 5.9. **(a)** A purely resistive branch b_j in the chosen path P_1^γ. **(b)** The result of transferring the current source onto this branch. **(c)** The Thevenin branch corresponding to the Norton branch of part (b).

If P_2^λ is another permissive path connecting n_1 and n_2, we obtain in the same way a branch-current-source vector $\mathbf{h}^{(2)}$ by transference, a branch-voltage-source vector $\mathbf{e}^{(2)} = -R\mathbf{h}^{(2)}$ by conversion, and a branch-current vector $\mathbf{i}^{(2)}$ by Theorem 5.2-8. Thus, $\mathbf{i}^{(2)} \in \mathcal{K}$, and

$$\langle R(\mathbf{h}^{(2)} + \mathbf{i}^{(2)}), \mathbf{s}\rangle \;=\; 0 \tag{5.26}$$

for all $\mathbf{s} \in \mathcal{K}$.

Assume that Condition 5.4-1 holds for the tips traversed by P_1^γ and P_2^λ. We are able to invoke Condition 3.5-1 and thereby Corollary 3.5-5 because any two nondisconnectable tips with one in P_1^γ and the other in P_2^λ will be shorted together according to Condition 5.4-1. Choose any branch of P_1^γ that is not in P_2^λ, and trace P_1^γ toward n_1. By Corollary 3.5-5, we will find a first node n_a met by both P_1^γ and P_2^λ. Upon tracing P_1^γ in the other direction, we will find another such first node n_b met by both paths. Thus, the flow of a current h along P_1^γ from n_a to n_b and then along P_2^λ from n_b to n_a is a loop current within $\mathbf{h}^{(1)} - \mathbf{h}^{(2)}$ on a loop L within $P_1^\gamma \cup P_2^\lambda$.

Now, $(P_1^\gamma \cup P_2^\lambda)\backslash L$ consists of four subpaths: a subpath P_{1a}^γ in P_1^γ between n_a and n_1, a subpath P_{2a}^λ in P_2^λ between n_a and n_1, a subpath P_{1b}^γ in P_1^γ between n_b and n_2, and a subpath P_{2b}^λ in P_2^λ between n_b and n_2. (These subpaths may be trivial if $n_a = n_1$ and/or $n_b = n_2$.) If P_{1a}^γ and P_{2a}^λ are not trivial, we can treat $P_{1a}^\gamma \cup P_{2a}^\lambda$ in the same way to find another loop in

$P_1^\gamma \cup P_2^\lambda$. Since P_1^γ and P_2^λ have only countably many branches, we can continue this process to find a finite or infinite sequence of loops in $P_1^\gamma \cup P_2^\lambda$, and the deletion of all those loops will yield a finite or infinite sequence of subpaths (possibly trivial), each residing in both P_1^γ and P_2^λ.

With a current of value h flowing in P_1^γ from n_1 to n_2 and flowing in P_2^λ from n_2 to n_1, we have that $\mathbf{h}^{(1)} - \mathbf{h}^{(2)}$ is the superposition of finitely or infinitely many loop currents. If infinitely many, the fact that P_1^γ and P_2^λ are both permissive implies that $\mathbf{h}^{(1)} - \mathbf{h}^{(2)}$ is the limit in $\mathcal{K}$ of those loop currents, where $\mathcal{K}$ is the space pertaining to $\mathbf{N}^\nu$, as before. All this shows that $\mathbf{h}^{(1)} - \mathbf{h}^{(2)} \in \mathcal{K}$.

Now, let us subtract (5.26) from (5.25). Thus, for all $\mathbf{s} \in \mathcal{K}$,

$$0 \;=\; \langle R(\mathbf{h}^{(1)} - \mathbf{h}^{(2)} + \mathbf{i}^{(1)} - \mathbf{i}^{(2)}, \mathbf{s}\rangle \;=\; (\mathbf{h}^{(1)} - \mathbf{h}^{(2)} + \mathbf{i}^{(1)} - \mathbf{i}^{(2)}, \mathbf{s}),$$

where the right-hand side is the inner product for $\mathcal{K}$. We can conclude that $\mathbf{h}^{(1)} - \mathbf{h}^{(2)} + \mathbf{i}^{(1)} - \mathbf{i}^{(2)} = 0$. Since $\mathbf{i}_o^{(1)} = \mathbf{i}^{(1)} + \mathbf{h}^{(1)}$ and $\mathbf{i}_o^{(2)} = \mathbf{i}^{(2)} + \mathbf{h}^{(2)}$, we have $\mathbf{i}_o^{(1)} = \mathbf{i}_o^{(2)}$. We are done.

Before stating the theorem (Theorem 5.4-2) we have established, let us summarize how the voltage-current regime for $\mathbf{N}^\nu$ has been obtained: With regard to a transfer of h onto a chosen permissive path P^γ, let $\mathbf{i}$ and $\mathbf{v}$ be the voltage vector and current vector dictated by Theorem 5.2-8 for the network obtained after transferring and converting h to get $\mathbf{e} = -R\mathbf{h}$ as indicated in Figure 5.8(c). Also, let

$$\mathbf{i}_o \;=\; \mathbf{i} + \mathbf{h} \tag{5.27}$$

and

$$\mathbf{v}_o \;=\; R(\mathbf{i} + \mathbf{h}). \tag{5.28}$$

Then, $\mathbf{i}_o$ and $\mathbf{v}_o$ are the current vector and the voltage vector for the ν-network $\mathbf{N}^\nu$.

Theorem 5.4-2. *Assume that the ν-network $\mathbf{N}^\nu$ is purely resistive and that its tips of ranks less than ν satisfy Condition 5.4-1. Append to $\mathbf{N}^\nu$ a single pure current source whose incident nodes are the terminal nodes of at least one permissive path within $\mathbf{N}^\nu$. Then, the voltage-current regime $\mathbf{v}_o, \mathbf{i}_o$ in $\mathbf{N}^\nu$, as given by (5.27) and (5.28), does not depend upon the choice of the permissive path.*

Henceforth, we specify $\mathbf{v}_o, \mathbf{i}_o$ as the voltage-current regime for the purely resistive ν-network $\mathbf{N}^\nu$ when $\mathbf{N}^\nu$ is excited by only one appended source and that source is a pure current source.

Another situation that will arise later on is a ν-network excited only by finitely many pure current sources. To obtain the voltage-current regime in this case, we can invoke the superposition principle. Thus, the desired voltage-current regime is the sum of the voltage-current regimes for the various sources considered individually.

Still another situation that we will meet is a ν-network $\mathbf{N}^\nu$ excited only by finitely many pure voltage sources, say, n of them. However, this situation will occur only when $\mathbf{N}^\nu$ has the following property: If the pure voltage sources are replaced by pure current sources, the resulting vector of n voltages at the terminals of the current sources will be given by an $n \times n$ nonsingular real matrix Z acting upon the vector of n current-source values. Therefore, we may convert the original voltage sources into current sources through the inverse matrix Z^{-1} and then may use Theorem 5.4-1 along with superposition to obtain the voltage-current regime for $\mathbf{N}^\nu$. All this will be explicated in greater detail later on.

(Before leaving this section, let us point out a correction for [35, page 98]. Rather than concluding as stated there that $\mathbf{i}^{(\nu)} + \mathbf{h}_a^{(\nu)} + \mathbf{h}_o$ is a member of $\mathcal{K}$, we should conclude that the difference between two such terms for different choices of ν is a member of $\mathcal{K}$. We can then continue as we have above with the argument establishing Theorem 5.4-2.)

5.5 Node Voltages

The electrical behavior of a 0-connected 0-network can be identified either by specifying its branch voltages or by specifying its node voltages with respect to some ground node. Either specification can be derived from the other one. Actually, the common practice is to use node voltages rather than branch voltages as the fundamental variables when examining conventional infinite networks in the context of discrete potential theory [27] or random walks on graphs [32].

However, this interchangeability between node voltages and branch voltages breaks down in general when the network is transfinite. As we have seen in Theorems 5.2-8 and 5.2-14, branch voltages will exist under quite broad conditions (Conditions 5.2-7 or Conditions 5.2-13), but it turns out that node voltages are not so readily available. Two difficulties may intervene by virtue of the fact that node voltages are algebraic sums of branch voltages along certain paths. In the first place, a voltage at a node n_0 transfinitely far away from a ground node n_g may fail to exist because the sum of voltages along every path between n_0 and n_g may diverge. Second, even when some of those sums exist, the voltage at n_0 may not be unique because different choices of the paths between n_0 and n_g may have different sums. Because of this, it can be argued that branch voltages are more fundamental than node voltages so far as transfinite electrical networks are concerned. After all, it is in the branches that power conversion takes place; nodes merely serve to short tips together. On the other hand, node voltages are at times essential and irreplaceable, as we shall see when we discuss transfinite discrete potential theory (Section 6.3) and transfinite random walks (Chapter 7). So, let us now define node voltages, examine transfinite networks wherein node voltages either fail to exist or are not unique, and finally establish conditions under which node voltages exist uniquely.

In this section $\mathbf{N}^\nu$ will denote a ν-connected ν-network satisfying Conditions 5.2-7. It is no restriction to have every branch in its Thevenin form. Thus, with our usual notations and orientations, $v_j = r_j i_j - e_j$ for the branch b_j; see Figure 5.3(a). Furthermore, the voltage-current regime for $\mathbf{N}^\nu$ will be the one designated by Theorem 5.2-8.

Choose a maximal node n_g and call n_g the *ground node*. Its *node voltage* u_g will always be set at 0 V. Also, let n_0 be any other maximal node of $\mathbf{N}^\nu$ and let P be a two-ended path terminating at n_g and n_0 and oriented from n_g to n_0. The index set Π for the branches of P is countable by virtue of the definitions of paths. As was defined in Section 5.2, P is called *permissive* if $\sum_{j\in\Pi} r_j < \infty$; a trivial path is automatically permissive.

With v_j being the voltage of the branch b_j as always, $-v_j$ will be called the *voltage rise* for b_j. The algebraic sum of the branch voltage rises along P from n_g to n_0 is

$$\sum_{(P)} \mp v_j \tag{5.29}$$

where the summation is over the branch-index set Π for P and the minus (plus) sign is used if b_j's orientation agrees (resp. disagrees) with the orientation of P. Equivalently, we can equate (5.29) to $\sum_{(P)} \pm v_j$ where now P is oriented from n_0 to n_g and the sign convention is reversed.

Lemma 5.5-1. *If the path P is permissive, then (5.29) converges absolutely.*

Proof. We estimate as in the proof of Theorem 5.3-4. For j restricted to Π, with $g_j = r_j^{-1}$, and by Schwarz's inequality, we may write

$$\sum |v_j| = \sum |r_j i_j - e_j| \le \sum \sqrt{r_j}|i_j|\sqrt{r_j} + \sum \sqrt{r_j}|e_j|\sqrt{g_j}$$

$$\leq \left(\sum r_j \sum i_j^2 r_j\right)^{1/2} + \left(\sum r_j \sum e_j^2 g_j\right)^{1/2} < \infty. \clubsuit$$

If there truly exists a permissive path P terminally incident to n_g and n_0, then the value (5.29) is defined to be the *node voltage* u_0 *at* n_0 *with respect to* n_g *along* P. We shall say also that n_0 *obtains the node voltage (5.29) with respect to* n_g *along* P. Naturally, any nonmaximal node is assigned the same node voltage as that of its maximal embracing node. It should be emphasized that according to this definition, node voltages are assigned only along permissive paths. For instance, (5.29) may converge even when P is not permissive, but nonetheless we will never use (5.29) to assign a node voltage along a nonpermissive path P. On the other hand, there may be more than one permissive path between n_g and n_0, and it may even happen that the node voltages assigned to n_0 along two such permissive paths may differ — as we shall see in Example 5.5-3.

Example 5.5-2. Let us illustrate the first difficulty regarding the existence of node voltages, the one whereby the algebraic sum of voltage rises along every path from the ground node n_g to a node n_0 is divergent. Perhaps the simplest example demonstrating this possibility is the 1-network of Figure 5.10(a). It is simply a 1-path $P^1 = \{n_g^0, P^0, n^1\}$, where P^0 is a one-ended 0-path starting at a ground 0-node n_g^0 and reaching a 1-node n^1. Every branch b_k $(k = 2, 3, 4, \ldots)$ of P^0 is in the Thevenin form of a linear resistor $r_k = k^{-1/2}$ in series with a voltage source $e_k = k^{-1}$ with polarity directed toward n^1. The sources satisfy the finite total isolated power condition $\sum_{k=2}^{\infty} e_k^2 g_k = \sum_{k=2}^{\infty} k^{-3/2}$, where $g_k = r_k^{-1} = k^{1/2}$. Thus, this 1-network does have a voltage-current regime dictated by Theorem 5.2-8. Moreover, the branch current i_k and branch voltage v_k are obviously $i_k = 0$ and $v_k = e_k$ for every branch. Consequently, the sum of the branch voltages diverges: $\sum_{k=2}^{\infty} v_k = \sum_{k=2}^{\infty} k^{-1} = \infty$. Hence, n^1 does not possess a (finite) node voltage with respect to the ground node n_g^0.

One might object that this is a trivial example since every branch current in P^1 is 0, making the branch voltages equal to the nonsummable branch source voltages. However, P^1 can be imbedded into a larger network in such a fashion that every one of its branches has a nonzero current, and yet its branch voltages still sum to infinity. This is so for the 1-network of Figure 5.10(b). In that network, two replicates of P^0 are connected at their corresponding nodes by resistors except at the first and last nodes, where they are connected by a source branch on the left and by a source branch in parallel with another resistor on the right. All of these (vertically drawn) resistors are 1 Ω in value. With $e_0 \neq 0$ and $\epsilon \neq 0$, every branch of this network carries a nonzero current. Indeed, we can use the superposition principal (Corollary 5.2-10) to determine the branch currents and branch voltages.

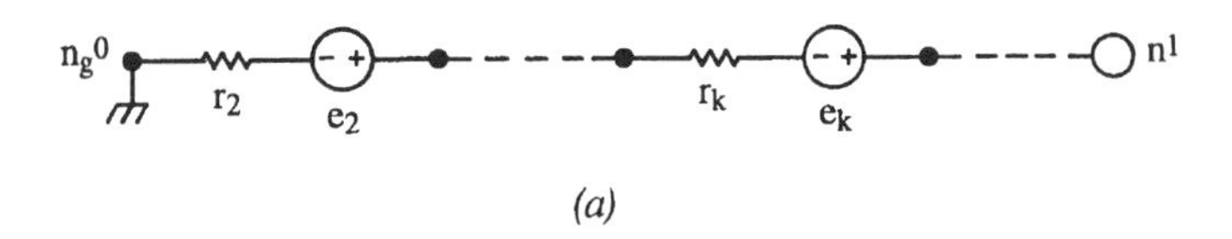

(a)

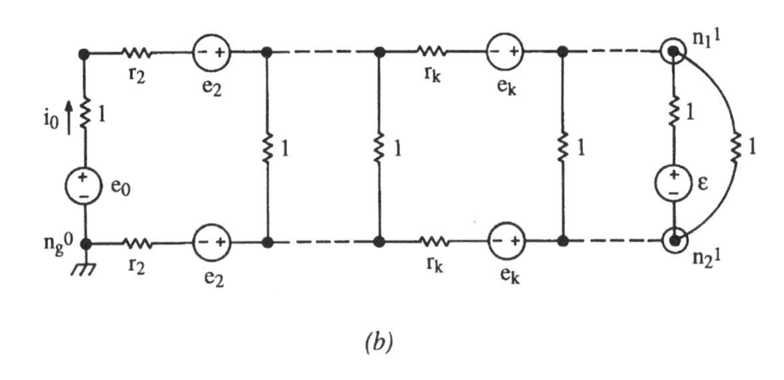

(b)

Figure 5.10. **(a)** A 1-network consisting of a single 1-path. Here, $r_k = k^{-1/2}$ and $e_k = k^{-1}$ for $k = 2, 3, 4, \ldots$. All branch currents are 0, and the 1-node n^1 does not have a node voltage with respect to the ground 0-node n_g^0. **(b)** Another 1-network in the form of a transfinite ladder. All vertical resistances have the value 1 Ω. The horizontal voltage sources and resistances are as in part (a). With $e_0 \neq 0$ and $\epsilon \neq 0$, every branch has a nonzero current, and yet neither n_1^1 nor n_2^1 has a node voltage with respect to n_g^0.

With $e_0 = \epsilon = 0$, the branch currents are induced by all the horizontal sources, and they are all 0 because of the balance between the upper and lower paths. It follows as before that the node voltages at n_1^1 and n_2^1 are both infinite — that is, they fail to exist as finite voltages.

As the next case, let $e_0 = 1$ and let all the other voltage sources be 0. Now, every branch to the left of the two 1-nodes carries a nonzero current. However, the two branches connected to those 1-nodes have zero currents because there is no permissive 1-loop that passes through those 1-nodes;

indeed, each such 1-loop must pass through an infinity of contiguous horizontal resistances, and these will sum to infinity. Because the total power dissipated in all the resistances is finite, we can conclude that the voltages on the vertical 1 Ω resistors tend to 0 as the 1-nodes are approached. In fact, the node voltages along the upper (or lower) horizontal path decrease (resp. increase) to the common voltage $(e_0 - i_0)/2$ at the 1-nodes. In short, n_1^1 and n_2^1 now have the same (finite) node voltage.

As the last case, let $\epsilon = 1$ and let all the other voltage sources be 0. Again because of the absence of permissive 1-loops, all the branches to the left of the 1-nodes carry zero currents, and the two branches incident to the 1-nodes carry nonzero currents, namely $\epsilon/2$.

Upon adding these voltage-current regimes, we see that every branch of the 1-network of Figure 5.10(b) has a nonzero current and that the 1-nodes n_1^1 and n_2^1 do not have node voltages with respect to ground. ♣

Example 5.5-3. The second difficulty about establishing node voltages concerns their possible nonuniqueness, that is, their possible dependence upon the choice of permissible paths. As an example of this, consider the 1-network of Figure 5.11, which consists of a sourceless lattice network and two appended source branches β_1 and β_2, each of which is a series connection of a 1 V source and a 1 Ω resistor. The resistance values in the lattice are $a_k = b_k = 1\ \Omega$ and $c_k = d_k = 2^{-k}\ \Omega$ for $k = 1, 2, 3, \ldots$. Let t_a^0, t_c^0, and t_d^0 be three 0-tips, each having a representative consisting only of a_k, c_k, and d_k branches, respectively. Then, let $n_a^1 = \{n_a^0, t_a^0\}$, $n_c^1 = \{n_c^0, t_c^0\}$, and $n_d^1 = \{n_d^0, t_d^0\}$, where n_a^0, n_c^0, and n_d^0 are three singleton 0-nodes containing respectively a $\vec{0}$-tip of β_1, a $\vec{0}$-tip of β_2, and a $\vec{0}$-tip of β_2. The other $\vec{0}$-tip of β_1 is contained in the 0-node n_1^0 as shown. Note that the permissive nondisconnectable tips t_c^0 and t_d^0 violate Condition 5.4-1.

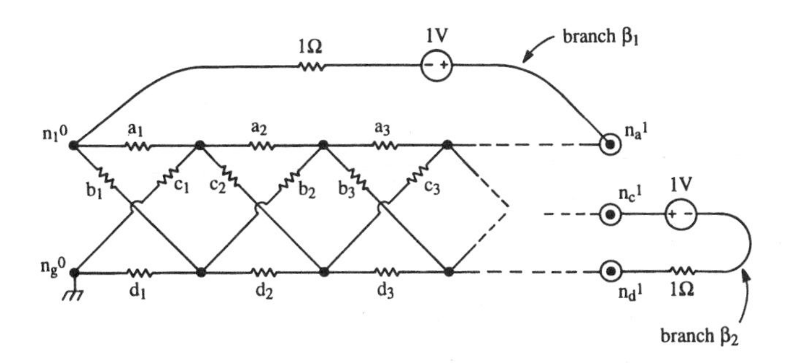

Figure 5.11. The 1-network discussed in Example 5.5-2.

Since $a_k = 1\ \Omega$ for all k, there is no permissive 1-loop that passes through β_1; thus, the current in β_1 is 0. The current in β_2 is 0 as well — but for a different reason. Even though each representative of t_c^0 and t_d^0 is permissive, there is no 1-loop that passes through β_2 because t_c^0 and t_d^0 are nondisconnectable, that is, any representative of t_c^0 and any representative of t_d^0 meet (infinitely often) and therefore cannot be parts of the same loop. Since there is no source within the lattice network, we can conclude that all the branches carry zero currents.

Now consider any possible node voltage at n_a^1. There are many permissive paths connecting ground n_g^0 and n_a^1, but they all pass through the source branch β_1 and are otherwise confined to the lattice. The voltage v_a^1 assigned to n_a^1 along any one of these paths is 1 V. To be sure, there are other paths that reach n_a^1 through t_a^0, but all of them are nonpermissive. (Had we allowed node voltages to be assigned along nonpermissive paths, v_a^1 would be 0 V for those paths.) In short, n_a^1 has the unique node voltage $v_a^1 = 1\ V$ because of the requirement that node voltages be assigned only along permissive paths.

As for the node voltage v_c^1 at n_c^1, there is a permissive path starting at n_g^0, passing along the c_k branches, and reaching n_c^1 through t_c^0. Thus, $v_c^1 = 0\ V$ for that path. There is another permissive path starting at n_g^0 and passing through the d_k branches, through n_d^1, and through the source branch β_2 to reach n_c^1 again. For that path, $v_c^1 = 1\ V$. We thus have nonuniqueness for the node voltage at n_c^1. Through a similar argument, the same conclusion can be drawn for n_d^1.

Here, too, we can construct another example, wherein the lattice branches carry nonzero currents, simply by inserting some voltage sources in some of the lattice branches. The source branches β_1 and β_2 would again have zero currents. Superposition would show that n_a^1 has a unique node voltage but n_c^1 and n_d^1 have nonunique node voltages. ♣

As these examples show, we are faced with the problem of finding conditions on the ν-network $\mathbf{N}^\nu$ under which unique node voltages exist with respect to a chosen ground node n_g. A solution is easily obtained from the results of the preceding section.

Theorem 5.5-4. *Assume that the tips of ranks less than ν in the ν-network $\mathbf{N}^\nu$ ($\nu \leq \omega$) satisfy Condition 5.4-1. Let n_g and n_0 be two nodes (of possibly different ranks) in $\mathbf{N}^\nu$, and let there be at least one permissive path* *connecting n_g and n_0. Then, n_0 has a unique node voltage with respect to n_g; that is, n_0 obtains the same node voltage with respect to n_g along all permissive paths between n_g and n_0.*

Proof. Assume there are two permissive two-ended paths P_1^γ and P_2^λ terminating at n_g and n_0. By the argument establishing Theorem 5.4-2, $P_1^\gamma \cup P_2^\lambda$ is the union of finitely or denumerably many loops and paths. The paths occur along the branches common to both P_1^γ and P_2^λ. By the permissivity of P_1^γ and P_2^λ, we can rearrange terms when algebraically summing voltages along P_1^γ and P_2^λ. Therefore, by that same permissivity and by Kirchhoff's voltage law applied to the loops, it follows that the node voltage assigned to n_0 along P_1^γ equals that obtained along P_2^λ. ♣

Corollary 5.5-5. *Assume that the tips of all ranks less than ν in the ν-network $\mathbf{N}^\nu$ ($\nu \leq \omega$) satisfy Condition 5.4-1. Also, assume that every two nodes of $\mathbf{N}^\nu$ are connected through at least one permissive path (i.e., assume that $\mathbf{N}^\nu$ is permissive). Choose a ground node n_g in $\mathbf{N}^\nu$ arbitrarily. Then, every node of $\mathbf{N}^\nu$ has a unique node voltage with respect to n_g.*

Finally, let us note in passing that Theorem 5.5-4 and Corollary 5.5-5 have been extended to a certain class of nonlinear transfinite networks [40].

Chapter 6

Permissively Finitely Structured Networks

The purpose of this chapter is to lay a foundation for a theory of random walks on transfinite networks, a subject that will be explored in Chapter 7. What is needed for this purpose is not only the uniqueness of the node voltages but also a maximum principle for them. The desired principle will assert that in a sourceless subsection every internal node voltage is less than the voltage at some bordering node of the subsection and greater than the voltage at another bordering node of the subsection — except in the case when all the node voltages within the subsection are identical. The uniqueness of the node voltages has already been established under the hypotheses of Theorem 5.5-4 and Corollary 5.5-5. The maximum principle is established in Section 6.3, and for this purpose we will impose stronger conditions than those used for uniqueness. In particular, we will assume that the network $\mathbf{N}^{\nu}$ is not only finitely structured but that the contraction paths can be chosen so that every one of them is permissive. These conditions will also suffice for our theory of transfinite random walks. Some ancillary results that are also established in this chapter are Kirchhoff's current law for a cut at an isolating set, the feasibility of exciting $\mathbf{N}^{\nu}$ by a pure voltage source, the boundedness of all the node voltages by the value of that voltage source, and another maximum principle expressed in terms of currents.

6.1 Permissively Finitely Structured ν-Networks

Recall that a path is called *permissive* if the sum of all the resistances in the path is finite. A trivial path is automatically permissive. As always, we restrict ν to $0 \leq \nu \leq \omega$.

Definition of a permissively finitely structured ν-network: A ν-network $\mathbf{N}^\nu$ will be called *permissively finitely structured* if the graph of $\mathbf{N}^\nu$ is finitely structured and if, for every β with $0 < \beta \leq \nu$ and for every maximal β-node n^β in $\mathbf{N}^\nu$, there is a contraction $\{\mathcal{W}_p\}_{p=1}^\infty$ to n^β, all of whose contraction paths are permissive.

From now on, we will always assume

Conditions 6.1-1. *The ν-network $\mathbf{N}^\nu$ is permissively finitely structured. Also, every branch of $\mathbf{N}^\nu$ has a positive resistance (and possibly a source as well).*

As was asserted by Lemma 4.5-1, $\mathbf{N}^\nu$ has no $\vec{\omega}$-node, and its rank cannot be $\vec{\omega}$.

Later on, we will be appending other sources to $\mathbf{N}^\nu$ but will denote the augmented network by $\mathbf{N}_a^\nu$; we will also use $\mathbf{N}_e^\nu$ in place of $\mathbf{N}_a^\nu$ when all the appended sources are pure voltage sources.

When the contraction paths for a contraction are all permissive, the contraction itself will be called *permissive*; whenever possible, we will always choose our contractions to be permissive.

Lemma 6.1-2. *Between every two nodes (of any ranks) in a permissively finitely structured network $\mathbf{N}^\nu$ ($\nu \leq \omega$), there is a permissive path that terminates at those two nodes.*

Proof. Let us assume at first that $\nu = \mu$, where μ is a natural number. Let n_a and n_b be two totally disjoint nodes. By the μ-connectedness of $\mathbf{N}^\mu$, there is a two-ended path P^α of rank α ($0 \leq \alpha \leq \mu$) that terminates at n_a and n_b. Thus, $\alpha - 1$ is the largest rank for all the tips traversed by P^α. If $\alpha = 0$, that path has only finitely many branches, and our conclusion follows. So, consider the case where $\alpha \geq 1$.

Let θ be the maximum rank among all the ranks of the maximal nodes that P^α meets. (θ might be larger than α.) Since $\mathbf{N}^\nu$ is finitely structured, its graph satisfies the hypothesis of Lemma 4.1-5, and therefore by conclusion (i) of that lemma P^α meets only finitely many maximal θ-nodes. We can choose an isolating set for each of those θ-nodes such that their corresponding arms are pairwise totally disjoint and such that neither n_a nor n_b is incident to an arm except when n_a or n_b is embraced by a θ-node. Consider the case where P^α meets a maximal θ-node n^θ through one or two nonelementary tips. Let $\mathbf{A}$ be the corresponding chosen arm, and $\mathcal{V}$ its base. Orient P^α from n_a to n_b. Then, there may be a first node $n_1 \in \mathcal{V}$ that P^α meets before meeting n^θ, a last node $n_2 \in \mathcal{V}$ that P^α meets after leaving n^θ, or both. In each case, replace that part of P^α between n_1 and n^θ or between n^θ and n_2 by a permissive path in $\mathbf{A}$ that terminates at n_1 and n^θ or at n^θ and n_2. This can be done because the contraction paths are permissive. The only other case to consider is when P^α meets n^θ only through elementary tips; now, nothing needs to be done.

Make such replacements when needed for all of the finitely many θ-nodes that P^α meets. Let P_r be the two-ended path resulting from the said replacements. Upon deleting the substituted paths from P_r, we are left with finitely many two-ended paths. For each of them, the maximum rank θ' among the ranks of all the maximal nodes that the path meets is less than θ (i.e., $\theta' \leq \theta - 1$). We can treat each one of those paths in the same way to obtain finitely many substituted permissive paths, whose deletions leave finitely many two-ended paths. For each of those, the maximum rank θ'' as described above is still lower (i.e., $\theta'' \leq \theta - 2$). Continuing this process, we are finally left with finitely many two-ended 0-paths, which are perforce permissive. Upon connecting together all of those 0-paths and the finitely many substituted paths in accordance with the tracing from n_a to n_b, we obtain a permissive path that terminates at those two nodes.

Since no finitely structured network has $\vec{\omega}$ as a rank, we are left with the case where $\nu = \omega$. In fact, to complete the proof we need only consider the case where $\theta = \omega$. Again we can choose an isolating set for each of the finitely many ω-nodes that P^α meets through one or two nonelementary tips — and can do so in such a fashion that the corresponding arms are pairwise totally disjoint and the same caveat concerning n_a and n_b holds as before. The part of P^α lying within each arm can be replaced by a permissive path. Moreover, the finitely many two-ended paths that remain after the parts of P^α lying within the arms are removed will have ranks that are natural numbers. We can therefore treat those finitely many two-ended paths as before to arrive at our conclusion again. ♣

We have already defined in Section 5.3 what is meant by "shorting a branch." We now wish to extend this idea, for, while manipulating networks, we will at times combine finitely many maximal nodes by *shorting* them. This is accomplished by creating a new node that embraces all the tips of all ranks that are embraced by the said nodes. The rank of the new node will be the maximum of the ranks of the nodes being shorted. With regard to the elementary tips, something more is done to complete the shorting: If any branch becomes a self-loop, it is removed (i.e., opened). Also, if parallel branches arise, they are combined by taking them in the Norton form and then adding conductances and current sources. Finally, if a branch arises that is not incident to an ordinary 0-node, that branch is replaced by two branches in series by introducing another ordinary 0-node, taking the original branch in the Thevenin form, and then assigning half of the original branch's resistance and voltage source to each of the two new branches. All this may be needed to maintain Conditions 4.3-1.

An important and easily checked result is

Lemma 6.1-3. *Let us short finitely many nodes (of any ranks) of a permissively finitely structured ν-network $\mathbf{N}^\nu$. Then, the resulting network is also a permissively finitely structured ν-network.*

6.2 Excitations by Pure Sources

The fundamental theory for voltage-current regimes in transfinite networks [35, Chapters 3 and 5], [37] takes pure voltage sources into account by transferring them into resistive branches. That transference can be accomplished only when one node of the voltage source is a maximal 0-node. For our purposes, we need a fundamental theory that encompasses pure voltage sources without transferring them. In particular, we wish to append a pure voltage source to $\mathbf{N}^\nu$ possibly at two nodes of ranks higher than 0. Consequently, our next objective is to generalize the fundamental theory accordingly. In doing so, we shall also generalize Kirchhoff's current law to make it applicable to a cut that isolates an α-node.

We continue to assume that the ν-network $\mathbf{N}^\nu$ satisfies Conditions 6.1-1. Index the branches of $\mathbf{N}^\nu$ by $j = 1, 2, 3, \ldots$. Let b_0 be a source branch and append it to any two nodes of $\mathbf{N}^\nu$, whatever be their ranks. Denote the resulting augmented network by $\mathbf{N}_a^\nu$. At this time, no restriction is placed on the kind of source branch b_0 may be; it may be a pure voltage source, a pure current source, or a source with a resistance. The current vector for $\mathbf{N}_a^\nu$ is denoted by $\mathbf{i} = (i_0, i_1, i_2, \ldots)$, where i_j is the current in branch b_j, $j = 0, 1, 2, \ldots$. We wish to construct a Hilbert space $\mathcal{K}_a$ of allowable branch-current vectors for $\mathbf{N}_a^\nu$, which is like the space $\mathcal{K}$ but without the branch b_0 contributing a term to the inner product for $\mathcal{K}_a$.

As the first step toward that objective, we consider Kirchhoff's current law applied to a cut $\mathbf{C}$ in $\mathbf{N}_a^\nu$ that isolates one of the maximal nodes n^α $(0 \leq \alpha \leq \nu)$ from all the other $(\alpha+)$-nodes in $\mathbf{N}_a^\nu$. In particular, if n^α is a maximal 0-node (i.e., $\alpha = 0$), we let $\mathbf{C}$ denote the set of all branches incident to n^α and we orient $\mathbf{C}$ away from n^α. However, if n^α is of higher rank (i.e., $0 < \alpha \leq \nu$), we let $\mathbf{C}$ denote a cut at an isolating set $\mathcal{W}$ for n^α (that is, $\mathbf{C}$ is a cut as defined at the end of Section 4.3), and we orient $\mathbf{C}$ away from $\mathcal{W}$ — effectively away from n^α again. In both cases, $\mathbf{C}$ is a finite set. If the source branch b_0 is incident to n^α, then $b_0 \in \mathbf{C}$. Kirchhoff's current law for $\mathbf{C}$ asserts that

$$\sum_{\mathbf{C}} \pm i_j = 0 \tag{6.1}$$

where i_j is the current in branch b_j, the summation is over the indices of the branches in $\mathbf{C}$, and the plus (minus) sign is used if the orientations of $\mathbf{C}$ and branch b_j agree (resp. disagree). We have yet to establish whether (6.1) holds.

The next step is to construct a Hilbert space $\mathcal{K}_a$ of current vectors $\mathbf{i} = (i_0, i_1, i_2, \ldots)$ in $\mathbf{N}_a^\nu$ for which the total power dissipation $\sum_{j=1}^\infty i_j^2 r_j$ within $\mathbf{N}^\nu$ is finite and also for which (6.1) holds regardless of the choice of $\mathbf{C}$. Let $\mathcal{I}_a$ be the set of all current vectors $\mathbf{i}$ for $\mathbf{N}_a^\nu$ such that $\sum_{j=1}^\infty i_j^2 r_j < \infty$. (In contrast to the space $\mathcal{I}$ discussed in Section 5.2, $\mathcal{I}_a$ cannot now be identified as an inner-product space; for instance, the nonzero vector $\mathbf{i}$ for which $i_0 = 1$ and $i_j = 0$, $j = 1, 2, 3, \ldots$, has zero power dissipation within $\mathbf{N}^\nu$.) Next, let $\mathcal{K}_a^0$ denote the span of all basic currents in $\mathcal{I}_a$.

Lemma 6.2-1. *Kirchhoff's current law (6.1) holds in* $\mathbf{N}_a^\nu$ *at every cut* $\mathbf{C}$ *whenever* $\mathbf{i} \in \mathcal{K}_a^0$.

Proof. Since there are only finitely many branches in $\mathbf{C}$, a loop can embrace branches of $\mathbf{C}$ at most finitely often. Moreover, any loop current passing through $\mathbf{C}$ contributes additive terms to the left-hand side of (6.1) an even number of times, positively for half of those times and negatively for the other half. Hence, its total contribution to the left-hand side of (6.1) is 0.

The same is true for any basic current i. Indeed, a basic current is an α-basic current for some α with $0 \leq \alpha \leq \nu$. Moreover, a 0-basic current is simply a 0-loop current. On the other hand, for $\alpha > 0$, an α-basic current $\mathbf{i}$ is a sum $\mathbf{i} = \sum \mathbf{i}_m$ of α-loop currents such that only finitely many of the $\mathbf{i}_m$ pass through any branch (with some other conditions imposed as well — see Section 5.2). This implies that only finitely many of the $\mathbf{i}_m$ pass through $\mathbf{C}$ because $\mathbf{C}$ is a finite set of branches. Since each $\mathbf{i}_m$ contributes 0 to the left-hand side of (6.1), $\mathbf{i}$ does, too.

We can now conclude that the same is true for every member of $\mathcal{K}_a^0$. ♣

We now define an inner product for $\mathcal{K}_a^0$ by

$$(\mathbf{i}, \mathbf{s}) = \sum_{j=1}^{\infty} i_j s_j r_j \tag{6.2}$$

where $\mathbf{i}, \mathbf{s} \in \mathcal{K}_a^0$. Even though (6.2) does not contain i_0 and s_0, it is nonetheless positive-definite. Indeed, $(\mathbf{i}, \mathbf{i}) \geq 0$ obviously. Moreover, if $(\mathbf{i}, \mathbf{i}) = \sum_{j=1}^{\infty} i_j^2 r_j = 0$, then $i_j = 0$ for all $j > 0$. Now, choose a cut $\mathbf{C}$ that isolates one of the two nodes to which b_0 is incident from the other one (that is, the isolating set $\mathcal{W}$ at which $\mathbf{C}$ is chosen separates the two nodes of b_0). Thus, b_0 is a member of $\mathbf{C}$, and by Lemma 6.2-1 we have

$$\pm i_0 = - \sum_{\mathbf{C}\backslash\{b_0\}} \pm i_j \tag{6.3}$$

where the summation is over the indices for the branches in $\mathbf{C}$ other than b_0 and the plus and minus signs are chosen as before. By (6.3), if $i_j = 0$ for all $j > 0$, then $i_0 = 0$, too. Therefore, $\mathbf{i} = 0$. Whence, the positive-definiteness of $(\mathbf{i}, \mathbf{s})$. The other inner-product axioms are also fulfilled.

Let $\mathcal{K}_a$ denote the closure of $\mathcal{K}_a^0$ in $\mathcal{I}_a$ under the norm $\|\mathbf{i}\| = (\mathbf{i}, \mathbf{i})^{\frac{1}{2}}$. Convergence under that norm implies branchwise convergence for every $j > 0$ (see Lemma 5.2-1). Since $\mathbf{C}\backslash\{b_0\}$ is a finite set, (6.3) implies branchwise convergence for $j = 0$ too. Thus, the members of $\mathcal{K}_a$ can be identified through branchwise convergence as current vectors in $\mathbf{N}_a^\nu$ satisfying (6.3). Moreover, it follows in a standard way (see, for example, [35, pages 63-64]) that $\mathcal{K}_a$ is complete and therefore is a Hilbert space with an inner product given by (6.2).

The branchwise convergence coupled with the finiteness of the set $\mathbf{C}$ allows us to extend Lemma 6.2-1 immediately:

Lemma 6.2-2. *Kirchhoff's current law (6.1) holds at every cut* $\mathbf{C}$ *whenever* $\mathbf{i} \in \mathcal{K}_a$.

Let $\mathcal{I}$ be the subset of $\mathcal{I}_a$ consisting of all $\mathbf{i} \in \mathcal{I}_a$ for which $i_0 = 0$. With (6.2) as the inner product, $\mathcal{I}$ is a Hilbert space. Let $\mathcal{K}$ be the corresponding subset of $\mathcal{K}_a$. (In fact, by deleting the 0-valued 0th component of each current vector, $\mathcal{I}$ and $\mathcal{K}$ can be identified with the spaces discussed in Section 5.2 for $\mathbf{N}^\nu$.) Now, augment $\mathbf{N}^\nu$ with an appended branch b_0 consisting of a pure current source h_0 to obtain the augmented network $\mathbf{N}_a^\nu$. Since there is at least one permissive path between any two nodes of $\mathbf{N}^\nu$ (Lemma 6.1-2), we can transfer h_0 to within $\mathbf{N}^\nu$ along a permissive path P to obtain a unique current vector $(i_1', i_2', \ldots)$ in the resulting network according to Theorem 5.2-14. (That is, in the resulting network the branch b_0 has been eliminated by removing it after the said transference is made.) We are now free to append the component $i_0' = 0$ to that current vector and will obtain thereby $\mathbf{i}' = (0, i_1', i_2', \ldots) \in \mathcal{K}$. Upon restoring b_0 and transferring the current source back to b_0, we obtain the corresponding current vector $\mathbf{i}$ for $\mathbf{N}_a^\nu$ as the superposition of $\mathbf{i}'$ and a loop current of value h_0 flowing around the loop $P \cup \{b_0\}$. Hence, $\mathbf{i}$ is a member of $\mathcal{K}_a$ and therefore satisfies Kirchhoff's current law according to Lemma 6.2-2. Moreover, $\mathbf{i}$ is independent of the choice of the permissive path P in $\mathbf{N}^\nu$ by which $\mathbf{i}$ was constructed (Theorem 5.4-2). In this way, a pure current source in b_0 generates a unique current vector $\mathbf{i} \in \mathcal{K}_a$.

We have yet to establish that any pure voltage source can be connected to any two nodes of $\mathbf{N}^\nu$. This would not be possible if $\mathbf{N}^\nu$ acted as a short or more generally as a pure voltage source between those nodes. However, as was just noted, we are permitted to connect a pure current source between those two nodes. We shall do so and then will show that $\mathbf{N}^\nu$ acts as a positive resistance between those nodes when its internal sources are removed (i.e., the current sources are opened and the voltage sources are shorted). This will justify the application of a pure voltage source to them. Actually, it is just as easy to show something more general; namely, at finitely many, say, K arbitrarily selected nodes, $\mathbf{N}^\nu$ behaves as a $(K-1)$-port with a nonsingular resistance matrix. We will need this more general version in Section 7.7.

So, remove the sources within $\mathbf{N}^\nu$ as stated. After this is done, $\mathbf{N}^\nu$ will be called *sourceless*. Arbitrarily select finitely many maximal nodes $n_1, \ldots, n_K$ of any ranks in $\mathbf{N}^\nu$ and connect pure current sources between them. (We view these sources as being external to $\mathbf{N}^\nu$.) Without loss of generality, we can take them to be $K-1$ current sources feeding the currents $h_2, \ldots, h_K$ from n_1 to $n_2, \ldots, n_K$, respectively. We designate n_1 as the ground node and obtain thereby the respective node voltages $u_2, \ldots, u_K$. In this way, $\mathbf{N}^\nu$ acts as an internally transfinite and sourceless, resistive $(K-1)$-port with n_1 as the common ground for the various ports . Moreover, $\mathbf{h} = (h_2, \ldots, h_K)$ is the imposed port-current vector, $\mathbf{u} = (u_2, \ldots, u_K)$ is the resulting port-voltage vector, and the mapping $Z : \mathbf{h} \mapsto \mathbf{u}$ is the $(K-1) \times (K-1)$ resistance matrix for this $(K-1)$-port. We will now show that Z is nonsingular. This will imply that any choice of the port-voltage vector $\mathbf{u}$ can be obtained by setting $\mathbf{h} = Z^{-1}\mathbf{u}$. This will also imply that any finite set of pure voltage sources can be appended to any nodes of $\mathbf{N}^\nu$ to obtain a unique voltage-current regime throughout $\mathbf{N}^\nu$ — even when the sources inside $\mathbf{N}^\nu$ are restored.

Lemma 6.2-3. *For the sourceless ν-network $\mathbf{N}^\nu$ satisfying Conditions 6.1-1, Z is symmetric and positive-definite and therefore nonsingular.*

Proof. The symmetry of Z follows from the reciprocity principle (Corollary 5.2-16) as follows. Let n_k and n_l be two of the nodes other than the ground node n_1 that are port terminals for Z. Choose a permissive path P_k from n_k to n_1 and another permissive path P_l from n_l to n_1. Let h be a current source of value 1 A and connect it from n_1 to n_k. We can transfer h onto P_k and then invoke Theorem 5.2-14 in order to get the voltage u_l at n_l as the algebraic sum of the voltages along P_l. Similarly, by applying h from n_1 to n_l and then transferring it onto P_l, we get the resulting voltage u_k at n_k as the algebraic sum of the voltages along P_k. Then, by (5.21) we get $u_l = u_k$. Hence, the lth-row, kth-column entry of Z is equal to the kth-row, lth-column entry.

We now prove that Z is positive-definite. Choose any vector $\mathbf{h} = (h_2, \ldots, h_K)$ of current sources h_k applied at the ports. For each n_k $(k = 2, \ldots, K)$, choose a cut $\mathbf{C}_k$ that isolates n_k from all the other port terminals. Thus, the source branch b_k for h_k is a member of $\mathbf{C}_k$, but the other source branches are not in $\mathbf{C}_k$. Hence, $\mathbf{C}_k = \mathbf{D}_k \cup \{b_k\}$, where $\mathbf{D}_k$ is the set of branches in $\mathbf{C}_k$ that lie within $\mathbf{N}^\nu$. According to the paragraph following Lemma 6.2-2, the current regime induced when h_k is acting alone is a member of $\mathcal{K}_a$ and therefore satisfies Kirchhoff's current law at $\mathbf{C}_k$; in this case, the algebraic sum of the currents in $\mathbf{D}_k$ is h_k flowing away from n_k. Similarly, when any other appended current sources is acting alone, that algebraic sum for $\mathbf{D}_k$ is 0. Consequently, by superposition, when all the current sources are acting simultaneously, the net current flowing through $\mathbf{D}_k$ away from n_k

is h_k. Therefore, there is at least one branch of $\mathbf{D}_k$ that carries a current of absolute value no less than $|h_k|/d_k$, where $d_k = \overline{\overline{\mathbf{D}_k}}$. Let $r_{min,k}$ be the least resistance among the branches of $\mathbf{D}_k$. Hence, the power dissipated in all the resistances of $\mathbf{D}_k$ is no less than $\delta_k h_k^2$, where $\delta_k = r_{min,k} d_k^{-2} > 0$. Hence, with a cut chosen for each of the nodes $n_2, \ldots, n_K$, we see that the power dissipated in all those cuts is no less than $\sum_{k=2}^{K} \delta_k h_k^2$. The last expression is positive if $\mathbf{h} \neq \mathbf{0}$.

Now, let $(\cdot, \cdot)_E$ be the inner product for $(K-1)$-dimensional Euclidean space. We now invoke Tellegen's equation in the form of (5.20). To do so, we transfer all the current sources into $\mathbf{N}^\nu$ along permissive paths. The sum of the voltages along each such path is one of the node voltages at a port terminal. Thus, with $\mathbf{w}$ set equal to the vector $\mathbf{v}$ of branch voltages, the left-hand side of (5.20) becomes the power $(\mathbf{u}, \mathbf{h})_E = (Z\mathbf{h}, \mathbf{h})_E$ supplied by the sources acting simultaneously. The right-hand side of (5.20) becomes the power dissipated in all the resistors of $\mathbf{N}^\nu$. Hence, $(Z\mathbf{h}, \mathbf{h})_E \geq \sum_{k=2}^{K} \delta_k h_k^2$, which proves that Z is positive-definite. ♣

When $K = 2$ so that only one pure current source is appended to $\mathbf{N}^\nu$, Z can be replaced by a real positive number z, which we shall refer to as the *input resistance* (also called the *driving-point resistance*) of $\mathbf{N}^\nu$ between the two nodes to which the source is appended. z is equal to the voltage between the those two nodes when the internal sources of $\mathbf{N}^\nu$ are removed and the applied pure current source is 1 A in value.

Let us return now to the case where $\mathbf{N}^\nu$ possibly has sources in its branches. As before, let one more source branch b_0 be appended to any two nodes n_1 and n_2 of $\mathbf{N}^\nu$ to obtain the augmented network $\mathbf{N}_a^\nu$. This is illustrated in Figure 6.1, where $\mathbf{N}^\nu$ is represented internally by a Thevenin equivalent branch — obtained as follows. When b_0 is opened, a voltage v_{oc} appears between n_1 and n_2 induced by the internal sources of $\mathbf{N}^\nu$. With b_0 restored and in the Thevenin form of a source e_0 in series with a nonnegative resistor $\rho \geq 0$, superposition dictates that the current i_0^ρ in the branch b_0 will induce an increment in the voltage between n_1 and n_2 due to the flow of i_0^ρ through the positive resistance z given by Lemma 6.2-3. (This is in fact Thevenin's theorem for $\mathbf{N}^\nu$ as viewed from n_1 and n_2.) Thus, with polarities chosen as in Figure 6.1, we have

$$e_0 - v_{oc} = (\rho + z) i_0^\rho. \tag{6.4}$$

Upon letting $\rho \to 0$, we obtain the following result.

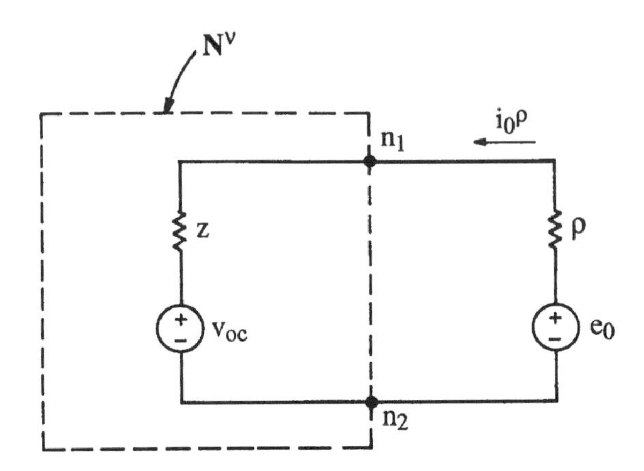

Figure 6.1. The network $\mathbf{N}^\nu$ with an externally appended source branch. $\mathbf{N}^\nu$ is represented internally by a Thevenin equivalent branch.

Theorem 6.2-4. *Let* $\mathbf{N}^\nu$ *satisfy Conditions 6.1-1, and let all its branches* b_j *(*$j = 1, 2, 3, \ldots$*) be in the Thevenin form. Let* b_0 *be a pure voltage source* e_0 *appended to any two nodes of* $\mathbf{N}^\nu$*, and let* $\mathbf{N}_a^\nu$ *be the resulting augmented network. Then, for* $\mathbf{N}_a^\nu$ *there is a unique* $\mathbf{i} \in \mathcal{K}_a$ *such that*

$$\sum_{j=0}^{\infty} e_j s_j = \sum_{j=1}^{\infty} r_j i_j s_j \tag{6.5}$$

for every $\mathbf{s} \in \mathcal{K}_a$.

Note. Observe that the summation on the left (right) of (6.5) is over all nonnegative (resp. positive) natural-number indices. Also, note that the fundamental Theorem 5.2-8 does not directly apply in this case because $r_0 = 0$ so that $e_0^2 g_0$ is infinite.

Proof. To prove this theorem, we will insert a resistance $\rho > 0$ in series with the voltage source e_0 within the branch b_0 to obtain the unique current vector $\mathbf{i}^\rho = (i_0^\rho, i_1^\rho, i_2^\rho, \ldots)$ dictated by Theorem 5.2-8, and then we will take $\rho \to 0$ to obtain (6.5) in the limit.

With ρ inserted as stated, Theorem 5.2-8 asserts that

$$\sum_{j=0}^{\infty} e_j s_j = \rho i_0^\rho s_0 + \sum_{j=1}^{\infty} r_j i_j^\rho s_j \tag{6.6}$$

for every $\mathbf{s} \in \mathcal{K}_a$. In this case, we also have (6.4). With λ being another positive value for the resistance inserted into b_0, (6.4) yields

$$i_0^\rho - i_0^\lambda = \frac{e_0 - v_{oc}}{\rho + z} - \frac{e_0 - v_{oc}}{\lambda + z} \rightarrow 0 \tag{6.7}$$

as $\rho, \lambda \rightarrow 0+$ independently. From (6.6) and (6.4), we obtain

$$\sum_{j=1}^{\infty} r_j (i_j^\rho - i_j^\lambda) s_j = (\lambda i_0^\lambda - \rho i_0^\rho) s_0 = z(i_0^\rho - i_0^\lambda) s_0. \tag{6.8}$$

Note now that both $\mathbf{i}^\rho$ and $\mathbf{i}^\lambda$ are members of $\mathcal{K}_a$. (For instance, $\mathbf{i}^\rho$ can be obtained by using i_0^ρ as a pure current source in b_0 in accordance with the paragraph following Lemma 6.2-2 again.) Also, recall that the norm $\|\mathbf{i}\|$ for any $\mathbf{i} \in \mathcal{K}_a$ is given by $\|\mathbf{i}\|^2 = \sum_{j=1}^{\infty} r_j i_j^2$. Consequently, we may set $s_j = i_j^\rho - i_j^\lambda$ in (6.8) and then invoke (6.7) to get

$$\|\mathbf{i}^\rho - \mathbf{i}^\lambda\|^2 = \sum_{j=1}^{\infty} r_j (i_j^\rho - i_j^\lambda)^2 = z(i_0^\rho - i_0^\lambda)^2 \rightarrow 0$$

as $\rho, \lambda \rightarrow 0+$ independently. Hence $\{\mathbf{i}^\rho : \rho > 0\}$ is a Cauchy directed set in the complete space $\mathcal{K}_a$ and therefore converges in $\mathcal{K}_a$ to an $\mathbf{i} \in \mathcal{K}_a$. Since the inner product of $\mathcal{K}_a$ is bicontinuous, we may pass to the limit in (6.6) to obtain (6.5).

$\mathbf{i}$ is uniquely determined by (6.5) because the right-hand side of (6.5) is the inner product $(\mathbf{i}, \mathbf{s})$ determined for all $\mathbf{s} \in \mathcal{K}_a$ by the left-hand side. ♣

Theorem 6.2-4 will be the fundamental theorem determining the voltage-current regime in $\mathbf{N}_a^\nu$ when only one pure voltage source e_0 is appended to $\mathbf{N}^\nu$. We may alter $\mathbf{N}^\nu$ by shorting a finite set of maximal nodes to obtain one terminal for e_0 and by shorting another finite set of maximal nodes (disjoint from the first set) to obtain the other terminal for e_0; this is permissible according to Lemma 6.1-3. That possibly altered network with e_0 appended will be denoted by $\mathbf{N}_e^\nu$. Furthermore, the voltage-current regime corresponding to finitely many pure voltage sources appended to the linear network $\mathbf{N}^\nu$ is determined from Theorem 6.2-4 by superposition.

Another result we shall need, this one in Section 7.5, arises from transfinite versions of Thomson's least power principle and Rayleigh's monotonicity law. We need only special cases of these. (A general development is given in [35, Sections 3.8 and 3.9], albeit for the case of 1-networks. Nevertheless, the discussion given there applies equally well to transfinite networks of higher ranks, but it does not subsume the version of Rayleigh's monotonicity law given by Lemma 6.2-6 below.)

Theorem 6.2-5. *Let* $\mathbf{N}^\nu$ *be a sourceless* ν*-network satisfying Conditions 6.1-1 and driven by a pure voltage source* e_0 *externally appended to two of its nodes. Let* $\mathbf{i} \in \mathcal{K}_a$ *be the resulting current vector with* i_0 *being the current in* e_0*. Let* $\mathbf{x} \in \mathcal{K}_a$ *be such that* $\mathbf{x} \neq \mathbf{i}$ *but* $x_0 = i_0$*. Then,* $\sum_{j=1}^{\infty} x_j^2 r_j > \sum_{j=1}^{\infty} i_j^2 r_j$*, where* $j = 1, 2, 3, \ldots$ *are the indices for the branches within* $\mathbf{N}^\nu$*.*

Proof. Set $\mathbf{x} = \mathbf{i} + \Delta\mathbf{i}$. Thus, $\Delta\mathbf{i} \in \mathcal{K}_a$ and $\Delta i_0 = 0$. Moreover, every component of the voltage-source vector $\mathbf{e}$ is 0 except for $e_0 \neq 0$. Choose $\mathbf{s} = \Delta\mathbf{i}$ in (6.5). This yields $\sum_{j=1}^{\infty} r_j i_j \Delta i_j = e_0 \Delta i_0 = 0$. Consider the following wherein the summations are over $j = 1, 2, 3, \ldots$.

$$\begin{aligned} \sum r_j x_j^2 &= \sum r_j (i_j + \Delta i_j)^2 \\ &= \sum r_j i_j^2 + 2 \sum r_j i_j \Delta i_j + \sum r_j \Delta i_j^2 \\ &= \sum r_j i_j^2 + \sum r_j \Delta i_j^2 \end{aligned}$$

Since $\Delta i_j \neq 0$ for at least some j and since $r_j > 0$ for all $j \geq 1$, our conclusion follows. ♣

With $\mathbf{N}^\nu$ still being a sourceless ν-network satisfying Conditions 6.1-1, assume that $\mathbf{N}^\nu$ is separated into two subnetworks $\mathbf{N}_1$ and $\mathbf{N}_2$ by a finite set $\mathcal{W}$ of maximal 0-nodes (as shown in Figure 6.2), and assume that each 0-node of $\mathcal{W}$ is of finite degree. Let $\mathbf{N}^\nu$ be excited by a pure voltage source e_0 in a source branch b_0 externally appended to two nodes n_1 and n_2 of $\mathbf{N}_1$ that are not in $\mathcal{W}$. We know from Lemma 6.2-3 that the input resistance z of $\mathbf{N}^\nu$ at n_1 and n_2 is a positive number $z = e_0/i_0$, where i_0 is the current in e_0. Let us assume furthermore that when all the branches of $\mathbf{N}_2$ are shorted, the network $\mathbf{N}'$ that $\mathbf{N}^\nu$ becomes is also a sourceless network satisfying Conditions 6.1-1 or possibly a finite 0-network. In fact, $\mathbf{N}'$ is $\mathbf{N}_1$ except for a single node that replaces $\mathcal{W}$. With $\mathbf{N}'$ excited by the external source e_0' applied to n_1 and n_2, the resulting input resistance z' is also a positive number $z' = e_0'/i_0'$, where i_0' is the current in e_0'. (The existence of a positive z' follows from Lemma 6.2-3 — except in the event that $\mathbf{N}'$ is a finite 0-network; in the latter case, the existence of $z' > 0$ follows from the theory of finite electrical networks.)

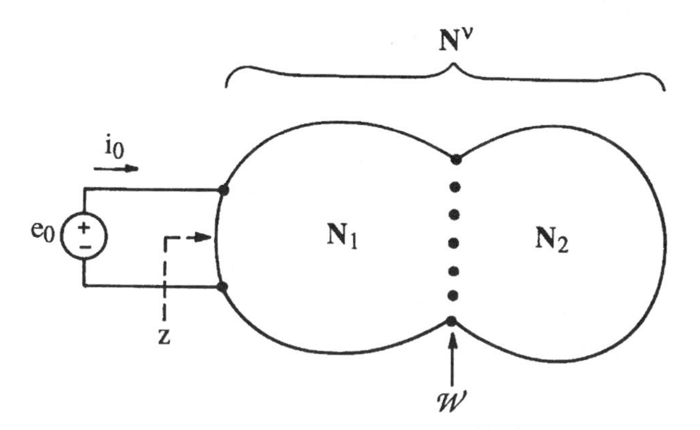

Figure 6.2. A sourceless transfinite $\mathbf{N}^\nu$ that is separated into two subnetworks $\mathbf{N}_1$ and $\mathbf{N}_2$ by a finite set $\mathcal{W}$ of maximal 0-nodes and is externally excited by a pure voltage source e_0 applied to two nodes n_1 and n_2 of $\mathbf{N}_1$. $z = e_0/i_0$ is the input resistance at n_1 and n_2. When all the branches of $\mathbf{N}_2$ are shorted, $\mathbf{N}^\nu$ becomes a smaller network $\mathbf{N}'$ obtained from $\mathbf{N}_1$ by shorting the nodes of $\mathcal{W}$.

Lemma 6.2-6. *Under the symbolism and assumptions just stated,* $z \geq z' > 0$.

Proof. Let $j = 1, 2, 3, \ldots$ be the branch indices for $\mathbf{N}^\nu$, and let r_j denote the branch resistances of $\mathbf{N}^\nu$. Also, set $r'_j = r_j$ if branch b_j is in $\mathbf{N}_1$, and set $r'_j = 0$ if branch b_j is in $\mathbf{N}_2$. Thus, $r_j = r'_j + \Delta r_j$, where $\Delta r_j = 0$ when b_j is in $\mathbf{N}_1$ and $\Delta r_j = r_j > 0$ when b_j is in $\mathbf{N}_2$. Let $\sum_{\mathbf{N}^\nu}$ be a summation over the branches of $\mathbf{N}^\nu$, and similarly for $\sum_{\mathbf{N}'} = \sum_{\mathbf{N}_1}$ and $\sum_{\mathbf{N}_2}$. Since $e_j = 0$ for $j = 1, 2, 3, \ldots$, Theorem 6.2-4 yields

$$z = \frac{e_0}{i_0} = \frac{1}{i_0^2}\sum_{\mathbf{N}^\nu} r_j i_j^2 = \frac{1}{i_0^2}\left(\sum_{\mathbf{N}_1} r'_j i_j^2 + \sum_{\mathbf{N}_2} \Delta r_j i_j^2\right). \tag{6.9}$$

Now, let b_0 denote the externally applied source branch. Also, let $\mathcal{K}_a$ (and $\mathcal{K}'_a$) be the usual solution space of current vectors for the network $\mathbf{N}^\nu \cup \{b_0\}$ (resp. $\mathbf{N}' \cup \{b_0\}$), and similarly for $\mathcal{K}_a^o$ (resp. $\mathcal{K}_a^{o\prime}$). The current vector $\mathbf{i}$ in $\mathbf{N}^\nu \cup \{b_0\}$ induced by e_0 is in $\mathcal{K}_a$. Also, the current vector $\mathbf{i}'$ in $\mathbf{N}' \cup \{b_0\}$ induced by e'_0 is in $\mathcal{K}'_a$. An important point is that the restriction of $\mathbf{i} \in \mathcal{K}_a$

to the branches of $\mathbf{N}'$ is a vector in $\mathcal{K}'_a$. This is because any loop current that passes between $\mathbf{N}_1$ and $\mathbf{N}_2$ is truncated into a loop current lying entirely within $\mathbf{N}'$ when the branches of $\mathbf{N}_2$ are shorted. Since there are only finitely many branches in $\mathbf{N}_1$ incident to the nodes of $\mathcal{W}$, any basic current in $\mathbf{N}^\nu$ becomes truncated into a basic current in $\mathbf{N}'$. Thus, these shortings and restrictions reduce any current vector in $\mathcal{K}^o_a$ to a current vector in $\mathcal{K}^{o\prime}_a$. It follows that they also reduce any current vector in $\mathcal{K}_a$ to a current vector in $\mathcal{K}'_a$ because, if a sequence in $\mathcal{K}_a$ converges, the sequence of restricted terms will converge in $\mathcal{K}'_a$.

Now, let us apply Theorem 6.2-5 to $\mathbf{N}' \cup \{b_0\}$. We have that the restriction of $\mathbf{i}$ to the branches of $\mathbf{N}' \cup \{b_0\}$ is in $\mathcal{K}'_a$, and therefore we can let $\mathbf{x}$ of Theorem 6.2-5 be that restriction. On the other hand, $\mathbf{i}' \in \mathcal{K}'_a$ is the current-vector solution for $\mathbf{N}' \cup \{b_0\}$ under the application of e'_0. We are free to adjust e'_0 to make $i'_0 = i_0$. Thus, we may write

$$\frac{1}{i_0^2}\sum_{\mathbf{N}_1} r'_j i_j^2 \;=\; \frac{1}{i_0^2}\sum_{\mathbf{N}'} r'_j i_j^2 \;\geq\; \frac{1}{i_0^2}\sum_{\mathbf{N}'} r'_j i_j'^2 \;=\; \frac{1}{i_0^2} e'_0 i'_0 \;=\; \frac{e'_0}{i'_0} \;=\; z'.$$

Upon comparing this to (6.9), we see that $z \geq z'$ because $\sum_{\mathbf{N}_2} \Delta r_j i_j^2 \geq 0$. ♣

6.3 Maximum Principles for Node Voltages and Branch Currents

Let $\mathbf{N}^0$ be a locally finite, connected, infinite 0-network whose branches are purely resistive (no branch sources), and let $\mathbf{N}^0$ carry a current that satisfies Kirchhoff's current law at every 0-node, Kirchhoff's voltage law around every 0-loop, and Ohm's law. (We are assuming that all sources are at infinity.) Upon choosing any node of $\mathbf{N}^0$ as ground and setting its voltage at 0, we obtain unique node voltages throughout $\mathbf{N}^0$. The maximum principle for those node voltages asserts that either they are all equal to 0 or there is no maximum value and no minimum value among them. If $\mathbf{N}^0$ were to be embedded as a $(1-)$-subsection in a larger, permissively finitely structured ν-network $\mathbf{N}^\nu$ ($\nu \geq 1$), this principle would assert that the maximum and minimum node voltages for that $(1-)$-subsection would have to occur at some of its bordering nodes. The objective of this section is to extend this principle to subsections of higher ranks in $\mathbf{N}^\nu$. We will also obtain an analogous maximum principle for branch currents.

Let $\mathbf{S}_b^{\beta-}$ be a $(\beta-)$-subsection of $\mathbf{N}_a^\nu$, where $\mathbf{N}_a^\nu$ is obtained by appending some sources to the network $\mathbf{N}^\nu$ satisfying Conditions 6.1-1. $\mathbf{S}_b^{\beta-}$ will be called *sourceless* if none of its branches is a source branch. Thus, the only way current can enter or leave $\mathbf{S}_b^{\beta-}$ is through its bordering nodes. Choose any node as ground. Since $\mathbf{S}_b^{\beta-}$ has only finitely many bordering $(\beta+)$-nodes, we can let u_{max} (and u_{min}) be the largest (and least) node voltage at those bordering $(\beta+)$-nodes. Consider the following properties. Part (a) is a transfinite generalization of the maximum principle.

Properties 6.3-1. *For all $\beta = 1, \ldots, \nu$ $(\nu \leq \omega, \nu \neq \vec{\omega})$, the following hold.*

(a) *There are exactly two possibilities for all the node voltages of any sourceless $(\beta-)$-subsection $\mathbf{S}_b^{\beta-}$ of $\mathbf{N}_a^\nu$.*

  (a1) *All those node voltages (for both internal and bordering nodes of $\mathbf{S}_b^{\beta-}$) have the same value.*

  (a2) *$\mathbf{S}_b^{\beta-}$ has at least two bordering $(\beta+)$-nodes with differing node voltages, and the internal node voltages of $\mathbf{S}_b^{\beta-}$ are strictly less than u_{max} and strictly larger than u_{min}.*

(b) *Let P^α $(\alpha < \beta)$ be any one-ended α-path in $\mathbf{S}_b^{\beta-}$ and let t^α be its α-tip. (P^α need not be permissive.) Let $\{m_1, m_2, \ldots\}$ be any sequence of nodes embraced by P^α that approaches t^α. Then, the node voltages at the m_l converge to the node voltage of the $(\alpha+1)$-node that embraces t^α.*

We will show that Properties 6.3-1 always hold. This will imply that those properties also hold for certain subnetworks that are not $(\beta-)$-subsections. In particular, let $\mathbf{N}_s^\alpha$ be a subnetwork of rank α of $\mathbf{N}_a^\nu$ with $\alpha < \nu$ such that $\mathbf{N}_s^\alpha$ meets $\mathbf{N}_a^\nu \backslash \mathbf{N}_s^\alpha$ at finitely many maximal nodes n_k $(k = 1, \ldots, K)$. Thus, the maximum rank among the nodes of $\mathbf{N}_s^\alpha$ is α, but the ranks of the n_k may be less than α. Choose any such $\beta > \alpha$. We can append to the n_k one-ended $(\beta-1)$-paths through their $(\beta-1)$-tips, with those paths being otherwise totally disjoint from $\mathbf{N}_a^\nu$ as well as from each other. All the currents in those paths will be 0, and moreover the voltage-current regime in $\mathbf{N}_a^\nu$ will not be disturbed by such appendages. Thus, $\mathbf{N}_s^\alpha$ becomes a $(\beta-)$-subsection. Our assertion that Properties 6.3-1 always hold now implies that they also hold for $\mathbf{N}_s^\alpha$, even before the appending of those $(\beta-1)$-paths.

To prove Properties 6.3-1, we start with a lemma.

Lemma 6.3-2. *Assume that Properties 6.3-1(a) hold. Let n^β be a maximal β-node ($\beta > 0$), whose incident ($\beta-$)-subsections are all sourceless. Assume also that at least one of those incident ($\beta-$)-subsections satisfies (a2). If the voltage u^β at n^β is no less than all the node voltages at the ($\beta+$)-nodes that are β-adjacent to n^β, then Kirchhoff's current law (6.1) is violated at some cut $\mathbf{C}$ for n^β.*

Proof. So far as Kirchhoff's current law is concerned, the ($\beta-$)-subsections incident to n^β that satisfy (a1) can be ignored because all their branch currents are 0. So, let us consider only those ($\beta-$)-subsections incident to n^β that satisfy (a2). In their union $\mathbf{U}$, choose a permissive contraction $\{\mathcal{W}_p\}_{p=1}^\infty$ to n^β. If n^β is 0-adjacent to one or more internal nodes of $\mathbf{U}$, then the voltages at those internal nodes are strictly less than the voltage u^β at n^β. Thus, the current leaving n^β through the branches incident to n^β will be positive. Let $\mathcal{V}_p$ be the base of $\mathcal{W}_p$. Since we may be dealing with only some of the ($\beta-$)-subsections incident to n^β, it is possible for $\mathcal{V}_p$ to be void. In that case, n^β must be 0-adjacent to internal nodes of $\mathbf{U}$, and the positive currents leaving n^β through the corresponding branches shows that Kirchhoff's current law is violated at n^β.

So assume that $\mathcal{V}_p$ is nonvoid. Therefore, it is nonvoid for every p, and we have $\overline{\overline{\mathcal{V}_p}} \leq m$, where m is finite and independent of p. Let $\mathbf{A}_p$ be the arm for $\mathcal{V}_p$ and set $\mathbf{M}_p = \mathbf{A}_p \backslash \mathbf{A}_{p+1}$. All the nodes of $\mathbf{M}_p$ are core nodes, and none of them is 0-adjacent to n^β. Moreover, for $p > 1$, $\mathcal{V}_p$ separates $\mathbf{M}_{p-1}$ and $\mathbf{M}_p$. Now, each contraction path for $\{\mathcal{W}_p\}_{p=1}^\infty$ is permissive, and the 0-node voltages at the $\mathcal{V}_p$ along any contraction path are strictly less than u^β (see Property 6.3-1(a2)) and converge to u^β (see Lemma 5.5-1). Since there are only finitely many contraction paths and since every node of $\mathcal{V}_p$ lies on a contraction path, it follows that we can choose two natural numbers p and q with $p < q$ such that the largest node voltage for $\mathcal{V}_p$ is less than the least node voltage for $\mathcal{V}_q$.

Set $\mathbf{M}_{p,q} = \bigcup_{k=p}^{q-1} \mathbf{M}_k$. We can generate the same voltage-current regime in $\mathbf{M}_{p,q}$ as it has in $\mathbf{N}_a^\nu$ by appending pure voltage sources to $\mathcal{V}_p \cup \mathcal{V}_q$ as follows: Let $n_{p,1}^0$ be a node of $\mathcal{V}_p$ with the largest node voltage $u_{p,1}$ for $\mathcal{V}_p$. Let $n_{p,k}^0$ be any other node of $\mathcal{V}_p$ and let $u_{p,k}$ be its voltage. Connect a pure voltage source of value $u_{p,1} - u_{p,k} \geq 0$ from $n_{p,k}^0$ to $n_{p,1}^0$ with its positive terminal at $n_{p,1}^0$. (That source will be a short if $u_{p,1} = u_{p,k}$.) Do this for all $n_{p,k}^0$ in $\mathcal{V}_p$. Similarly, connect a pure voltage source from a node

$n_{q,1}^0$ of $\mathcal{V}_q$ with the least node voltage $u_{q,1}$ for $\mathcal{V}_q$ to each of the other nodes of $\mathcal{V}_q$ to establish their relative node voltages at the values they have in $\mathbf{N}_a^\nu$. Finally, for the same purpose, connect a pure voltage source $e_{p,q}$ of value $u_{q,1} - u_{p,1} > 0$ from $n_{p,1}^0$ to $n_{q,1}^0$, positive terminal at $n_{q,1}^0$. All these appended voltages are nonnegative, and moreover $e_{p,q}$ is positive. That these sources produce the same regime in $\mathbf{M}_{p,q}$ as $\mathbf{M}_{p,q}$ has in $\mathbf{N}_a^\nu$ follows from the fact that otherwise the difference between the two regimes would be a regime dissipating power in $\mathbf{M}_{p,q}$ with no sources applied to $\mathbf{M}_{p,q}$ — an impossibility.

We shall now argue by superposition. Assume that $e_{p,q}$ is acting alone (i.e., all other voltage sources set equal to zero). Then, $\mathcal{V}_p$ (and $\mathcal{V}_q$) is shorted into a single maximal 0-node n_p (and n_q). The resulting network $\mathbf{M}'_{p,q}$ has the pure voltage source $e_{p,q}$ connected between n_p and n_q.

We are free to assume that n_p and n_q are of rank β. (As was explained above, we can append one-ended $(\beta-1)$-paths to n_p and n_q through $(\beta-1)$-tips without disturbing the voltage-current regime in $\mathbf{M}'_{p,q}$.) All the other nodes of $\mathbf{M}'_{p,q}$ have ranks less than β. In this way, $\mathbf{M}'_{p,q}$ can be viewed as a union of finitely many $(\beta-)$-subsections.

So, by Property 6.3-1(a2), all nodes of $\mathbf{M}'_{p,q}$ other than n_p and n_q have node voltages strictly larger than that at n_p and strictly less than that at n_q. Let $\mathbf{C}'_q$ be the set of branches of $\mathbf{M}_{p,q}$ that are incident to $\mathcal{V}_q$. Orient those branches away from $\mathcal{V}_q$. It follows that in $\mathbf{M}'_{p,q}$ the current in each branch of $\mathbf{C}'_q$ is positive (i.e., is nonzero and directed away from n_q).

Next, let the voltage source e connected between $n_{p,k}^0$ and $n_{p,1}^0$ act alone. Virtually the same argument leads to a similar conclusion: The currents in the branches of $\mathbf{C}'_q$ are nonnegative.

On the other hand, let e' be the voltage source connected between the nodes $n_{q,1}^0$ and $n_{q,k}^0$ of $\mathcal{V}_q$. When e' is acting alone and is not 0, we have positive currents flowing away from $n_{q,k}^0$ through the branches of $\mathbf{C}'_q$ incident to $n_{q,k}^0$ and positive currents flowing toward the other nodes of $\mathcal{V}_q$ through the remaining branches of $\mathbf{C}'_q$ as well as toward the nodes of $\mathcal{V}_p$ through the branches of $\mathbf{M}_{p,q}$ incident to $\mathcal{V}_p$. By Kirchhoff's current law applied at the shorts imposed to make e' act alone (that is, at the nodes of e' with the said shorts in place), the algebraic sum of the currents in the branches of $\mathbf{C}'_q$ measured away from $\mathcal{V}_q$ is nonnegative.

By superposition, the algebraic sum of the currents in $\mathbf{C}'_q$ is positive. Altogether then, we can now conclude that, whether or not the bases $\mathcal{V}_p$ are void, Kirchhoff's current law will be violated at some cut $\mathbf{C}$ for n^β. ♣

We will use an inductive argument to establish Properties 6.3-1. We could start with the maximum principle for infinite 0-networks as described in the first paragraph of this section. But we won't. Instead, we will start at an even earlier stage and will thereby establish that conventional maximum principle as well. This will be accomplished by taking each branch to be a more primitive kind of subsection, namely a $(0-)$-subsection having no internal nodes, exactly two $(0-)$-tips, and two 0-nodes as its only bordering nodes. In fact, we can view each branch as being $(0-)$-connected to itself but not $(0-)$-connected to any other branch. In this way, the branch satisfies the definition of a subsection in a trivial way. Moreover, all of Properties 6.3-1 hold for that branch. In particular, part (b) holds vacuously.

To proceed, we will first argue the case where β is a natural number. Assume that Properties 6.3-1 hold for the natural number β replaced by $\beta-1$ and for all ranks lower than $\beta-1$ as well. (Thus, if $\beta=1$, we have that Properties 6.3-1 hold for $\beta-1 = 0$, which means we are dealing with a $(0-)$-subsection, that is, a branch.) We will prove that Properties 6.3-1 also hold for β.

Clearly, Properties (a1) and (a2) are mutually exclusive for any arbitrarily chosen $(\beta-)$-subsection $\mathbf{S}_b^{\beta-}$. Assuming (a1) does not hold, we shall prove that (a2) must hold. We can again ignore any subsection of rank less than $\beta-1$ and satisfying (a1) because all of its branches are null branches — i.e., all its branch voltages and currents are zero.

Let α be the internal rank of $\mathbf{S}_b^{\beta-}$, that is, the largest rank among all the internal nodes of $\mathbf{S}_b^{\beta-}$. Thus, $\alpha<\beta$. If $\alpha<\beta-1$, then $\mathbf{S}_b^{\beta-}$ is also a $((\beta-1)-)$-subsection, and therefore (a2) holds by the inductive hypothesis (or trivially if $\beta=1$, for then we have a $(0-)$-subsection, namely a single branch). So, consider the case where $\alpha=\beta-1$. Some or all of the $(\alpha-)$-subsections in $\mathbf{N}_a^{\nu}$ partition $\mathbf{S}_b^{\beta-}$, and Properties 6.3-1 (with β replaced by α) hold for each of them — by the inductive hypothesis again. In fact, (a2) holds for at least one of them since (a1) does not hold for $\mathbf{S}_b^{\beta-}$.

Let n_a^{α} be any arbitrarily chosen internal α-node of $\mathbf{S}_b^{\beta-}$. There will be another node n_b, which is either an internal α-node of $\mathbf{S}_b^{\beta-}$ or a bordering $(\beta+)$-node of $\mathbf{S}_b^{\beta-}$, such that the voltage at n_b differs from the voltage u_a^{α} of n_a^{α}. We can choose an α-path that terminates at n_a^{α}, reaches n_b, and lies in the core of $\mathbf{S}_b^{\beta-}$, except possibly for a last branch incident to n_b (Lemma 4.1-3(iv)). Upon tracing along that path starting from n_a^{α}, we will find an internal α-node n_0^{α} (possibly n_a^{α} itself) with the same voltage as n_a^{α} but α-adjacent either to n_b or to an internal α-node with a voltage different from

u_a^α. Let $\mathbf{S}_u$ be the union of all the $(\alpha-)$-subsections that are incident to those two α-adjacent nodes with differing voltages. By virtue of Condition 4.3-1 $\mathbf{S}_u$ is a finite union and has only finitely many incident $(\alpha+)$-nodes. By the inductive hypothesis again, (a2) holds for each member of $\mathbf{S}_u$. Let $n_x^{\alpha+}$ be an $(\alpha+)$-node incident to $\mathbf{S}_u$ with the largest voltage $u_x^{\alpha+}$ as compared to the other $(\alpha+)$-nodes incident to $\mathbf{S}_u$. We have $u_x^{\alpha+} \geq u_a^\alpha$. If $n_x^{\alpha+}$ is a bordering node of $\mathbf{S}_b^{\beta-}$, then either $u_x^{\alpha+} > u_a^\alpha = u_0^\alpha$ or $u_x^{\alpha+} = u_a^\alpha = u_0^\alpha$. In the latter case, the internal node n_0^α will be α-adjacent to another $(\alpha+)$-node with a voltage less than u_0^α. All this implies that either there is a bordering node of $\mathbf{S}_b^{\beta-}$ with a voltage larger than u_a^α or there is an internal α-node n_1^α (possibly n_0^α itself) with $u_1^\alpha \geq u_a^\alpha$ and with n_1^α incident to an $(\alpha-)$-subsection whose internal node voltages are strictly less than u_1^α.

Now, consider the set $\mathcal{N}_1^{\alpha+}$ of all the $(\alpha+)$-nodes that are α-adjacent to n_1^α. At least one node of $\mathcal{N}_1^{\alpha+}$ must have a voltage larger than u_1^α, for otherwise Kirchhoff's current law would be violated at a cut for n_1^α according to Lemma 6.3-2 (with β replaced by $\alpha = \beta - 1$). Thus, either we have a bordering node for $\mathbf{S}_b^{\beta-}$ in $\mathcal{N}_1^{\alpha+}$ with a voltage larger than u_1^α or, failing that, we can select an internal α-node $n_2^\alpha \in \mathcal{N}_1^{\alpha+}$ with a voltage u_2^α no less than the voltages at all the other α-nodes α-adjacent to n_1^α and with $u_2^\alpha > u_1^\alpha \geq u_a^\alpha$. Note also that, since n_1^α and n_2^α are α-adjacent, there is an $(\alpha-)$-path that reaches n_1^α and n_2^α, lies within an $(\alpha-)$-subsection, and reaches no other α-node (Lemma 4.2-3).

In the case where we have selected n_2^α, we can invoke the inductive hypothesis (a2) and Lemma 6.3-2 again to deduce that either there is a bordering node with a voltage larger than u_2^α or there is an internal α-node n_3^α with the following properties: n_3^α is α-adjacent to n_2^α but not to n_1^α; $u_3^\alpha > u_2^\alpha > u_1^\alpha \geq u_a^\alpha$; u_3^α is the largest voltage for all the α-nodes that are α-adjacent to n_2^α. Here, too, we can connect n_2^α to n_3^α by an $(\alpha-)$-path as before. This yields an α-path from n_1^α to n_3^α.

Further repetitions of this argument generate an α-path P^α whose α-node voltages are no less than u_a^α and are strictly increasing when ordered according to a tracing that starts at n_1^α. Either P^α is two-ended and terminates at a bordering node of $\mathbf{S}_b^{\beta-}$ with a voltage larger than u_a^α, or it is one-ended. In the latter case, we know from Lemma 4.5-4 that P^α will eventually lie in every arbitrarily chosen arm for the bordering $(\beta+)$-node $n^{\beta+}$ $(\beta > \alpha)$ that embraces the α-tip of P^α. However, we cannot yet assert that the α-node voltages of P^α converge to the voltage $u^{\beta+}$ of $n^{\beta+}$ because P^α may not be permissive. Nor can we use Property 6.3-1(b) yet because our inductive hypothesis has it that Property 6.3-1(b) holds for β replaced by $\beta - 1$, that is, for an α less than $\beta - 1$. We need that property for $\alpha = \beta - 1$.

So, as our next step, we show that Property 6.3-1(b) holds for $\alpha = \beta - 1$. Choose a permissive contraction $\{\mathcal{W}_p\}_{p=1}^{\infty}$ to $n^{\beta+}$ and let $\{\mathbf{A}_p\}_{p=1}^{\infty}$ be the corresponding sequence of arms. By Lemma 4.5-4, P^α eventually lies in $\mathbf{A}_1$. Let $\mathbf{M}_p = \mathbf{A}_p \backslash \mathbf{A}_{p+1}$. By the definition of a contraction (see Section 4.4), $\mathbf{M}_p$ is not void. We also have that, for $p > 1$, $\mathcal{V}_p$ separates $\mathbf{M}_{p-1}$ and $\mathbf{M}_p$. Moreover, $\mathbf{M}_p$ has only finitely many α-nodes (perhaps, none at all) according to Lemma 4.5-5.

As in the proof of Lemma 6.3-2, $\mathbf{M}_p$ can be viewed as being a finite union of $(\beta-)$-subsections in a $(\beta+)$-network with the bordering nodes of $\mathbf{M}_p$ being $(\beta+)$-nodes that embrace the nodes of $\mathcal{V}_p \cup \mathcal{V}_{p+1}$. Indeed, we can append finitely many pure voltage sources to the nodes of $\mathcal{V}_p \cup \mathcal{V}_{p+1}$ to produce the same voltage-current regime in $\mathbf{M}_p$ as it has as a part of $\mathbf{N}_a^\nu$. Moreover, we can append finitely many one-ended $(\beta - 1)$-paths to the nodes of $\mathcal{V}_p \cup \mathcal{V}_{p+1}$ through their $(\beta - 1)$-tips with those paths being otherwise totally disjoint from each other and from $\mathcal{M}_p$. The appending of those paths will not alter the voltage-current regime in $\mathbf{M}_p$ and will yield a larger, finitely structured $(\beta+)$-network.

The argument we have already constructed for $\mathbf{S}_b^{\beta-}$ can now be applied to $\mathbf{M}_p$. Since $\mathbf{M}_p$ has only finitely many α-nodes, this will lead to a two-ended α-path that must terminate at a node of $\mathcal{V}_p \cup \mathcal{V}_{p+1}$, and the conclusion is that every α-node voltage for $\mathbf{M}_p$ is no larger than the largest voltage for the nodes of $\mathcal{V}_p \cup \mathcal{V}_{p+1}$. By a similar argument "no larger" and "largest" can be replaced by "no less" and "least".

Now, let $u_{max,p}$ and $u_{min,p}$ be the largest and least voltage for the nodes of $\mathcal{V}_p$, respectively. There are only finitely many contraction paths for the chosen contraction, each being permissive, and every node of $\mathcal{V}_p$ lies on a permissive contraction path. It follows from Lemma 5.5-1 that the voltages at the various $\mathcal{V}_p$ along any contraction path converge to the voltage $u^{\beta+}$ at $n^{\beta+}$ and that therefore $u_{max,p} \to u^{\beta+}$ and $u_{min,p} \to u^{\beta+}$ as $p \to \infty$.

As a consequence of the last two paragraphs, the α-node voltages along the path P^α converge to $u^{\beta+}$. This conclusion extends immediately to the voltages along any sequence of nodes embraced by P_α and approaching $n^{\beta+}$. Indeed, all but finitely many of those nodes will lie in $\cup_{p=1}^{\infty} \mathbf{M}_p$. If one of those nodes n^δ is of rank δ ($0 \leq \delta < \alpha$) and lies in $\mathbf{M}_p$, its voltage u^δ will be no larger (no less) than the largest (least) bordering voltage for the $(\alpha-)$-subsection in which n^δ resides — according to our inductive hypothesis again. (In the same way as before, we can view the nodes of $\mathcal{V}_p \cup \mathcal{V}_{p+1}$ as being α-nodes.) By what we have shown above, this in turn implies that

$$\min(u_{min,p-1}, u_{min,p}) \leq u^\delta \leq \max(u_{max,p-1}, u_{max,p}).$$

Our asserted extension follows.

Altogether then, we have established inductively that Property 6.3-1(b) holds for $\alpha = \beta - 1$.

We can now complete our proof of Property 6.3-1(a2) (for the case where β is a natural number) under the assumption that (a1) does not hold. We now have it that, in the event that P^α is a one-ended α-path, the α-node voltages along P^α converge to the node voltage $u^{\beta+}$ at $n^{\beta+}$. Consequently, $u^{\beta+} > u_a^\alpha$. Since n_a^α was arbitrarily chosen as an internal α-node of $\mathbf{S}_b^{\beta-}$, it now follows that the voltages at every internal α-node is strictly less than the maximum voltage u_{max} for the bordering $(\beta+)$-nodes of $\mathbf{S}_b^{\beta-}$. By the inductive hypothesis, this is also true for every internal node, whatever be its rank. A similar argument allows us to replace "strictly less" by "strictly greater" and "maximum voltage u_{max}" by "minimum voltage u_{min}." Whence Property 6.3-1(a2).

The next theorem summarizes what has so far been established. Remember that $\mathbf{N}^\nu$ is permissively finitely structured with positive resistances in all of its branches (i.e., Conditions 6.1-1 are assumed) and that $\mathbf{N}_a^\nu$ is $\mathbf{N}^\nu$ with finitely many (possibly pure) sources appended. If the appended sources are pure voltage sources, the voltage-current regime is determined by Theorem 6.2.4 along with superposition if there are more than one source.

Theorem 6.3-3. *For any natural number β with $\beta \leq \nu$, let $\mathbf{S}_b^{\beta-}$ be a $(\beta-)$-subsection of $\mathbf{N}_a^\nu$ and let it be sourceless (thus, none of its internal nodes is incident to a source branch of $\mathbf{N}_a^\nu$). Then, Properties 6.3-1 hold for it.*

Let us now consider the case where $\beta = \omega = \nu$. We assume that, among all the tips with which the subsection $\mathbf{S}_b^{\omega-}$ reaches its bordering nodes, at least one of them is an $\vec{\omega}$-tip. (Otherwise $\mathbf{S}_b^{\omega-}$ would also be an $(\alpha-)$-subsection for some natural number α and Theorem 6.3-3 would then apply.)

Let n_k^ω $(k = 1, \ldots, K)$ be the bordering nodes of $\mathbf{S}_b^{\omega-}$. Choose in $\mathbf{S}_b^{\omega-}$ a permissive contraction $\{\mathcal{W}_{k,p}\}_{p=1}^\infty$ for each n_k^ω that $\mathbf{S}_b^{\omega-}$ reaches through a nonelementary tip, and let $\{\mathbf{A}_{k,p}\}_{p=1}^\infty$ be the set of corresponding arms.

These arms are void if $\mathbf{S}_b^{\omega-}$ is incident to $n_k^{\omega-}$ only through elementary tips. Set $\mathbf{F}_p = \mathbf{S}_b^{\omega-} \backslash \bigcup_{k=1}^K \mathbf{A}_{k,p}$; thus, $\mathbf{F}_p$ is the subnetwork of $\mathbf{S}_b^{\omega-}$ induced by all the branches of $\mathbf{S}_b^{\omega-}$ that are not in any $\mathbf{A}_{k,p}$. Now, there will be some natural number $\mu_p - 1$ that uniformly bounds all the ranks of all the nodes of $\mathbf{F}_p$. Indeed, if this were not so, $\mathbf{F}_p$ would have to be an $\vec{\omega}$-network since it has nodes of arbitrarily large natural-number ranks but no ω-nodes. Moreover, since $\mathbf{F}_p$ is locally finitely structured, it has no $\vec{\omega}$-nodes. Therefore, by Theorem 4.2-7, it would have at least one $\vec{\omega}$-path. The $\vec{\omega}$-tip of that path would then have to be embraced by an ω-node, in contradiction to the fact that $\mathbf{F}_p$ has no ω-node.

We can convert each node in each base $\mathcal{V}_{k,p}$ into a μ_p-node by appending a one-ended $(\mu_p - 1)$-path through its $(\mu_p - 1)$-tip to that node and making that path otherwise totally disjoint from $\mathbf{N}_a^\nu$. We do so. This makes each $\mathbf{F}_p$ be a (μ_p-)-subsection $\mathbf{F}_p^{\mu_p-}$. These appended paths can be ignored so far as the electrical behavior of $\mathbf{S}_b^{\omega-}$ is concerned because all the branches of those paths are a null branches. Moreover, we are free to choose the μ_p $(p = 1, 2, \ldots)$ strictly increasing. The $\mathbf{F}_p^{\mu_p-}$ fill out $\mathbf{S}_b^{\omega-}$ in the sense that $\mathbf{F}_q^{\mu_q-} \subset \mathbf{F}_p^{\mu_p-}$ for $q < p$ and $\mathbf{S}_b^{\omega-} = \bigcup_{p=1}^\infty \mathbf{F}_p^{\mu_p-}$.

Let us now assume that (a1) of Conditions 6.3-1 does not hold for $\mathbf{S}_b^{\omega-}$. Therefore, it will not hold for some $\mathbf{F}_q^{\mu_q-}$, nor will it hold for every $\mathbf{F}_p^{\mu_p-}$ where $p > q$. By Theorem 6.3-3, we have for each $p > q$ that the maximum voltage u_p of the node voltages for $\mathbf{F}_p^{\mu_p-}$ occurs at one or more of the bordering nodes of $\mathbf{F}_p^{\mu_p-}$. These bordering nodes are base nodes for the contraction. Also, the u_p will be nondecreasing as $p \to \infty$. Moreover, because the contraction paths are all permissive, the voltages along any contraction path converge to the voltage at the bordering node of $\mathcal{S}_b^{\omega-}$ that path reaches. Since the u_p occur at base nodes and therefore on contraction paths, and since there are only finitely many such paths, the maximum voltage u_p for $\mathbf{F}_p^{\mu_p-}$ converges as $p \to \infty$ to the maximum voltage u among the bordering nodes of $\mathbf{S}_b^{\omega-}$. Finally, since, for each p, u_p is strictly larger than the internal-node voltages for $\mathbf{F}_p^{\mu_p-}$ according to Theorem 6.3-3, u is in turn strictly larger than every internal-node voltage for $\mathbf{S}_b^{\omega-}$. This establishes Property 6.3-1(a) for $\mathbf{S}_b^{\omega-}$.

As for Property 6.3-1(b), choose any bordering node $n_k^{\omega+}$ that $\mathbf{S}_b^{\omega-}$ reaches through a nonelementary tip. Also, choose a permissive contraction $\{\mathcal{W}_{k,p}\}_{p=1}^\infty$ in $\mathbf{S}_b^{\omega-}$ for $n_k^{\omega+}$, and let $\{\mathbf{A}_{k,p}\}_{p=1}^\infty$ be the corresponding arms. For k fixed and $q > p$, set $\mathbf{M}_{p,q} = \mathbf{A}_{k,p} \backslash \mathbf{A}_{k,q}$. By appending one-ended paths to the base nodes of $\mathcal{V}_{k,p}$ and $\mathcal{V}_{k,q}$, we can view $\mathbf{M}_{p,q}$ as a ($\mu-$)-subsection for some $\mu < \vec{\omega}$. This allows us to apply Theorem 6.3-3 to $\mathbf{M}_{p,q}$ to conclude that the internal-node voltages of $\mathbf{M}_{p,q}$ are bounded above and below by the maximum and minimum of the node voltages at $\mathcal{V}_{k,p} \cup \mathcal{V}_{k,q}$. Since the latter voltages converge to the voltage at $n_k^{\omega+}$, so too do those of the $\mathbf{M}_{p,q}$ as $p, q \to \infty$. This shows that the node voltages along any path that reaches $n_k^{\omega+}$ through $\mathbf{S}_b^{\omega-}$ converge to the voltage at $n_k^{\omega+}$.

In summary, we have the following when $\beta = \omega = \nu$.

Theorem 6.3-4 *Let $\mathbf{S}_b^{\omega-}$ be a sourceless ($\omega-$)-subsection in $\mathbf{N}_a^{\omega}$. Then, Properties 6.3-1 hold for it.*

If $\mathbf{S}_b^{\beta-}$ has only one bordering node, then the only possibility is (a1):

Corollary 6.3-5. *Under the hypothesis of either Theorem 6.3-3 or Theorem 6.3-4, assume also that $\mathbf{S}_b^{\beta-}$ has only one incident ($\beta+$)-node $n^{\beta+}$. Then, the node voltages for $\mathbf{S}_b^{\beta-}$ are all the same, namely, $u^{\beta+}$.*

Corollary 6.3-6. *Let $\mathbf{N}_s$ be any connected subnetwork of $\mathbf{N}_a^{\nu}$ that meets its complement $\mathbf{N}_a^{\nu} \backslash \mathbf{N}_s$ with only finitely many maximal nodes n_k ($k = 1, \ldots, K$). Assume furthermore that $\mathbf{N}_s$ is sourceless, that is, no branch of $\mathbf{N}_s$ is a source branch. Then, Properties 6.3-1 hold for $\mathbf{N}_s$. If in addition $K = 1$, then the node voltages of $\mathbf{N}_s$ are all the same.*

Proof. Assume at first that ν is a natural number μ. As we have done previously, we can append one-sided μ-paths to the n_k in such a fashion that the μ-paths are incident to the n_k through their μ-tips and are otherwise totally disjoint from each other and from $\mathbf{N}_a^{\mu}$. $\mathbf{N}_a^{\mu}$ along with the appended branches becomes a ($\mu+1$)-network, and each n_k becomes the exceptional element of a ($\mu + 1$)-node $n_k^{\mu+1} = \{n_k, t_k^{\mu}\}$, where t_k^{μ} is the μ-tip of the kth appended path. Because of the finite structuring of $\mathbf{N}_a^{\mu}$, $\mathbf{N}_s$ meets each $n_k^{\mu+1}$ with no more than finitely many contraction paths. This implies that $\mathbf{N}_s$, as a subnetwork of the ($\mu + 1$)-network, is the union of no more than finitely many (($\mu + 1$)$-$)-subsections. Therefore, we may apply Theorem 6.3-3 and Corollary 6.3-5 to each of those (($\mu+1$)$-$)-subsections to obtain our conclusions.

For the case where $\nu = \omega$, we now append one-ended ω-paths to the n_k to get ($\omega + 1$)-nodes $n_k^{\omega+1}$. This then requires an application of Theorem 6.3-3 and Corollary 6.3-5 to an ($\omega + 1$)-network. Although we have not explicitly established that theorem and corollary for ($\omega + 1$)-networks, this can be done just by repeating the arguments for the case where β is a natural number. ♣

Corollary 6.3-7. *Assume now that* $\mathbf{N}^\nu$ *is sourceless. Let* $\mathbf{N}^\nu_e$ *now denote* $\mathbf{N}^\nu_a$ *with exactly one source — a pure voltage source appended to any two nodes. Let that source's value be 1 V and let the negative terminal of the source be ground with a node voltage of 0. Then, the voltage at every node of* $\mathbf{N}^\nu_e$ *is no less than 0 and no larger than 1.*

Proof. This is a special case of Corollary 6.3-6. $\mathbf{N}_s$ is now $\mathbf{N}^\nu$ (that is, upon deleting the source from $\mathbf{N}^\nu_e$, we get $\mathbf{N}_s$). Also, there are now only two n_k nodes, the maximal nodes to which the source is appended. ♣

The next corollary sharpens the last one.

Corollary 6.3-8. *Assume the hypothesis of Corollary 6.3-7. Let* n_e *and* n_g *be the two maximal nodes to which the 1 V source of Corollary 6.3-7 is appended, and let* n_g *be the ground node. Also, let* n_0 *be another maximal node of* $\mathbf{N}^\nu_e$ *different from* n_e *and* n_g*.*

(i) *If there is a path* P *in* $\mathbf{N}^\nu_e$ *that meets* n_0 *and* n_g *but does not meet* n_e*, then* $u_0 < 1$.

(ii) *If there is a path* P *in* $\mathbf{N}^\nu_e$ *that meets* n_0 *and* n_e *but does not meet* n_g*, then* $u_0 > 0$.

Proof. We can now consider together all the cases where $0 \leq \nu \leq \omega$, $\nu \neq \vec{\omega}$. Under the hypothesis of (i), suppose $u_0 = 1$. All the node voltages are no larger than 1 according to Corollary 6.3-7. Upon tracing the path P starting at n_0, we will meet a node with a voltage less than 1. Two cases arise:

Case 1. We meet a maximal 0-node n^0_1 *with a voltage equal to 1 and adjacent to a 0-node with a voltage less than 1.* By Kirchhoff's current law, there must be another 0-node in $\mathbf{N}^\nu_e$ that is 0-adjacent to n^0_1 and has a voltage greater than 1. This violates Corollary 6.3-7.

Case 2. We meet a maximal δ*-node* n^δ_1 *with* $0 < \delta \leq \nu$, $u^\delta_1 = 1$, *and* n^δ_1 *incident to a* $(\delta-)$*-subsection having at least some of its internal or bordering node voltages less than 1.* By Property 6.3-1(a2), at least one of the bordering node voltages of that $(\delta-)$-subsection will be less than 1.

By Lemma 6.3-2, if Kirchhoff's current law is to be satisfied at every cut that isolates n_1^δ, there must be another $(\delta+)$-node that is δ-adjacent to n_1^δ and has a voltage larger than 1. Again Corollary 6.3-7 is violated.

These two are the only possible cases. Hence, our supposition is false, and (i) is true. (ii) is established similarly. ♣

Example 6.3-9. By virtue of Corollary 6.3-7, we are now able to extend the classical triangle inequality for input resistances to any transfinite sourceless network $\mathbf{N}^\nu$ satisfying Conditions 6.1-1. As was shown by Lemma 6.2-3, $\mathbf{N}^\nu$ behaves as a positive resistance between any two of its nodes. Let n_a, n_b, and n_c be three maximal nodes of $\mathbf{N}^\nu$. We apply a pure current source of value 1 A to $\mathbf{N}^\nu$ in two ways. The first way: The source extracts 1A from n_a and injects 1A into n_b. This produces a voltage difference v_{ba} between n_b and n_a with positive polarity at n_b, as well as a voltage difference v_{ca} between n_c and n_a with positive polarity at n_c (v_{ca} may be 0). Since $\mathbf{N}^\nu$ behaves as a positive resistance between n_a and n_b, the same voltage v_{cb} between n_c and n_b will be produced by the application of a pure voltage source of value v_{ba} between n_b and n_a with the stated polarity. So, by Corollary 6.3-7 and the linearity of $\mathbf{N}^\nu$, we have $0 \leq v_{ca} \leq v_{ba}$. The second way: The source extracts 1 A from n_b and injects 1 A into n_c. With "w" now denoting voltage differences, we have by a similar reasoning that $0 \leq w_{ca} \leq w_{cb}$. Furthermore, with two current sources acting in these two ways simultaneously, we have in effect a single current source as follows: 1 A extracted from n_a, 1 A injected into n_c, and 0 A source current at n_b. Let d_{ca} denote the voltage difference between n_c and n_a under the last application of a current source. Then, by superposition, $d_{ca} = w_{ca} + v_{ca} \leq w_{cb} + v_{ba}$. By definition, the input resistance between any two nodes of $\mathbf{N}^\nu$ is the voltage produced between them when a 1 A source is impressed upon them. Letting "r" denote those input resistances, we thus have the *triangle inequality* for the input resistances between any three nodes of the transfinite network $\mathbf{N}^\nu$:

$$r_{ca} \leq r_{cb} + r_{ba}. \quad ♣$$

Finally, the argument used in establishing Theorem 6.2-4, when combined with the reciprocity and maximum principles yields a result that we shall need in Section 7.5. It is a maximum principle expressed in terms of currents.

Theorem 6.3-10. *Let* $\mathbf{N}^\nu$ *be sourceless and satisfy Conditions 6.1-1. Let* h_a *be a pure current source appended to two maximal nodes* n_1 *and* n_2

of $\mathbf{N}^\nu$, *and let a short be connected between another two maximal nodes* n_3 *and* n_4 *of* $\mathbf{N}^\nu$. *Assume that at least one (but not necessarily both) of* n_1 *and* n_2 *is different from both* n_3 *and* n_4. *Let* i_b *be the current in the short between* n_3 *and* n_4 *induced by* h_a. *Then,* $|i_b| \leq |h_a|$.

Note. Were n_1 to coincide with n_3 and were n_2 to coincide with n_4, we would immediately have $|i_b| = |h_a|$ because the input resistance z of $\mathbf{N}^\nu$ between n_1 and n_2 is positive.

Also, note that the current in the short between n_3 and n_4 can be determined as follows. Before n_3 and n_4 are shorted, choose a cut $\mathbf{C}_3$ for n_3 that isolates n_3 from n_4 and choose another cut $\mathbf{C}_4$ for n_4 that isolates n_4 from n_3. Orient cut branches toward their respective nodes n_3 and n_4. Then, after n_3 and n_4 are shorted, $\mathbf{C}_3 \cup \mathbf{C}_4$ — with all resulting self-loops opened — is a cut for the single node combined from n_3 and n_4. The currents in the branches of $\mathbf{C}_3 \cup \mathbf{C}_4$ are determined by Theorem 5.4-2 whereby h_a is transferred along a permissive path connecting n_1 and n_2. Then, the subtraction of the algebraic sum of the branch currents in $\mathbf{C}_4$ from the algebraic sum of the branch currents in $\mathbf{C}_3$ yields the current in the short flowing from n_3 to n_4.

Proof. In place of the short, let us append to n_3 and n_4 a source branch b_b in the Thevenin form having a positive resistor ρ in series with a voltage source e_b. We can make a Thevenin-to-Norton transformation to convert b_b into a parallel combination of ρ and a current source $k_b^\rho = -e_b/\rho$. As for the current source h_a, we can transfer it along a permissive path connecting n_1 and n_2. The network is now in a form appropriate for the application of Corollary 5.2-16. Consider two cases.

Case 1. $h_a = 0$ and $k_b^\rho \neq 0$: Let w_a^ρ be the algebraic sum of the voltages along P induced by k_b^ρ.

Case 2. $h_a \neq 0$ and $k_b^\rho = 0$: Now, b_b is a purely resistive branch with resistance $\rho > 0$. Let v_b^ρ and i_b^ρ be the branch voltage and branch current in b_b induced by the current sources of value h_b along P. In this case, $i_b^\rho = v_b^\rho/\rho$.

By the reciprocity principle (Corollary 5.2-16), $w_a^\rho h_a = v_b^\rho k_b^\rho$. Therefore, $w_a^\rho h_a = v_b^\rho(-e_b/\rho) = -i_b^\rho e_b$. By the argument used in the proof of Theorem 6.2-4, all currents and voltages converge as $\rho \to 0+$. Thus, $w_a^\rho \to w_a$ and $i_b^\rho \to i_b$. Consequently, $w_a h_a = -i_b e_b$. By the maximum principle in the form of Corollary 6.3-7, the condition $e_b = 1$ implies that $|w_a| \leq 1$ because w_a is the difference between two node voltages induced by e_b. It now follows that $|h_a| \geq |i_b|$. ♣

Chapter 7

Transfinite Random Walks

The idea of a random walk on a graph is an idea of considerable importance in probability theory and has a variety of applications in the sciences as well as in other branches of mathematics. The recent survey of Woess [32] provides an excellent overview of the subject along with an extended bibliography. However, all that theory has been restricted to walks on conventional (finite and infinite) graphs. The recent advent of transfinite graphs invites an examination of random walks upon them. This immediately raises the question of how a random walk might wander through a transfinite node and thereby progress from one 0-section to another.

Our solution to this problem is based upon the fact that a large part of conventional random-walk theory uses the so-called "nearest-neighbor rule," which specifies probabilities of transitions between adjacent 0-nodes. That rule has an interpretation in terms of electrical networks, whereby the probabilities of transitions from a given node to its adjacent nodes are proportional to the conductances incident to that node, with the result that other probabilities concerning the random walk are given in terms of node voltages, in particular, by means of a rule established by Nash-Williams [19]. Because of this, our transfinite generalization of electrical network theory provides a means — an extension of the Nash-Williams rule — for transfinitely generalizing random walks governed by the nearest-neighbor rule.

As with our other transfinite generalizations, a variety of complications beset this one too. Such for example is the fact that almost all random walks on a transfinite graph are confined to single 0-sections and only an exceedingly small fraction of them do succeed in penetrating through transfinite nodes into adjacent 0-sections. This motivates the idea of a "roving" random walk, one that does pass through transfinite nodes. (Actually, this very problem also arises when conventional random walks are interpreted in terms of electrical networks; see Section 7.3.)

Our theory is restricted to transfinite walks on permissively finitely structured networks. This allows us to invoke the maximum principle for node voltages, which will be essential to our analysis. That structure also enables a recursive development of a theory for transfinite random walks through an inductive argument that first shows how a random walk on a $(\beta-)$-subsection may reach a bordering $(\beta+)$-node, then how the walk may leave a β-node, and finally how it may wander among the β-nodes of a $((\beta+1)-)$-subsection.

The final result relates random walks on transfinite networks to random walks on ordinary networks in the following way. Given any arbitrarily chosen finite set of nodes in the transfinite network, the probabilities of transitions between those nodes are shown to be describable by an irreducible and reversible Markov chain. (Such Markov chains are surveyed in Appendix C.) It is an established fact that the latter represents a random walk on a finite network. Thus, the wanderings of a random walk among the finitely many chosen nodes of the transfinite network is mimicked by a random walk on the finite network.

Throughout this chapter, the following will always be assumed.

Condition 7.0-1. *The ν-network $\mathbf{N}^\nu$ is sourceless (i.e., no branch of $\mathbf{N}^\nu$ has a source) and is permissively finitely structured with $1 \leq \nu \leq \omega$, $\nu \neq \vec{\omega}$.*

Thus, every branch of $\mathbf{N}^\nu$ is a positive resistor. $\mathbf{N}^\nu$ will be excited by appending source branches to it.

7.1 The Nash-Williams Rule

Let $\mathbf{N}^0$ be a sourceless, 0-connected, locally finite 0-network in which every branch b_j has a positive finite conductance g_j. Assume that $\mathbf{N}^0$ has no self-loops. A *walk* W^0 on $\mathbf{N}^0$ is an alternating sequence

$$W^0 \;=\; \{\ldots, n_m, b_m, n_{m+1}, b_{m+1}, \ldots\} \qquad (7.1)$$

of 0-nodes and branches such that each branch b_m is incident to the two nodes n_m and n_{m+1} adjacent to it in the sequence. Unlike a 0-path, the nodes and branches may appear more than once in the sequence — except that nodes that are adjacent in the sequence are different because there are no self-loops. However, like a path, the walk W^0 may be two-ended, one-ended, or endless. Whenever W^0 terminates on either end, it is required that it terminate at a node. Thus, a 0-path is a special case of a 0-walk. The *orientation* of W^0 is the direction conforming with increasing values of the index m, and the opposite direction is the *reverse orientation* of W^0.

A *random 0-walk* on $\mathbf{N}^0$ is a 0-walk on $\mathbf{N}^0$ in which the choices of the branches are uncertain and governed by some probabilistic rule. For us, that will be the *nearest-neighbor rule*, defined as follows. Given the 0-node n_m, let J_m be the index set for the branches incident to n_m. Consider a random walk that has progressed up to the node n_m. Then, the probability that the next branch in the random walk is b_j $(j \in J_m)$ is by definition $g_j / \sum_{i \in J_m} g_i$. It will be more convenient for us to restate this in terms of the node sequence in (7.1). Let K_m be the index set of all the 0-nodes that are 0-adjacent to n_m. Also, let g_{mk} $(k \in K_m)$ now denote the conductances incident to n_m; thus, $g_{mk} = g_{km}$. Then, the probability P_{mk} that the next node after n_m in the random walk is n_k $(k \in K_m)$ is

$$P_{mk} = \frac{g_{mk}}{\sum_{l \in K_m} g_{ml}}. \tag{7.2}$$

Thus, $\sum_{k \in K_m} P_{mk} = 1$. Furthermore, we set $P_{mk} = 0$ whenever the node n_k is not adjacent to n_m (i.e., $k \notin K_m$). We also set $P_{kk} = 0$ to reflect the fact that there are no self-loops in $\mathbf{N}^0$. The P_{mk} are called *transition probabilities* or, more specifically, *one-step transition probabilities*, and there is one such probability for each ordered pair of nodes in $\mathbf{N}^0$. In general, $P_{mk} \neq P_{km}$. Note that $\sum_{k \in K} P_{mk} = 1$, where now m is fixed and K is the index set for all the nodes in $\mathbf{N}^0$.

The P_{mk} can be measured electrically in terms of node voltages. With the walk at n_m and with n_k $(k \in K_m)$ being a node 0-adjacent to n_m, let the node voltage at n_k be 1 V, let the voltage at n_m float (i.e., be determined by Kirchhoff's laws and Ohm's law), and let the node voltage at all other nodes n_l $(l \in K, l \neq m, l \neq k)$ be 0 V. This yields a voltage at n_m which turns out to be equal to P_{mk} as given by (7.2). This fact can be shown through a simple manipulation of Kirchhoff's laws and Ohm's law.

Thus, the nearest-neighbor rule yields a Markov chain, whose state space is the set of nodes in $\mathbf{N}^0$ and whose transition probabilities are given by (7.2). Actually, this is a special kind of Markov chain, one that is irreducible and reversible (see Appendices C3 and C5). Such a chain is also called a *random walk*, but it too is a special kind of random walk, one that is governed by the nearest-neighbor rule. It will be convenient at times to speak of a *random walker* Ψ that performs the random walk. Thus, Ψ is an entity that wanders among the nodes of $\mathbf{N}^0$; at each node it casts a die reflecting the probabilities (7.2) to choose thereby the next branch and node in its walk.

Nash-Williams has derived from the nearest-neighbor rule a law based on node voltages that specifies the probability that Ψ will meet one set of nodes before it meets another disjoint set of nodes [19]. Since it is so essential to our needs, let us now present his analysis.

Henceforth, we assume that Ψ wanders in $\mathbf{N}^0$ according to the nearest-neighbor rule (7.2). Let $\mathcal{N}$ be the node set of $\mathbf{N}^0$, let $\mathcal{N}_f$ be any nonvoid finite set of nodes in $\mathbf{N}^0$, and let $\mathcal{N}_t = \mathcal{N} \backslash \mathcal{N}_f$ be the set of nodes in $\mathcal{N}$ that are not in $\mathcal{N}_f$. Also, let n be any node of $\mathcal{N}_f$. Furthermore, let $Q(n, \mathcal{N}_t)$ denote the probability that the random walker Ψ, after starting from n, will meet $\mathcal{N}_t$ within at most finitely many steps, and let $Q_k(n, \mathcal{N}_t)$ denote the probability that Ψ will do so in k or fewer steps.

Lemma 7.1-1. $Q(n, \mathcal{N}_t) = 1$.

Note. This lemma asserts that it is a certainty that Ψ, after starting from n, will meet $\mathcal{N}_t$ within finitely many steps.

Proof. Let n be any node of $\mathcal{N}_f$. Let $\delta(n, \mathcal{N}_t)$ denote the distance from n to $\mathcal{N}_t$, that is, the length of the shortest path from n to $\mathcal{N}_t$, where the length of a path is the number of branches in it. Set

$$d = \max_{n \in \mathcal{N}_f} \delta(n, \mathcal{N}_t)$$

and

$$Q_{min} = \min_{n \in \mathcal{N}_f} Q_d(n, \mathcal{N}_t).$$

Q_{min} is a lower bound on the probability that Ψ, after starting from some unspecified node of $\mathcal{N}_f$, will meet $\mathcal{N}_t$ in d or fewer steps. Therefore, $1 - Q_{min}$ is an upper bound on the probability that Ψ, after starting as stated, will not meet $\mathcal{N}_t$ in d or fewer steps.

For any given $n \in \mathcal{N}_f$, we can find a path P from n to $\mathcal{N}_t$ whose length is no larger than d. By the nearest-neighbor rule (7.2), for each node n_a of P, there is a positive probability that Ψ will proceed from n_a to the next node in P in one step. It follows that there is a positive probability that Ψ

will proceed along P to meet $\mathcal{N}_t$ in d or fewer steps. Since $\mathcal{N}_f$ is a finite set, we can infer that $Q_{min} > 0$.

Now, kd is a positive natural number, and $1 - Q_{kd}(n, \mathcal{N}_t)$ is the probability that Ψ, after starting from $n \in \mathcal{N}_f$, will not meet $\mathcal{N}_t$ in kd or fewer steps. Consequently,

$$0 \leq 1 - Q_{kd}(n, \mathcal{N}_t) \leq (1 - Q_{min})^k \rightarrow 0$$

as $k \rightarrow \infty$. This shows that the probability of Ψ never meeting $\mathcal{N}_t$ after starting from n is 0. Hence, $Q(n, \mathcal{N}_t) = 1$. ♣

As the next step, we partition $\mathcal{N}_t$ into two nonvoid subsets $\mathcal{N}_e$ and $\mathcal{N}_g$; thus, $\mathcal{N}_t = \mathcal{N}_e \cup \mathcal{N}_g$ and $\mathcal{N}_e \cap \mathcal{N}_g = \emptyset$. We are interested in the probability $P(n)$ that Ψ, after starting from $n \in \mathcal{N}_f$, will meet $\mathcal{N}_e$ before meeting $\mathcal{N}_g$. To be more explicit, we shall also denote this probability $P(n)$ by

$$Prob(\mathrm{s}n, \mathrm{r}\mathcal{N}_e, \mathrm{b}\mathcal{N}_g). \tag{7.3}$$

By allowing n to vary throughout $\mathcal{N}_f$, we hereby have a real number $P(n)$ assigned to every node $n \in \mathcal{N}_f$, and those numbers satisfy $0 \leq P(n) \leq 1$. Furthermore, let us set $P(n) = 1$ if $n \in \mathcal{N}_e$ and $P(n) = 0$ if $n \in \mathcal{N}_g$. We will interpret these numbers as node voltages. These in turn determine all the branch voltages in the usual way; that is, if branch b_j is incident away from node n_a and incident toward node n_b, then its branch voltage v_j is given by $v_j = P(n_a) - P(n_b)$. According to Lemma 5.1-1, these branch voltages v_j satisfy Kirchhoff's voltage law around every loop in $\mathbf{N}^0$. Furthermore, the v_j determine branch currents i_j through Ohm's law: $i_j = g_j v_j$. Thus, the current in any branch incident to two nodes of $\mathcal{N}_e$ or to two nodes of $\mathcal{N}_g$ is 0.

We may think of all the nodes of $\mathcal{N}_e$ as being effectively shorted together into a single node n_e by virtue of their common node voltage of 1 V, and similarly the nodes of $\mathcal{N}_g$ can be thought of as a single node n_g with a voltage of 0 V. Furthermore, we can take it that there is a single pure voltage source of 1 V incident to n_e and n_g, which establishes their node voltages. The result is a finite network $\mathbf{N}_e^0$; indeed, since the node set $\mathcal{N}_f$ is finite and $\mathbf{N}^0$ is locally finite, only finitely many branches are incident to n_e (and similarly for n_g).

Lemma 7.1-2. *The branch currents in the finite network* $\mathbf{N}_e^0$ *satisfy Kirchhoff's current law at every node.*

Proof. Let n_m be any node of $\mathcal{N}_f$ and let $\{n_k\}_{k \in K_m}$ be the finite set of all nodes 0-adjacent to n_m, as before. It is a certainty that Ψ, after starting from n_m, will meet a node n_k $(k \in K_m)$ in one step. Moreover, these one-step transitions are mutually exclusive events. We have defined the node voltage $P(n)$ to be the probability (7.3). Therefore, by conditional probabilities and the nearest-neighbor rule (7.2), we have

$$P(n_m) \;=\; \sum_{k \in K_m} \left(\frac{g_{mk}}{\sum_{l \in K_m} g_{ml}} P(n_k) \right). \tag{7.4}$$

(Some of the nodes n_k in the right-hand side may be in $\mathcal{N}_e$ or $\mathcal{N}_g$, but this equation holds nonetheless.) Now, the algebraic sum of the currents leaving n_m is

$$P(n_m) \sum_{l \in K_m} g_{ml} \;-\; \sum_{k \in K_m} g_{mk} P(n_k). \tag{7.5}$$

Upon substituting (7.4) into (7.5), we see that (7.5) is equal to 0.

We have yet to consider the two nodes n_e and n_g. Let us write out Kirchhoff's current law at every node of $\mathcal{N}_f$ as an algebraic sum of currents leaving each such node. Upon adding together all those finitely many sums of finitely many terms, we get another sum which also equals 0. Now, the current in any branch incident to two nodes of $\mathcal{N}_f$ will appear twice in that latter sum, once with a plus sign and once with a minus sign. Hence, it cancels out. On the other hand, the current in any branch incident to a node of $\mathcal{N}_f$ and to n_e or n_g appears only once in that latter sum. It follows that the algebraic sum of all the currents flowing from n_e to nodes of $\mathcal{N}_f$ is equal to the algebraic sum of all the currents flowing from $\mathcal{N}_f$ to n_g. In addition there may be branches incident to n_e and n_g, but the current leaving n_e in such a branch equals the current entering n_g through that branch. We can identify the sum of all the currents leaving n_e as the current in the source branch, and this in turn is equal to the sum of all the currents entering n_g. Altogether then, Kirchhoff's current law is satisfied at n_e, at n_g, and at every node of $\mathcal{N}_f$; that is, it is satisfied throughout $\mathbf{N}^0_e$. ♣

All this implies that we can determine the probabilities $P(n)$ (that is, (7.3)) electrically by applying Kirchhoff's laws and Ohm's law to $\mathbf{N}^0_e$ and solving for its node voltages, with n_g taken as ground. This is Nash-Williams' rule. With regard to our original infinite, locally finite, 0-connected, sourceless network $\mathbf{N}^0$ with no self-loops, with $\mathcal{N}_f$ being any finite set of nodes in $\mathbf{N}^0$, and with $\mathcal{N}_e \cup \mathcal{N}_f \cup \mathcal{N}_g$ being a partition of the node set of $\mathbf{N}^0$, as before, we can restate this conclusion as follows.

Theorem 7.1-4. (The Nash-Williams Rule.) *For any node n of $\mathcal{N}_f$, the probability (7.3) that a random walker Ψ, after starting from n, will meet $\mathcal{N}_e$ before meeting $\mathcal{N}_g$ is the voltage at n when $\mathcal{N}_e$ is held at 1 V and $\mathcal{N}_g$ is held at 0 V. That voltage is determined by Kirchhoff's laws and Ohm's law.*

7.2 Transfinite Walks

We start by defining a 0-walk on the finitely structured ν-network $\mathbf{N}^\nu$. We have already done so for a 0-walk on a 0-network, but let us repeat that definition in order to add one more condition and to introduce some more terminology.

Definition of a 0-walk: A *0-walk* W^0 is a walk of the conventional sort as defined in Section 7.1; it is an alternating sequence of 0-nodes n_m^0 and branches b_m:

$$W^0 = \{\ldots, n_m^0, b_m, n_{m+1}^0, b_{m+1}, \ldots\} \tag{7.6}$$

such that the following are satisfied.

(a) For each m, b_m is incident to both n_m^0 and n_{m+1}^0.

(b) If the sequence terminates on either side, it terminates at a 0-node.

(c) Other than the terminal 0-nodes, the 0-nodes of (7.6) are maximal. (However, the terminal 0-nodes may be embraced by nodes of higher ranks.)

That the nonterminal nodes of (7.6) are maximal implies that W^0 is restricted to the interior of a 0-subsection except possibly terminally. Furthermore, since there are no self-loops, the adjacent 0-nodes n_m^0 and n_{m+1}^0 are different for each m. Except for this, the elements of W^0 may repeat.

Let us define some terminology regarding a 0-walk W^0. W^0 is called *nontrivial* if it has at least one branch. The *orientation* (*reverse orientation*) of a nontrivial 0-walk W^0 is the direction of increasing (decreasing) indices m. We say that W^0 *embraces* itself, all its elements, all elementary tips embraced by its nodes, and all its subsequences that are 0-walks by themselves. W^0 may either be *finite* or, synonymously, *two-ended* with two terminal nodes, *one-ended* with exactly one terminal node, or *endless* without any terminal node. When W^0 has a terminal node, we say that W^0 *stops at* (*starts at*) its terminal node on the right (on the left). We also say that W^0 *leaves* a terminal node if it starts at that node. The branches in W^0 induce a subgraph, which we shall refer to as the *subgraph traversed by* W^0 or simply as the *subgraph of* W^0. We say that W^0 *meets* a tip, a node, a branch, or another subgraph if its subgraph does so. (Remember that "meeting" is a stronger concept than "reaching"; see Section 2.1.) Thus, W^0 *meets* each of its elements and *passes through* each of its elements other than any terminal node. However, a 0-walk cannot pass through a γ-node n^γ $(\gamma > 0)$; it can only "reach" n^γ either by terminating at a 0-node embraced by n^γ or by proceeding infinitely through an arm for n^γ.

To be more precise about this last idea of "reaching n^γ through an arm," let us choose a contraction $\{\mathcal{W}_p\}_{p=1}^{\infty}$ to a maximal γ-node n^γ and let $\{\mathbf{A}_p\}_{p=1}^{\infty}$ be the corresponding sequence of arms. Also, let us denote one-ended parts of a one-ended or endless 0-walk W^0 by

$$W^0_{m,\infty} = \{n^0_m, b_m, n^0_{m+1}, b_{m+1}, \ldots\}$$

and

$$W^0_{-\infty,m} = \{\ldots, b_{m-2}, n^0_{m-1}, b_{m-1}, n^0_m\}.$$

(Here, too, elements may repeat.) We say that W^0 *stops at* (alternatively, *starts at* or *leaves*) n^γ either if W^0 terminates on the right (resp. on the left) at a 0-node embraced by n^γ or if, for every natural number q, there is an m depending on q such that $W^0_{m,\infty}$ (resp. $W^0_{-\infty,m}$) remains within $\mathbf{A}_q$. In both cases, we also say that W^0 *reaches* n^γ. In the former case, W^0 also meets n^γ. In the latter case, we say that W^0 *eventually lies in* $\mathbf{A}_q$ and *reaches* n^γ *through* $\mathbf{A}_q$.

This definition of "reaching through an arm" does not depend upon the choice of the contraction. Indeed, let $\{\mathbf{A}_p\}_{p=1}^{\infty}$ and $\{\mathbf{A}'_p\}_{p=1}^{\infty}$ be two sequences of arms corresponding to two choices of the contraction. Then, given any q, we can find an r such that $\mathbf{A}'_r \subset \mathbf{A}_q$. This is so because of Condition 4.4-1(a) and the fact that there are only finitely many branches incident to the base $\mathcal{V}_q$ of $\mathbf{A}_q$. Thus, we can choose r so large that none of those branches is in $\mathbf{A}'_r$. But this implies that all of $\mathbf{A}'_r$ is in $\mathbf{A}_q$, for otherwise $\mathcal{W}_q$ would not separate n^γ from the complement $\tilde{\mathbf{A}}_q$ of $\mathbf{A}_q$. Hence, to insure that W^0 eventually lies in $\mathbf{A}_q$, we need merely ascertain that it eventually lies in $\mathbf{A}'_r$ for some sufficiently large r — from which comes the asserted independence from the choice of the contraction.

Example 7.2-1. Consider an endless 0-path P^0 that reaches two 1-nodes n_1^1 and n_2^1, as illustrated in Figure 7.1. Consider also a one-ended 0-walk that progresses back and forth along P^0 in an expanding fashion as follows. (We need merely list the sequence of nodes in order to specify W_a^0.)

$$\begin{aligned} W_a^0 \;=\; & \{n_0^0, n_1^0, n_0^0, n_{-1}^0, n_0^0, n_1^0, n_2^0, n_1^0, \ldots, n_{-1}^0, n_{-2}^0, n_{-1}^0, \ldots, \\ & n_2^0, n_3^0, n_2^0, \ldots, n_{-2}^0, n_{-3}^0, n_{-2}^0, \ldots, \\ & n_m^0, n_{m+1}^0, n_m^0, \ldots, n_{-m}^0, n_{-m-1}^0, n_{-m}^0, \ldots\} \end{aligned}$$

The subgraph of W_a^0 is P^0. Were we to mimic our definition of "meeting" when defining "reaching" for a walk, we would say that W_a^0 "reaches" both n_1^1 and n_2^1 because P^0 does so. However, we have not and will not define the "reaching" of a walk this way but will instead restrict ourselves to the stricter definition given above. Thus, W_a^1 does not reach either n_1^1 or n_2^1 because it does not eventually lie entirely within any given arm for a contraction to n_1^1 or to n_2^1. More generally, a 0-walk cannot reach a γ-node ($\gamma > 0$) if it keeps returning to some 0-node infinitely often before reaching that γ-node.

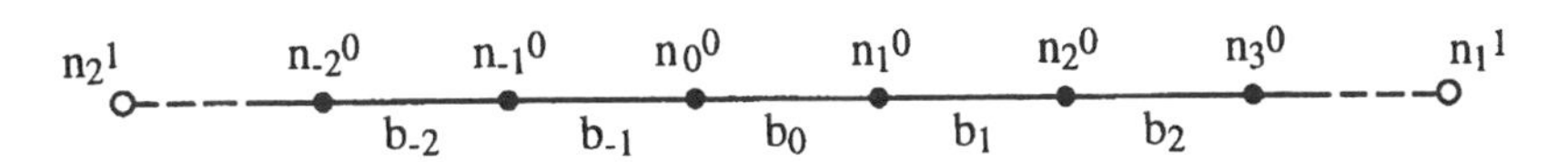

Figure 7.1. An endless 0-path P^0 that reaches the 1-nodes n_1^1 and n_2^1.

(It will follow from Rule 7.3-1 below and Corollary 6.3-8(i) that the probability of a random walker entering a given arm $\mathcal{A}_p$, reaching a base $\mathcal{V}_q$ ($q > p$), and then returning to the base $\mathcal{V}_p$ without reaching the 1-node n^1 incident to that arm is less than 1. Therefore, the probability of it repeating such a transition infinitely often is 0. W_a^0 is such a wholly improbable walk.)

On the other hand, consider a 0-walk W_b^0 that progresses back and forth along P^0 in a more restricted fashion by taking two steps forward and one step backward as follows:

$$\begin{aligned} W_b^0 \;=\; & \{n_0^0, n_1^0, n_2^0, n_1^0, n_2^0, n_3^0, n_2^0, n_3^0, n_4^0, n_3^0, \ldots, \\ & n_{m-1}^0, n_m^0, n_{m+1}^0, n_m^0, n_{m+1}^0, n_{m+2}^0, n_{m+1}^0, \ldots\}. \end{aligned}$$

This 0-walk does eventually lie within every arm for a contraction to n_1^1. Thus, W_b^0 does reach n_1^1 (but not n_2^1). ♣

With 0-walks in hand, we can define β-walks recursively, where now β is any positive natural number. ($\vec{\omega}$-walks and ω-walks will be defined subsequently.)

Definition of a β-walk: A *β-walk* W^β is a two-ended or one-ended or endless alternating sequence:

$$W^\beta = \{\ldots, n_m^\beta, W_m^{\beta-}, n_{m+1}^\beta, W_{m+1}^{\beta-}, \ldots\} \tag{7.7}$$

where, for every m, the following conditions hold:

(a) n_m^β is a maximal β-node — except that if (7.7) terminates on the left or on the right, the terminal element is a node of rank β or less and need not be maximal (that is, the terminal node may be embraced by a node of any higher rank, perhaps higher than β). Moreover, there is at least one β-node.

(b) $W_m^{\beta-}$ is a "nontrivial" α-walk, where $0 \leq \alpha < \beta$ with α depending in general on m. By *nontrivial* we mean that the α-walk "embraces" at least one branch. ("Embracing" is defined below and carries the same meaning it always has.)

(c) Finally, $W_m^{\beta-}$ "stops at" the node on its right and "starts at" the node on its left.

To complete this recursive definition of a β-walk, we have to define "stops at" and "starts at" for a β-walk. This has been done already for a 0-walk.

First of all, the *orientation* (*reverse orientation*) of W^β is in the direction of increasing (decreasing) indices m. If W^β terminates on the right (left), we say that W^β *stops at* (*starts at* or *leaves*) its terminal node. But, there is another way "stopping" or "starting" can occur: We define one-ended portions of a one-ended or endless β-walk W^β as follows:

$$W_{m,\infty}^\beta = \{n_m^\beta, W_m^{\beta-}, n_{m+1}^\beta, W_{m+1}^{\beta-}, \ldots\},$$

$$W_{-\infty,m}^\beta = \{\ldots, W_{m-2}^{\beta-}, n_{m-1}^\beta, W_{m-1}^{\beta-}, n_m^\beta\}.$$

Choose a contraction $\{\mathcal{W}_p\}_{p=1}^\infty$ to some maximal γ-node n^γ, where $\gamma > \beta$, and let $\{\mathbf{A}_p\}_{p=1}^\infty$ be the corresponding sequence of arms. We say that W^β

stops at (*starts at* or *leaves*) n^γ either if $\mathcal{W}^\beta$ terminates on the right (left) at some node embraced by n^γ or if, for every natural number q, there is an m depending on q such that $W^\beta_{m,\infty}$ (resp. $W^\beta_{-\infty,m}$) remains within $\mathbf{A}_q$. In the latter case, we say that for both stopping and starting at n^γ, W^β *eventually lies in* $\mathbf{A}_q$ and *reaches* n^γ *through* $\mathbf{A}_q$.

Finally, if a walk stops at or starts at a node n, we also say that it does so at any node that embraces or is embraced by n. This completes our recursive definition of a β-walk.

The same argument as that used for 0-walks shows that this definition of stopping and starting does not depend upon the choice of the contraction $\{\mathcal{W}_p\}_{p=1}^\infty$.

For a simple example of these ideas, refer to Example 7.2-1. The 0-walk W^0_b can be extended into the 1-walk $\{n^0_0, W^0_b, n^1_1\}$; this walk stops at n^1_1 and starts at n^0_0. However, the 0-walk W^0_a cannot be extended into a 1-walk since it never reaches any 1-node.

Note that a β-walk W^β is perforce restricted to the interior of a $((\beta+1)-)$-subsection except possibly terminally. This is because every nonterminal β-node n^β_m in (7.7) is maximal and moreover, by our recursive definition, because every other node embraced by W^β is also not embraced by a $((\beta+1)+)$-node except possibly for the terminal nodes of W^β.

Some more terminology: W^β is said to *embrace* itself, all its elements, all the elements embraced by its elements, and so forth down to all the elementary tips embraced by its embraced 0-nodes. It will also *embrace* walks that are subsequences of (7.7) or walks that are subsequences of embraced walks of lower ranks. Thus, "embrace" has the same meaning for walks as it does for paths and for transfinite graphs in general.

The branches of a β-walk W^β induce a *subgraph traversed by* W^β — or synonymously, the *subgraph of* W^β. We say that W^β *meets* a tip, a node, a branch, or another subgraph if its subgraph does so. "*Passes through*" is used also in place of "meets" if "stopping at" and "starting at" do not apply. Thus, a β-walk cannot pass through a γ-node n^γ if $\gamma > \beta$; it can only stop or start at n^γ. Furthermore, the β-walk (7.7) is said to perform a *one-step* β*-transition* from n^β_m to n^β_{m+1}.

As our last task in this section, we define $\vec{\omega}$-walks and ω-walks in a finitely structured ω-network $\mathbf{N}^\omega$. (Remember that there are no finitely structured $\vec{\omega}$-networks and no $\vec{\omega}$-nodes in a finitely structured ω-network, according to Lemma 4.5-1.)

Definition of an $\vec{\omega}$-walk. A *one-ended $\vec{\omega}$-walk* $W^{\vec{\omega}}$ is a one-ended alternating sequence of the form

$$W^{\vec{\omega}} = \{n_0^{\beta_0}, W_0^{\beta_1-}, n_1^{\beta_1}, W_1^{\beta_2-}, n_2^{\beta_2}, W_2^{\beta_3-}, \ldots\} \tag{7.8}$$

or of the form

$$W^{\vec{\omega}} = \{\ldots, W_{-3}^{\beta_{-3}-}, n_{-2}^{\beta_{-2}}, W_{-2}^{\beta_{-2}-}, n_{-1}^{\beta_{-1}}, W_{-1}^{\beta_{-1}-}, n_0^{\beta_0}\} \tag{7.9}$$

where, for every index $m = \ldots, -1, 0, 1, \ldots$, the following conditions hold:

(a) The β_m are natural numbers with $\beta_m < \beta_{m+1}$ for all $m = 0, 1, 2, \ldots$ and $\beta_m > \beta_{m+1}$ for all $m = -1, -2, -3, \ldots$.

(b) The nodes $n_m^{\beta_m}$ are maximal β_m-nodes except possibly for $n_0^{\beta_0}$, which need not be maximal.

(c) The $W_m^{\beta_m-}$ are nontrivial α_m-walks, where $0 \le \alpha_m < \beta_m$, that reach $n_m^{\beta_m}$ and $n_{m+1}^{\beta_{m+1}}$. (For example, in (7.9) $W_{-1}^{\beta_{-1}-}$ starts at $n_{-1}^{\beta_{-1}}$ and stops at $n_0^{\beta_0}$.)

An *endless $\vec{\omega}$-walk* is a two-way infinite sequence consisting of (7.9) followed by (7.8) with coincident $n_0^{\beta_0}$.

Because of (b), an $\vec{\omega}$-walk is perforce restricted to an $(\omega-)$-subsection. Also, $W^{\vec{\omega}}$ is automatically *nontrivial* in the sense that each walk $W_m^{\beta_m-}$ is nontrivial.

"Orientation," "stopping," "starting," and "reaching" are defined as before: An $\vec{\omega}$-walk is *oriented* in the direction of increasing indices m. Next, let $W_{m,\infty}^{\vec{\omega}}$ be a one-ended portion of (7.8) of the form

$$W_{m,\infty}^{\vec{\omega}} = \{n_m^{\beta_m}, W_m^{\beta_{m+1}-}, n_{m+1}^{\beta_{m+1}}, W_{m+1}^{\beta_{m+2}-}, \ldots\},$$

where $m \ge 0$, and let $W_{-\infty,-m}^{\vec{\omega}}$ be a one-ended portion of (7.9) of the form

$$W_{-\infty,-m}^{\vec{\omega}} = \{\ldots, W_{-m-2}^{\beta_{-m-2}-}, n_{-m-1}^{\beta_{-m-1}}, W_{-m-1}^{\beta_{-m-1}-}, n_{-m}^{\beta_{-m}}\},$$

where $-m \le 0$. Also, let n^ω be an ω-node. Choose a contraction to n^ω and let $\{\mathcal{A}_p\}_{p=1}^{\infty}$ be its sequence of arms. We shall say that an $\vec{\omega}$-walk $W^{\vec{\omega}}$ *stops at* (*starts at* or *leaves*) n^ω either if it terminates on the right (left) at some node n^μ embraced by n^ω, where μ is a natural number, or if for every

q there exists an m such that $W^{\vec{\omega}}_{m,\infty}$ (resp. $W^{\vec{\omega}}_{-\infty,-m}$) is contained in $\mathcal{A}_q$. In either case, we say that $W^{\vec{\omega}}$ *reaches* n^{ω}.

The ideas of "the subgraph of $W^{\vec{\omega}}$," "embracing," and "meeting" are defined exactly as before.

We turn now to an "ω-walk" in an ω-network $\mathbf{N}^{\omega}$.

Definition of an ω-walk: An *ω-walk* W^{ω} is a (two-ended, one-ended, or endless) alternating sequence of the form

$$W^{\omega} = \{\ldots, n^{\omega}_m, W^{\omega-}_m, n^{\omega}_{m+1}, W^{\omega-}_{m+1}, \ldots\} \qquad (7.10)$$

where the following hold:

(a) Each n^{ω}_m is an ω-node (perforce maximal) except possibly when (7.10) terminates, in which case the terminal element is a node of any rank and possibly nonmaximal. Moreover, there is at least one ω-node in W^{ω}.

(b) Each $W^{\omega-}_m$ is a nontrivial walk whose rank is either a natural number or $\vec{\omega}$.

(c) Each $W^{\omega-}_m$ stops at the node on its right and starts at the node on its left.

"Orientation," "stopping," "starting," "leaving," "reaching," "embracing," "meeting," "nontrivial," "traversed subgraphs," and "one-step ω-transitions" are defined exactly as before; just replace β by ω in the definition for a walk with a natural-number rank β.

Another fact that is of some importance concerns how a β-walk W^{β} (now, $1 \leq \beta \leq \omega$) may reach a node n^{γ}_r from some starting node n^{δ}_s. W^{β} cannot do so if it keeps returning to n^{δ}_s infinitely often before it reaches n^{γ}_r. To say this in another way, we have

Lemma 7.2-2. *Let n^{γ}_r and n^{δ}_s be two maximal nodes. A β-walk W^{β} $(1 \leq \beta \leq \omega)$ can reach n^{γ}_r after starting from n^{δ}_s only if it meets n^{δ}_s at most finitely many times before reaching n^{γ}_r for the first time.*

Proof. Choose a contraction to n^{γ}_r. Then, there is an arm $\mathbf{A}_q$ for n^{γ}_r that does not meet n^{δ}_s. If W^{β} keeps returning to n^{δ}_s infinitely often, there will be no m for which $W^{\beta}_{m,\infty}$ remains within $\mathbf{A}_q$. This implies our conclusion. ♣

This lemma extends the idea illustrated in Example 7.2-1.

7.3 Transfinite Random Walks — The General Idea

Our objective throughout the rest of this work is to establish a theory for transfinite random walks that wander among nodes of various ranks in a permissively finitely structured ν-network $\mathbf{N}^\nu$. As in Section 7.1, it will be convenient at times to speak in terms of a *random walker* Ψ that performs the random walk on $\mathbf{N}^\nu$. We assume throughout that Ψ adheres to the nearest-neighbor rule (7.2) when leaving a maximal 0-node n_0^0.

Our theory for transfinite random walks will be established recursively. A first step of generalization, namely from random 0-walks to random 1-walks, is made in the following way. We start with a truncation of a 0-subsection $\mathbf{S}_b^0$ in the ν-network N^ν. That truncation is along isolating sets in $\mathbf{S}_b^0$, one such set for each bordering $(1+)$-node of $\mathbf{S}_b^0$. The Nash-Williams rule (Theorem 7.1-4), which is a consequence of the nearest-neighbor rule, gives relative probabilities of reaching those isolating sets for a random walker Ψ starting at some node within the truncation. Then, by replacing the isolating sets by contractions, we expand the truncation to fill out $\mathbf{S}_b^0$ — and through a limiting process obtain relative probabilities of transitions to the bordering $(1+)$-nodes of $\mathbf{S}_b^0$. Next, through a similar limiting process, we obtain relative probabilities of transitions from a 1-node n_s^1 to the $(1+)$-nodes that are 1-adjacent to n_s^1. The latter comprise a generalization of the nearest-neighbor rule through an increase in rank from 0 to 1. Finally, by examining a certain Markov chain corresponding to a random 1-walk, we are able to extend the Nash-Williams rule to the wandering of Ψ through a truncation of a 1-subsection. As a particular case, we have an extension of that rule for the wandering of Ψ in a permissively finitely structured 1-network.

We shall recursively extend all this to higher ranks of random walks by defining the probabilities of transitions among nodes of higher ranks in a fashion consistent with the definitions for the lower ranks. We will show that the defined probabilities arise as limiting cases of prior generalizations. However, this does not eliminate the need for those definitions because transfinite random walks whose transition probabilities are not such continuous extensions may be conceivable. In short, we are using the nearest-neighbor rule coupled with the Nash-Williams rule as the paradigm for our transfinite random walks and are thereby basing those walks on the theory of transfinite electrical networks.

There is a critical assumption that we shall always impose, namely that Ψ *roves*. This will mean that Ψ never stops at any node n_s but keeps wandering, and that every walk starting at any chosen node n_s and reaching any other chosen node is a possible walk for Ψ to traverse as Ψ passes through n_s. Thus, a roving Ψ can reach any node from any other node. Whether the probability of doing so is positive is another matter; it may be 0. This may appear to be contradictory at this point of our discussion, for how can something be possible if its probability is 0? The reason for — and the feasibility of — this condition of roving is explained below after we discuss how that 0 probability can arise.

As in Section 7.1, it will be convenient to use a certain concise notation for relative probabilities of transitions. Consider a random walker Ψ roving through $\mathbf{N}^{\nu}$. We say that Ψ *reaches a node set* $\mathcal{N}_t$ if Ψ reaches any node of $\mathcal{N}_t$. Let $\mathcal{N}_e$ and $\mathcal{N}_g$ be two disjoint finite nonvoid sets of maximal nodes in $\mathbf{N}^{\nu}$, and let n_s be another maximal node of $\mathbf{N}^{\nu}$ that is not in $\mathcal{N}_e \cup \mathcal{N}_g$. Then,

$$Prob(\mathrm{s}n_s, \mathrm{r}\mathcal{N}_e, \mathrm{b}\mathcal{N}_g) \tag{7.11}$$

will denote the probability that Ψ, having started at n_s, will reach $\mathcal{N}_e$ before reaching $\mathcal{N}_g$. There is a tacit condition regarding (7.11), namely that Ψ truly reaches $\mathcal{N}_e \cup \mathcal{N}_g$. That Ψ roves makes this possible. But more is involved. When we write the relative transition probability (7.11), it is understood that we are comparing only and all those walks that, after starting from n_s, do reach $\mathcal{N}_e \cup \mathcal{N}_g$ (despite the fact that the probability of Ψ reaching $\mathcal{N}_e \cup \mathcal{N}_g$ may well be 0 if we expand our examination to all walks that start from n_s). This too is further explained below.

Here at last is the generalization of the Nash-Williams rule that will govern the wanderings of Ψ on $\mathbf{N}^{\nu}$.

Rule 7.3-1. *Let $\mathcal{N}_e$ and $\mathcal{N}_g$ be two disjoint finite nonvoid sets of maximal nodes such that $\mathcal{N}_e \cup \mathcal{N}_g$ separates a subgraph $\mathbf{T}$ of $\mathbf{N}^{\nu}$ from $\mathbf{N}^{\nu}\backslash\mathbf{T}$. Let n_s be a node of $\mathbf{T}$ with $n_s \notin (\mathcal{N}_e \cup \mathcal{N}_g)$. Then, the probability (7.11) of Ψ reaching $\mathcal{N}_e$ before reaching $\mathcal{N}_g$, given that Ψ starts at n_s and reaches $\mathcal{N}_e \cup \mathcal{N}_g$, is the node voltage u_s at n_s when $\mathcal{N}_e$ is held at 1 V and $\mathcal{N}_g$ is held at 0 V.*

As a special case, we can let $\mathbf{T}$ be all of $\mathbf{N}^{\nu}$, so that $\mathbf{N}^{\nu}\backslash\mathbf{T}$ is void. In this case, $\mathcal{N}_e$ and $\mathcal{N}_g$ can be any disjoint finite nonvoid sets of maximal nodes.

Let us now be more precise about what we mean by a "transition." A *transition from* n_s *to a node* n *of a finite node set* $\mathcal{N}_t$ is the set of all walks that start at n_s and reach n before reaching $\mathcal{N}_t \backslash \{n\}$. Rule 7.3-1 assigns probabilities to each such transition (replace $\mathcal{N}_e$ by $\{n\}$ and $\mathcal{N}_g$ by $\mathcal{N}_t \backslash \{n\}$). That assignment is Markovian in that it does not depend upon the prior wanderings of Ψ before it starts from n_s. Next, a *transition from* n_s *to* $\mathcal{N}_t$ is the union of all transitions from n_s to the nodes of $\mathcal{N}_t$. It follows that (7.11) is the sum of the probabilities of transitions to the individual nodes of $\mathcal{N}_e$, given that Ψ makes a transition from n_s to $\mathcal{N}_e \cup \mathcal{N}_g$.

Furthermore, let us check the consistency of Rule 7.3-1 under the following circumstance. Let n_s, $\mathcal{N}_e$, $\mathcal{N}_g$, and $\mathbf{T}$ be as stated in Rule 7.3-1, and let $\mathcal{M}$ be another finite set of maximal nodes in T that separates n_s from $\mathcal{N}_e \cup \mathcal{N}_g$ with $n_s \notin \mathcal{M}$ and $\mathcal{M} \cap (\mathcal{N}_e \cup \mathcal{N}_g) = \emptyset$. Assume that Ψ, after starting from n_s, reaches $\mathcal{N}_e \cup \mathcal{N}_g$. Then, Ψ must meet at least one node of $\mathcal{M}$ before reaching $\mathcal{N}_e \cup \mathcal{N}_g$. Let m_i $(i = 1, \ldots, I)$ be the nodes of $\mathcal{M}$. By conditional probabilities, we should have

$$Prob(\mathrm{s}n_s, \mathrm{r}\mathcal{N}_e, \mathrm{b}\mathcal{N}_g) = \sum_{i=1}^{I} Prob(\mathrm{s}n_s, \mathrm{r}m_i, \mathrm{b}(\mathcal{M}\backslash\{m_i\}))Prob(\mathrm{s}m_i, \mathrm{r}\mathcal{N}_e, \mathrm{b}\mathcal{N}_g). \quad (7.12)$$

This equation can be established electrically.

Let u_s (and u_{m_i}) be the voltage at n_s (respectively, at m_i) when $\mathcal{N}_e$ is held at 1 V and $\mathcal{N}_g$ is held at 0 V. Also, let $v_s(i)$ be the voltage at n_s when m_i is held at 1 V and $\mathcal{M}\backslash\{m_i\}$ is held at 0 V. By the superposition principle for electrical networks,

$$u_s = \sum_{i=1}^{I} v_s(i) u_{m_i}. \quad (7.13)$$

By Rule 7.3-1, the voltages in (7.13) correspond to the probabilities in (7.12). This verifies (7.12).

We now turn to the reason why the condition of roving is being imposed upon Ψ. Assume that Ψ is at a maximal β-node n_s^β where $\beta > 0$. Can Ψ leave n_s^β? The answer is "no" — in the following probabilistic sense. Let us assume that the Rule 7.3-1 still governs the wanderings of Ψ as it starts from n_s^β. Then, given any other node, the probability that Ψ will reach that other node before returning to n_s^β is 0.

To show this, choose a permissive contraction to n_s^β. Let $\mathcal{X}_p$ and $\mathcal{X}_q$ be two conjoining sets for that contraction, let $\mathcal{V}_p$ and $\mathcal{V}_q$ be the corresponding bases, and let $p < q$. Also, let $\mathcal{D}$ be the set of nodes that are 0-adjacent to n_s^β. Ψ can leave n_s^β in two ways:

Case 1. Ψ *leaves* n_s^β *along an incident branch.* We are seeking the probability that Ψ will reach $\mathcal{D}$ before reaching $\mathcal{V}_q$ for any q—assuming, of course, that there is at least one branch incident to n_s^β so that $\mathcal{D}$ is nonvoid. By a formal application of Rule 7.3-1, $Prob(\mathrm{s}n_s^\beta, \mathrm{r}\mathcal{D}, \mathrm{b}\mathcal{V}_q)$ is the voltage u_s^β at n_s^β when $\mathcal{D}$ is held at 1 V and $\mathcal{V}_q$ is held at 0 V. By the voltage-divider rule, $u_s^\beta = R_q/(R_d + R_q)$, where R_q is the resistance of the arm $\mathbf{A}_q$ between n_s^β and a short at $\mathcal{V}_q$ and R_d is the parallel resistance of the branches incident to n_s^β. Since the contraction paths in $\mathbf{A}_q$ are permissive, R_q is finite, and moreover $R_q \to 0$ as $q \to \infty$. Therefore, $u_s^\beta \to 0$ as $q \to \infty$. The heuristic interpretation of this is that the fraction representing those walks that leave n_s^β through a branch incident to n_s^β, as compared to all the walks leaving n_s^β along the arm $\mathbf{A}_q$, tends to 0 as $q \to \infty$.

Case 2. Ψ *leaves* n_s^β *along the arm* $\mathbf{A}_q$. That is, Ψ reaches $\mathcal{V}_q$ for some sufficiently large q greater than p before returning to n_s^β. ($\mathcal{D}$ may or may not be void in this case.) Let $n_{q,i}^0$ be any node of $\mathcal{V}_q$. With Ψ starting from $n_{q,i}^0$, how probable is it that Ψ reaches $\mathcal{V}_p$ before returning to n_s^β? To answer this, we formally apply Rule 7.3-1 with $n_s = n_{q,i}^0$, $\mathcal{N}_e = \mathcal{V}_p$, and $\mathcal{N}_g = \{n_s^\beta\}$. This assumes that Ψ does reach $\mathcal{V}_p \cup \{n_s^\beta\}$. Since there is a part of a contraction path that connects $n_{q,i}^0$ to a node of $\mathcal{V}_p$ and does not meet n_s^β, Corollary 6.3-8 asserts that $Prob(\mathrm{s}n_{q,i}^0, \mathrm{r}\mathcal{V}_p, \mathrm{b}n_s^\beta)$ is positive. The only other possibility is that Ψ wanders indefinitely without ever reaching $\mathcal{V}_p$ or n_s^β; in this case it is a true statement that Ψ does not reach $\mathcal{V}_p$ before returning to n_s^β.

We now argue that $Prob(\mathrm{s}n_{q,i}^0, \mathrm{r}\mathcal{V}_p, \mathrm{b}n_s^\beta)$ tends to 0 as $q \to \infty$ (for instance, $n_{q,i}^0$ can shift along a contraction path to approach n_s^β). That probability is the voltage $u_{q,i}$ at $n_{q,i}^0$ when $\mathcal{V}_p$ is held at 1 V and n_s^β is held at 0 V. But, by Theorems 6.3-3 and 6.3-4 and Property 6.3-1(b), $u_{q,i} \to 0$ as $q \to \infty$. The heuristic interpretation of this is that the fraction representing those walks that start from $n_{q,i}^0$ and reach $\mathcal{V}_p$ before reaching n_s^β, as compared to the walks that start from $n_{q,i}^0$ and reach n_s^β before reaching $\mathcal{V}_p$, tends to 0 as $q \to \infty$.

Both cases taken together mean that only a vanishingly small fraction of the random walks that start at n_s^β will reach $\mathcal{X}_p$ without first returning to n_s^β, whatever be p. This in turn implies the following: Once Ψ reaches a

β-node n_s^β, the probability that Ψ will reach any other maximal node before returning to n_s^β is 0. Moreover, a 0-walk is defined as a sequence, and therefore there is no 0-walk in which Ψ returns to n_s^β infinitely many times and then reaches $\mathcal{X}_p$. Furthermore, consider the case where $\mathcal{X}_p$ is replaced by a γ-node n^γ $(\gamma > 0)$. Our arguments concerning probabilities hold just as well in this case, for we can choose $\mathcal{X}_p$ to isolate n^γ from n_s^β. Now, for Ψ to reach n^γ along a walk of rank greater than 0, it cannot keep returning to n_s^β infinitely many times before reaching n^γ (Lemma 7.2-2). All of this implies that the probability of Ψ ever reaching a node other than n_s^β is 0. That is, Ψ is stuck at n_s^β — at least in a probabilistic sense.

However, this does not mean that there are no walks starting at n_s^β and reaching $\mathcal{V}_p$ or $\mathcal{D}$. It simply means that we are dealing with the exceptional case when we consider such walks. Such walks are always available to a roving Ψ. In short, we are free to restrict our attention to walks like that and to discard all others. We can then assign relative probabilities for transitions from a β-node to other nodes in accordance with Rule 7.3-1.

Let us expand upon this matter, for it may still be disconcerting that we are dealing with an event whose probability is 0. Is it true that "an event whose probability is 0 does not happen"? As a rough analogy, consider the interval between 0 and 1 and let its probability measure be Lebesgue measure. Then, the probability assigned to the set of rational numbers is 0, but we are not forced to conclude that rational numbers "do not happen." For the purposes of a subsequent analysis, we are free to restrict our attention to the set of rational numbers between 0 and 1 and to replace Lebesgue measure by a new probability measure whereby that set of rational numbers has probability 1. This is exactly what is happening when we assert that "Ψ roves." In particular, given a finite set $\mathcal{N}_t$ of maximal nodes with $n_s^\beta \notin \mathcal{N}_t$, we are discarding those walks that, after leaving n_s^β, never reach $\mathcal{N}_t$. Thus, we are making it a certainty that Ψ will reach $\mathcal{N}_t$. Under this restriction, we will show in subsequent sections that Rule 7.3-1 extends the nearest-neighbor rule to transfinite random walks in a continuous fashion.

Actually, we claim even more, namely that our imposition of the roving condition is nothing other than what has been tacitly assumed in the conventional theory of random walks on 0-networks under the nearest-neighbor rule whenever the electrical analog has been used. To see this, consider a random walk on a sourceless 0-network $\mathbf{N}^0$ under the nearest-neighbor rule. Then, the multistep transition probabilities are governed by the Nash-Williams rule, which treats the random walk as an electrical phenomenon.

Now, a physical resistor r is often a medium of resistive material, in which case it can be divided into a series circuit of many resistors summing to r. In fact, a mathematical model of that resistor can be divided into an endless path having the same total resistance r. The replacement of every resistor in $\mathbf{N}^0$ by such an endless path changes $\mathbf{N}^0$ into a 1-network $\mathbf{N}^1$, and the original random walk on $\mathbf{N}^0$ becomes a random 1-walk on $\mathbf{N}^1$ — but only if Ψ roves. Otherwise, the original random walk on $\mathbf{N}^0$ will remain stuck at a 0-node if we view each resistor as a continuous distribution of resistive material. Thus, the nearest-neighbor rule tacitly imposes a roving condition on Ψ. It throws away every walk that starts down a resistor but never reaches the adjacent 0-node.

(Let us mention in passing that there is another approach to the condition of roving [41]. When Ψ leaves n_s^β and reaches a finite set $\mathcal{N}_t$ of $(\beta+)$-nodes, it may return to n_s^β several times before reaching $\mathcal{N}_t$ (see Lemma 7.2-2 again). We can simply ignore those returns and consider a transition from n_s^β to $\mathcal{N}_t$ that does not return to n_s^β until possibly after it reaches $\mathcal{N}_t$. Walks fulfilling this condition may be called "β-roving." But, this then leads to a hierarchy of roving walks of increasing ranks, and the resulting discussion becomes more complicated than the approach adopted in this book.)

Our recursive procedure for extending Rule 7.3-1 from lower ranks to higher ranks can be summarized as follows. We start with any truncation of a 0-subsection, in which case Rule 7.3-1 is simply the Nash-Williams rule (Theorem 7.1-4). That rule can be extended continuously to any truncation of a 1-subsection through a specific procedure E, as was indicated in the second paragraph of this section. Then, given any rank β ($1 \leq \beta \leq \omega, \beta \neq \vec{\omega}$), we assume that Rule 7.3-1 holds for every $(\beta-)$-subsection. Upon applying the procedure E, we extend Rule 7.3-1 to any arbitrary β-subsection. This is the content of Sections 7.4 through 7.7 below, and procedure E is completely specified therein. Actually, the first step of extension from rank 0 to rank $\beta = 1$ is exactly the same as that for the general case from the ranks $\beta-$ to rank β, and therefore it need not be made separately — just set $\beta = 1$ in procedure E to get it. Moreover, procedure E works just as well when $\beta = \omega$ as it does when β is a natural number. All this establishes Rule 7.3-1 for every truncation of every β-subsection and, in fact, for the entire network $\mathbf{N}^\nu$. With the case $\beta = \omega$ in hand, we can proceed in exactly the same way to extend Rule 7.3-1 to still higher ranks, but we stop at $\beta = \omega$.

Finally, before leaving this section, let us discuss a difficulty concerning transfinite random walks as temporal processes. Conventional random walks usually are viewed as discrete time series whose individual steps occur at uniformly displaced instants of time. On the other hand, a transfinite random walk entails in general infinitely many steps. Were we to assign the same transit time to each step, the random walker might never complete a transition between nodes of higher ranks within a finite time interval. We can suggest three ways of handling this matter.

First of all, we could simply ignore it. There is no precept dictating that a transfinite walk must be a temporal process. We can expunge time and view transfinite random walks as being nontemporal abstractions.

Second, we could allow uniform time steps were we to admit transfinite transit times. This might be rationalized by the following argument. In dynamic analyses, it is not unusual to allow time to pass to infinity in order to determine final states. So, if time can be stretched out infinitely, why not transfinitely? (After all, this has been done with ordinals.) In fact, we can borrow some nonstandard analysis [23] (for introductory treatments, see [12] or [16]) to sum all the time steps into a hyperreal transit time.

A third way of imbedding transfinite random walks into time is as follows. Let us make the time of passage through any branch proportional to — perhaps equal to — the resistance of the branch. If a walk happens to be a permissive path, then the time taken by that walk will be finite. Moreover, by virtue of Lemma 6.1-2, every transition between two given nodes of arbitrary ranks will contain at least one permissive path and therefore at least one walk of finite duration. However, in general there will be other walks in that transition that are of infinite duration. In any case, there will be a finite infimum for the durations of all the walks in the transition. We may therefore assign that infimum as the "duration infimum" of a transition and try to work with that quantity in order to examine the temporal behavior of transfinite random walks.

7.4 Reaching a Bordering Node

We now establish Rule 7.3-1 for the case where the random walker Ψ starts from an internal node of a subsection and reaches a bordering node of that subsection. The analysis is the same whatever be the rank of the subsection. So, let β be any rank such that $1 \leq \beta \leq \nu \leq \omega, \beta \neq \vec{\omega}$. Consider any $(\beta-)$-subsection $\mathbf{S}_b^{\beta-}$ in $\mathbf{N}^\nu$ and let α be the internal rank of $\mathbf{S}_b^{\beta-}$. Thus, $\mathbf{S}_b^{\beta-} = \mathbf{S}_b^\alpha$. We wish to obtain relative probabilities of transitions from an internal maximal node n_s of $\mathbf{S}_b^\alpha$ to the various $(\beta+)$-nodes $n_k^{\beta+}$ $(k = 1, \ldots, K)$ bordering $\mathbf{S}_b^\alpha$ — and possibly to other internal nodes of $\mathbf{S}_b^\alpha$ as well. Figure 7.2 illustrates our symbolism. Let $n_k^{\beta+}$ $(k = 1, \ldots, K)$ be the bordering nodes of $\mathbf{S}_b^\alpha$. Choose a permissive contraction $\{\mathcal{W}_{k,p_k}\}_{p_k=1}^\infty$ within $\mathbf{S}_b^\alpha$ for each $n_k^{\beta+}$. We define $\mathbf{F}(p_1, \ldots, p_K)$ as the subnetwork of $\mathbf{S}_b^\alpha$ induced by all branches of $\mathbf{S}_b^\alpha$ that do not reside in the arms $\mathbf{A}_{1,p_1}, \ldots, \mathbf{A}_{K,p_K}$ corresponding to the isolating sets $\mathcal{W}_{1,p_1}, \ldots, \mathcal{W}_{K,p_K}$. This also is illustrated in Figure 7.2. For sufficiently large $p_1, \ldots, p_K$, the rank of $\mathbf{F}(p_1, \ldots, p_K)$ will be α too (see Lemma 4.4-2). $\mathbf{F}(p_1, \ldots, p_K)$ may be incident to a bordering node of $\mathbf{S}_b^\alpha$ through (at most finitely many) branches. Furthermore, $\mathbf{F}(p_1, \ldots, p_K)$ will be α-connected so long as the arms $\mathbf{A}_{k,p_k}$ are chosen small enough (i.e., the p_k are chosen large enough). Indeed, if this were not so, some branches of some cuts for the bordering nodes of $\mathbf{S}_b^\alpha$ would not be connected through paths that do not meet those bordering nodes — in violation of Lemma 4.1-3(iv).

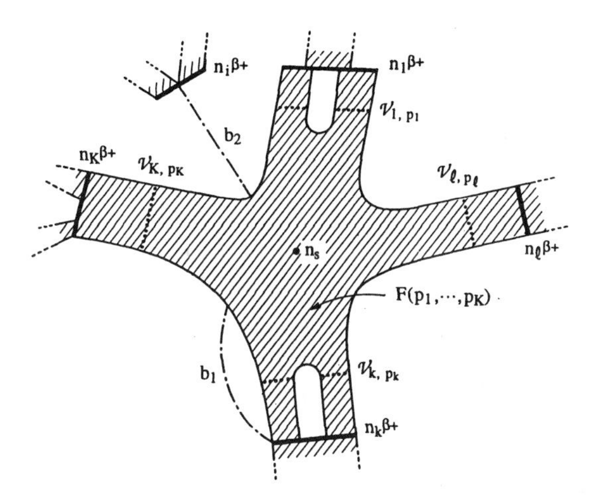

Figure 7.2. An α-subsection $\mathbf{S}_b^\alpha$. The heavy lines denote the $(\beta+)$-nodes bordering $\mathbf{S}_b^\alpha$. The dot-dash lines denote branches b_1 and b_2 of $\mathbf{S}_b^\alpha$ incident to bordering $(\beta+)$-nodes. $\mathcal{V}_{k,p_k}$ denotes and arm base in $\mathbf{S}_b^\alpha$ corresponding to the $(\beta+)$-node $n_k^{\beta+}$; it is indicated by a line of dots. $\mathbf{F}(p_1, \ldots, p_k)$ is the subnetwork within the boundary consisting of the $\mathcal{V}_{k,p_k}$ $(k = 1, \ldots, K)$; $\mathbf{F}(p_1, \ldots, p_K)$ contains the branches b_1 and b_2. ($\mathcal{W}_{k,p_k}$ is $\mathcal{V}_{k,p_k} \cup \{n_k^0\}$, where n_k^0 is the 0-node embraced by $n_k^{\beta+}$ and incident to $\mathbf{S}_b^\alpha$. In general, either $\mathcal{V}_{k,p_k}$ may be void or n_k^0 may be absent, but not both.)

When the p_k are chosen so large that $\mathbf{F}(p_1, \ldots, p_K)$ is of rank α and is α-connected, we call $\mathbf{F}(p_1, \ldots, p_K)$ a *truncation of* $\mathbf{S}_b^\alpha$.

Note also that the truncation $\mathbf{F}(p_1, \ldots, p_K)$ will have only finitely many α-nodes, for otherwise it would be incident to an $(\alpha+1)$-node according to Theorem 4.2-4; the latter possibility was eliminated by the deletion of the arms $\mathbf{A}_{k,p_k}$. Altogether then, we have it that $\mathbf{F}(p_1, \ldots, p_K)$ is permissively finitely structured as an α-network.

In order to recursively generalize the Nash-Williams rule, let us now assume that Rule 7.3-1 has been established for the case where $\mathbf{T}, \mathcal{N}_e, \mathcal{N}_g$, and n_s are chosen as follows. $\mathbf{T}$ is the truncation $\mathbf{F}(p_1, \ldots, p_K)$ of $\mathcal{S}_b^\alpha$. $\mathcal{N}_e$ and $\mathcal{N}_g$ are two nonvoid, disjoint, finite sets of maximal nodes in $\mathbf{F}(p_1, \ldots, p_K)$ such that, for each k, $\mathcal{W}_{k,p_k}$ lies entirely within $\mathcal{N}_e$ or entirely within $\mathcal{N}_g$ — and stays there as $p_k \rightarrow \infty$. Thus, all the $\mathcal{W}_{k,p_k}$ lie in $\mathcal{N}_e \cup \mathcal{N}_g$, but $\mathcal{N}_e \cup \mathcal{N}_g$ may contain other nodes as well. Finally, n_s is a maximal node of $\mathbf{F}(p_1, \ldots, p_K)$ with $n_s \notin (\mathcal{N}_e \cup \mathcal{N}_g)$. (When $\beta = 1$, that is, when $\alpha = 0$, Rule 7.3-1 is the Nash-Williams rule (Theorem 7.1-4) for these choices of $\mathbf{T}, \mathcal{N}_e, \mathcal{N}_g$, and n_s.) We also assume that Ψ roves; thus, Ψ does not stop at any node and can reach $\mathcal{N}_e \cup \mathcal{N}_g$.

We wish to extend Rule 7.3-1 by expanding the truncation $\mathbf{F}(p_1, \ldots, p_K)$ to fill out $\mathcal{S}_b^\alpha$. To this end, let $\mathcal{M}_e$ be the set obtained by deleting from $\mathcal{N}_e$ all nodes of each set $\mathcal{W}_{k,p_k}$ contained in $\mathcal{N}_e$ and adding in their place $n_k^{\beta+}$. Also, construct $\mathcal{M}_g$ from $\mathcal{N}_g$ in the same way. (All the $\mathcal{W}_{k,p_k}$ may be in $\mathcal{N}_e$, in which case $\mathcal{M}_g = \mathcal{N}_g$; similarly, $\mathcal{M}_e = \mathcal{N}_e$ if all the $\mathcal{W}_{k,p_k}$ are in $\mathcal{N}_g$.) Also, replace $\mathbf{F}(p_1, \ldots, p_K)$ by $\mathbf{S}_b^\alpha$. Thus, n_s can be any internal maximal node of $\mathcal{S}_b^\alpha$. Let u_s be the voltage at n_s when $\mathcal{M}_e$ is held at 1 V and $\mathcal{M}_g$ is held at 0 V. Through a formal application of Rule 7.3-1, we have $u_s = Prob(\mathrm{s}n_s, \mathrm{r}\mathcal{M}_e, \mathrm{b}\mathcal{M}_g)$, a result we wish to obtain through a limiting process.

In the next lemma, we send $p_k \rightarrow \infty$ for every k. It is understood that $\mathcal{N}_e \cup \mathcal{N}_g$ is adjusted accordingly; that is, the nodes of $\mathcal{M}_e \cup \mathcal{M}_g$ that are not bordering nodes of $\mathbf{S}_b^\alpha$ are fixed nodes of $\mathcal{N}_e \cup \mathcal{N}_g$, and all other nodes of $\mathcal{N}_e \cup \mathcal{N}_g$ are those of the $\mathcal{W}_{k,p_k}$; as the p_k increase, the nodes in the bases $\mathcal{V}_{k,p_k}$ are changed accordingly. Furthermore, let us denote the voltage v_s at n_s, when $\mathcal{N}_e$ is held at 1 V and $\mathcal{N}_g$ is held at 0 V, by $v_s(p_1, \ldots, p_K)$ in order to display its dependence upon the p_k.

Lemma 7.4-1. *$v_s(p_1, \ldots, p_K)$ converges to u_s as the $p_1, \ldots, p_K$ tend to infinity independently.*

Proof. For each $k = 1, \ldots, K$, let $n^0_{k,p_k,i}$ denote the ith node in $\mathcal{W}_{k,p_k}$, and let $u_{k,p_k,i}$ denote the corresponding node voltage resulting from 1 V at $\mathcal{M}_e$ and 0 V at $\mathcal{M}_g$. Corollary 6.3-7 as applied to $\mathbf{S}^{\alpha}_{b}$ implies that $0 \leq u_{k,p_k,i} \leq 1$ for all k and i. By superposition, $v_s(p_1, \ldots, p_K) - u_s$ is the voltage at n_s resulting from the following application of node voltages for all k and i: $n^0_{k,p_k,i}$ is held at $1 - u_{k,p_k,i}$ if $n^{\beta+}_k \in \mathcal{M}_e$, and $n^0_{k,p_k,i}$ is held at $-u_{k,p_k,i}$ if $n^{\beta+}_k \in \mathcal{M}_g$. All other nodes of $\mathcal{M}_e$ and $\mathcal{M}_g$ are held at 0 V.

Now, let u_{max} be the maximum of all the voltages $1 - u_{k,p_k,i}$ at the nodes $n^0_{k,p_k,i}$ in the $\mathcal{W}_{k,p_k}$ corresponding to all the $n^{\beta+}_k \in \mathcal{M}_e$, and let u_{min} be the minimum of all the voltages $-u_{k,p_k,i}$ at the nodes $n^0_{k,p_k,i}$ in the $\mathcal{W}_{k,p_k}$ corresponding to all the $n^{\beta+}_k \in \mathcal{M}_g$. We have $u_{max} \geq 0$ and $u_{min} \leq 0$. Moreover, u_{max} and u_{min} depend in general on $p_1, \ldots, p_K$. Since $\mathbf{F}(p_1, \ldots, p_K)$ is a linear network, Corollary 6.3-7 as applied to $\mathbf{F}(p_1, \ldots, p_K)$ also implies that

$$u_{min} \leq v_s(p_1, \ldots, p_K) - u_s \leq u_{max}.$$

Recall that for each fixed k but varying p_k, the cardinalities of the $\mathcal{W}_{k,p_k}$ are uniformly bounded (in fact, are no larger than $m+1$, where m is the number of contraction paths for the chosen contraction to $n^{\beta+}_k$). Hence, by Property 6.3-1(b), which holds in this case according to Theorems 6.3-3 and 6.3-4, we have that $u_{min} \to 0$ and $u_{max} \to 0$ as the $p_1, \ldots p_K$ tend to infinity independently. ♣

The last lemma immediately yields the following extension of Rule 7.3-1.

Theorem 7.4-2. *Assume that Rule 7.3-1 holds for every truncation $\mathbf{F}(p_1, \ldots, p_K)$ of the α-subsection S^{α}_{b}. As the $p_k \to \infty$, the voltages of Rule 7.3-1 extend continuously from those of $\mathbf{F}(p_1, \ldots, p_K)$ to those of $\mathbf{S}^{\alpha}_{b}$ and thereby yield the following result for a corresponding extension of the probabilities of Rule 7.3-1: Let n_s be an internal maximal node of $\mathbf{S}^{\alpha}_{b}$, and let $\mathcal{M}_e$ and $\mathcal{M}_g$ be nonvoid, disjoint, finite sets of maximal nodes with $\mathcal{M}_e \cup \mathcal{M}_g$ containing all the maximal bordering nodes of $\mathbf{S}^{\alpha}_{b}$ and with $n_s \notin (\mathcal{M}_e \cup \mathcal{M}_g)$. Then, for a roving Ψ and given that Ψ reaches $\mathcal{M}_e \cup \mathcal{M}_g$, $Prob(\mathrm{s}n_s, \mathrm{r}\mathcal{M}_e, \mathrm{b}\mathcal{M}_g)$ is the voltage at n_s when $\mathcal{M}_e$ is held at 1 V and $\mathcal{M}_g$ is held at 0 V.*

We take this continuous extension of Rule 7.3-1 as the definition of the probabilities of transitions from an internal node of an α-subsection $\mathbf{S}_b^\alpha$ to the bordering $(\beta+)$-nodes of $\mathbf{S}_b^\alpha$ and possibly to other nodes as well.

Let us now consider the probability of Ψ reaching $\mathcal{M}_e \cup \mathcal{M}_g$. We shall call an α-subsection $\mathbf{S}_b^\alpha$ *transient* if the following two conditions are satisfied:

(a) For any arbitrarily chosen maximal 0-node n_s^0 of $\mathbf{S}_b^\alpha$, there is a positive probability that the roving Ψ, after starting from n_s^0, will reach a bordering node of $\mathbf{S}_b^\alpha$ before returning to n_s^0.

(b) For any arbitrarily chosen maximal δ-node n_s^δ $(0 < \delta \leq \alpha)$ of $\mathbf{S}_b^\alpha$ and for every conjoining set $\mathcal{X}$ for n_s^δ, there is at least one node n_a^0 of $\mathcal{X}$ with a positive probability that the roving Ψ, having started from n_a^0, will reach a bordering node of $\mathbf{S}_b^\alpha$ before reaching n_s^δ.

Part (a) of this definition subsumes the conventional definition of transiency for an infinite 0-connected 0-network $\mathbf{N}^0$. Indeed, that definition demands that there be a positive probability that Ψ, after starting from n_s^0, will never return to n_s^0 [27, page 17]. Now, we can append a bordering node to $\mathbf{N}^0$ by making it the set of all 0-tips for $\mathbf{N}^0$. Then, a walk that starts at n_s^0 and reaches that bordering node before returning to n_s^0 will pass through infinitely many branches to get to that bordering node and thereby "will never return to n_s^0" in conventional terms.

With regard to part (b), note that since Ψ roves, it will continue on from n_s^δ to reach some other maximal node of $\mathbf{S}_b^\alpha$. But, since $\mathbf{S}_b^\alpha$ is finitely structured, this implies that Ψ will certainly reach some conjoining set $\mathcal{X}$ for n_s^0 in $\mathbf{S}_b^\alpha$. In this case, however, the said positive probability can get arbitrarily small if the base $\mathcal{V}$ of $\mathcal{X}$ is arbitrarily close to n_s^δ. In other words, Ψ can escape from n_s^δ if it roves; however, the same cannot be said for a nonroving Ψ at n_s^δ — that Ψ may get stuck at n_s^δ in the sense that the probability of Ψ leaving n_s^δ is 0.

Theorem 7.4-3. *Assume that Ψ roves. Then, every α-subsection $\mathbf{S}_b^\alpha$ of $\mathbf{N}^\nu$ is transient.*

Proof. Let n_s be any internal maximal node of $\mathbf{S}_b^\alpha$, and assume that Ψ starts at n_s. If n_s is a maximal 0-node, then by the nearest-neighbor rule and for each node n_a that is 0-adjacent to n_s, the probability is positive that Ψ will reach n_a in one 0-step. If n_a is embraced by an $((\alpha+1)+)$-node, then the theorem follows immediately. So, assume the latter is not the case for all nodes 0-adjacent to n_s when n_s is a maximal 0-node.

Next, let n_s be a maximal δ-node $(1 \leq \delta \leq \alpha)$. As was pointed out above, the roving of Ψ implies that Ψ will continue on from n_s to reach some conjoining set $\mathcal{X}$ for n_s in $\mathcal{S}_b^\alpha$. Choose $n_a \in \mathcal{X}$ arbitrarily. There is a path (possibly just a single branch) that meets n_s and n_a and does not meet $\mathcal{X}\backslash\{n_a\}$. Hence, by Rule 7.3-1 and by Corollary 6.3-8, $Prob(\mathrm{s}n_s, \mathrm{r}n_a, \mathrm{b}\mathcal{X}\backslash\{n_a\})$ is positive, whatever be the choice of $n_a \in \mathcal{X}$.

Now, consider both cases where either n_s is an maximal 0-node with $\mathcal{X}$ being the set of nodes adjacent to n_s or n_s is a maximal δ-node $(1 \leq \delta \leq \alpha)$ with $\mathcal{X}$ being a conjoining set for n_s. In order to show that $\mathbf{S}_b^\alpha$ is transient, we need only show that $Prob(\mathrm{s}n_a, \mathrm{r}\mathcal{N}_b, \mathrm{b}n_s)$ is positive, where now $\mathcal{N}_b$ is taken to be the set of bordering nodes of $\mathbf{S}_b^\alpha$. By Rule 7.3-1 again — as extended by Theorem 7.4-2, the latter can be accomplished by showing that for some choice of $n_a \in \mathcal{X}$, the voltage at n_a is positive when n_s is held at 0 V and when all the bordering nodes of $\mathbf{S}_b^\alpha$ are held at 1 V. But this follows from Corollary 6.3-8 because there is a path that connects a node of $\mathcal{X}$ to a bordering node and does not meet n_s. ♣

7.5 Leaving a β-Node

Having examined probabilities as Ψ reaches a bordering $(\beta+)$-node, let us now consider probabilities as the roving Ψ leaves a maximal β-node n_s^β. Again the arguments are the same for all ranks β $(1 \leq \beta \leq \omega, \beta \neq \vec{\omega})$. This time, let $\mathcal{N}_e$ and $\mathcal{N}_g$ be two nonvoid, disjoint, finite sets of maximal $(\beta-)$-nodes such that $\mathcal{N}_e \cup \mathcal{N}_g$ separates n_s^β from all other $(\beta+)$-nodes and $n_s^\beta \notin \mathcal{N}_e \cup \mathcal{N}_g$. (See Figure 7.3.) Since $\mathbf{N}^\nu$ is finitely structured, we can choose a permissive contraction $\{\mathcal{W}_p\}_{p=1}^\infty$ to n_s^β such that no node of $\mathcal{N}_e \cup \mathcal{N}_g$ lies in the first arm $\mathbf{A}_1$ and thereby in any arm $\mathbf{A}_p$ of that contraction.

Next, let $\mathbf{M}$ be the subnetwork induced by all the branches that are not separated from n_s^β by $\mathcal{N}_e \cup \mathcal{N}_g$. (In Figure 7.3, $\mathbf{M}$ is the subnetwork lying within the ring of nodes comprising $\mathcal{N}_e \cup \mathcal{N}_g$.) The only node of $\mathbf{M}$ whose rank is no less than β is n_s^β. We wish to know $Prob(\mathrm{s}n_s^\beta, \mathrm{r}\mathcal{N}_e, \mathrm{b}\mathcal{N}_g)$. However, we cannot as yet use Rule 7.3-1 because our recursive hypothesis with respect to Rule 7.3-1 is presently assumed only for a starting node n_s of rank less than β. What we can do, however, is first short the arm $\mathbf{A}_p$ to obtain a single maximal 0-node n_p^0 and then examine $Prob(\mathrm{s}n_p^0, \mathrm{r}\mathcal{N}_e, \mathrm{b}\mathcal{N}_g)$. As $p \to \infty$, the $\mathcal{V}_p$ contract to n_s^β, and hopefully the corresponding $Prob(\mathrm{s}n_p^0,$

r$\mathcal{N}_e$, b$\mathcal{N}_g$) converge (they will). The limit can then be taken as the definition of $Prob(\text{s}n_s^\beta, \text{r}\mathcal{N}_e, \text{b}\mathcal{N}_g)$.

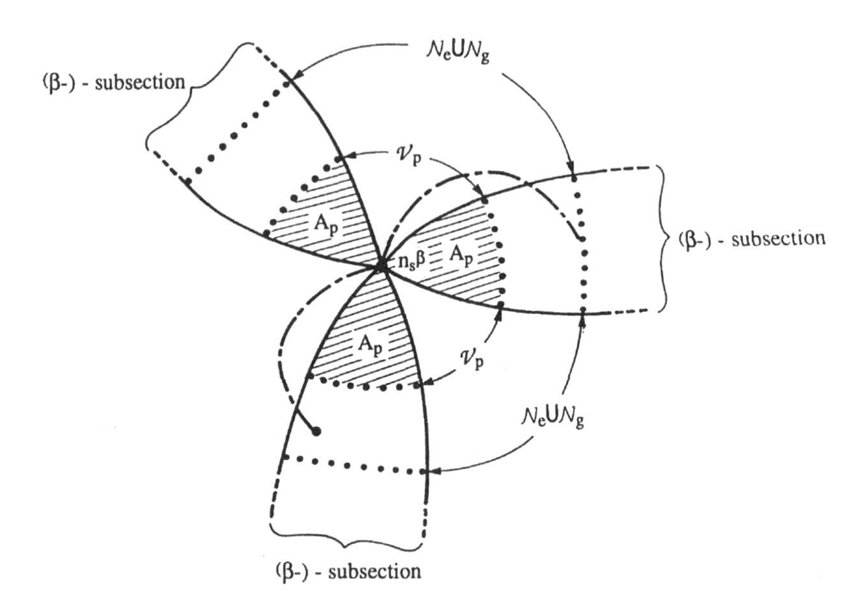

Figure 7.3. A β-node n_s^β with its incident $(\beta-)$-subsections. Branches incident to n_s^β are denoted by dot-dash lines. The base $\mathcal{V}_p$ is denoted by a ring of dots. The isolating set $\mathcal{W}_p$ is $\mathcal{V}_p$ along with the 0-node embraced by n_s^β if that 0-node exists. The arm $\mathbf{A}_p$ is shown cross-hatched. $\mathcal{N}_e \cup \mathcal{N}_g$ is also shown by a ring of dots. $\mathcal{N}_e \cup \mathcal{N}_g$ does not meet $\mathbf{A}_p$ and separates $\mathcal{W}_p$ from all the $(\beta+)$-nodes other than n_s^β; moreover, $n_s^\beta \notin \mathcal{N}_e \cup \mathcal{N}_g$.

To proceed, for our recursive argument, we now assume that Rule 7.3-1 has been established for a roving Ψ for all ranks less than β. Remember that for a 0-network Rule 7.3-1 is the Nash-Williams rule (Theorem 7.1-4). Let $\mathbf{M}_p$ be the network obtained from $\mathbf{M}$ by replacing every branch of $\mathbf{A}_p$ by a short. That shorting creates a maximal 0-node n_p^0 and eliminates n_s^β. Thus, the internal rank of $\mathbf{M}_p$ is α, where $\alpha < \beta$. So, we can apply Rule 7.3-1 to $\mathbf{M}_p$ with $n_s = n_p^0$ to get $Prob(\text{s}n_p^0, \text{r}\mathcal{N}_e, \text{b}\mathcal{N}_g)$ as the voltage v_p at n_p^0 when $\mathcal{N}_e$ is held at 1 V and $\mathcal{N}_g$ is held at 0 V. On the other hand, we can formally apply Rule 7.3-1 to $\mathbf{M}$ to get $Prob(\text{s}n_s^\beta, \text{r}\mathcal{N}_e, \text{b}\mathcal{N}_g)$ as the voltage u_s^β at n_s^β under the same excitation. In this section, we will show that $v_p \rightarrow u_s^\beta$ as $p \rightarrow \infty$. This will prove that Rule 7.3-1 extends continuously to a formula for determining $Prob(\text{s}n_s^\beta, \text{r}\mathcal{N}_e, \text{b}\mathcal{N}_g)$ with respect to $\mathbf{M}$.

The needed excitation can be produced by a pure 1-V voltage source $e_0 = 1$ in a branch b_0 appended to $\mathbf{M}$ or $\mathbf{M}_p$; b_0 is connected from a short at $\mathcal{N}_g$ to a short at $\mathcal{N}_e$. Let $\mathbf{M}_a$ (and $\mathbf{M}_{pa}$) be the result of so appending b_0 to $\mathbf{M}_p$ (and $\mathbf{M}_{pa}$). In accordance with Theorem 6.2-4, we obtain a unique voltage-current regime $\mathbf{v}, \mathbf{i}$ in $\mathbf{M}_a$ and another one $\mathbf{v}_p, \mathbf{i}_p$ in $\mathbf{M}_{pa}$. We have $\mathbf{i} \in \mathcal{K}_a$ (and $\mathbf{i}_p \in \mathcal{K}_{pa}$), where $\mathcal{K}_a$ (and $\mathcal{K}_{pa}$) is the Hilbert space indicated in Theorem 6.2-4 with respect to the network $\mathbf{M_a}$ (and $\mathbf{M}_{pa}$) in place of $\mathbf{N}_a^\nu$.

We can extend the current vector $\mathbf{i}_p \in \mathcal{K}_{pa}$ for $\mathbf{M}_{pa}$ into a current vector $\mathbf{i}_p^e$ for $\mathbf{M}_a$ as follows. On $\mathbf{M}_{pa}$, $\mathbf{i}_p^e$ and $\mathbf{i}_p$ agree. Next, let $\mathbf{C}_p$ be the cut at $\mathcal{W}_p$. $\mathbf{C}_p$ resides in $\mathbf{M}_{pa}$. Let n_k be any node of $\mathcal{V}_p$, and let P_k be that part of a contraction path that connects n_k to n_s^β. (If more than one contraction path passes through n_k, choose one of them for P_k.) Let $c_{p,k}$ be the algebraic sum of the currents in the branches of $\mathbf{C}_p$ that are incident to n_k. (Measure those currents as directed toward n_k.) Assign to P_k the current flow of value $c_{p,k}$ directed toward n_s^β. Do the same thing for every node of $\mathcal{V}_p$ to obtain a current flow from each node of $\mathcal{V}_p$ along a contraction path to n_s^β. In $\mathbf{M}_{pa}$, $\mathbf{C}_p$ is simply the finite set of branches incident to the maximal 0-node n_p^0. By virtue of Lemma 6.2-2, Kirchhoff's current law is satisfied at $\mathbf{C}_p$, and it follows that the algebraic sum of all the flows $c_{p,k}$ is 0. As for those branches of $\mathbf{A}_p$ that are not in any P_k, let their currents be 0. In this way we extend $\mathbf{i}_p \in \mathcal{K}_{pa}$ into a current vector $\mathbf{i}_p^e$ for $\mathbf{M}_a$. We will prove below that $\mathbf{i}_p^e \in \mathcal{K}_a$.

On the other hand, $\mathbf{v}_p$ is the branch-voltage vector for $\mathbf{M}_{pa}$ dictated by Theorem 6.2-4; that is, $v_{p0} = -e_0$ and, when $j \neq 0$ and the branch b_j lies in $\mathbf{M}_{pa}$, $v_{p,j} = r_j i_{p,j}$. We extend $\mathbf{v}_p$ into a voltage vector $\mathbf{v}_p^e$ for $\mathbf{M}_a$ simply by assigning 0 V as the branch voltage for each branch in $\mathbf{A}_p$. Note that for the regime $\mathbf{v}_p^e, \mathbf{i}_p^e$, Ohm's law is not satisfied along the paths P_k, but for our purposes this will be of no concern.

Given any branch-voltage vector $\mathbf{w}$ and any current vector $\mathbf{s}$ for $\mathbf{M}_a$, we define a coupling of $\mathbf{w}$ and $\mathbf{s}$ by $\langle \mathbf{w}, \mathbf{s} \rangle = \sum w_j s_j$, where the summation is over the indices of the branches in $\mathbf{M}_a$. Our next objective is to show that

$$\langle \mathbf{v} - \mathbf{v}_p^e, \mathbf{i} - \mathbf{i}_p^e \rangle \;=\; 0. \tag{7.14}$$

That $\langle \mathbf{v}, \mathbf{i} \rangle = 0$ follows directly from the application of Theorem 6.2-4 to $\mathbf{M}_a$; see (6.5). Similarly, upon applying that theorem to $\mathbf{M}_{pa}$, we get $\langle \mathbf{v}_p, \mathbf{i}_p \rangle = 0$. Since $\mathbf{v}_p^e$ vanishes throughout $\mathbf{A}_p$, we also have $\langle \mathbf{v}_p^e, \mathbf{i}_p^e \rangle = 0$.

We now wish to show that

$$\langle \mathbf{v}, \mathbf{i}_p^e \rangle = 0. \tag{7.15}$$

For this purpose, we use the space $\mathcal{K}_a$ indicated in Theorem 6.2-4, but now with $\mathbf{N}_a^\nu$ replaced by $\mathbf{M}_a$. If we can show that $\mathbf{i}_p^e \in \mathcal{K}_a$, then (7.15) will follow from (6.5) with $\mathbf{s} = \mathbf{i}_p^e$.

Lemma 7.5-1. $\mathbf{i}_p^e \in \mathcal{K}_a$.

Proof. Upon applying Theorem 6.2-4 for $\mathbf{M}_{pa}$, we obtain the current vector $\mathbf{i}_p \in \mathcal{K}_{pa}$; $\mathbf{i}_p^e$ extends $\mathbf{i}_p$ as described above. Assume at first that $\mathbf{i}_p \in \mathcal{K}_{pa}^0$, where $\mathcal{K}_{pa}^0$ is the span of basic currents in $\mathbf{M}_{pa}$. Thus, the flow of $\mathbf{i}_p$ through the single maximal 0-node n_p^0, obtained by shorting $\mathbf{A}_p$ and thereby $\mathcal{W}_p$, can be represented by a finite number of loop currents in $\mathbf{M}_{pa}$; thus, the currents in the branches of the cut $\mathbf{C}_p$ arise from the superposition of finitely many loop currents. If such a loop current passes along a short between two different nodes of $\mathcal{W}_p$, that loop current can be taken to flow — not along that short — but instead along one or two contraction paths between $\mathcal{W}_p$ and n_s^β, and thereby from one node of $\mathcal{W}_p$ through n_s^β to another node of $\mathcal{W}_p$. (Possibly n_s^β embraces a 0-node of $\mathcal{W}_p$, in which case just one contraction path might be traversed.) With such an alteration for every one of the said loop currents, we will obtain $\mathbf{i}_p^e$. Moreover, this procedure will convert any basic current for $\mathbf{M}_{pa}$ into a basic current for $\mathbf{M}_a$. We can conclude that $\mathbf{i}_p^e \in \mathcal{K}_a^0$.

Next, assume that $\mathbf{i}_p \in \mathcal{K}_{pa}$. The flow through the aforementioned short (i.e., through the maximal 0-node n_p^0) can again be represented by finitely many loop currents in $\mathbf{M}_{pa}$. Once again, each such loop current can be extended into a loop current passing through n_s^β via one or two contraction paths. The result is an extension $\mathbf{i}_p^e$ of $\mathbf{i}_p \in \mathcal{K}_{pa}$.

Moreover, $\mathbf{i}_p$ is the limit in $\mathcal{K}_{pa}$ of a sequence of current vectors in $\mathcal{K}_{pa}^0$, each of which can be extended in the same way into a member of $\mathcal{K}_a^0$. Furthermore, convergence in $\mathcal{K}_{pa}$ implies branchwise convergence. Since the current $c_{p,k}$ in each path P_k is the algebraic sum of the currents in the finitely many cut-branches that are incident to the node $n_k \in \mathcal{V}_p$ at which P_k terminates, we get branchwise convergence on P_k as well.

We now argue that, since each P_k is permissive, we get convergence on P_k in accordance with the norm of $\mathcal{K}_a$. Indeed, corresponding to a sequence $\{\mathbf{i}_{p,l}\}_{l=1}^\infty$ of vectors in $\mathcal{K}_{pa}^0$ that converges to $\mathbf{i}_p \in \mathcal{K}_{pa}$, we have a sequence

$\{\mathbf{i}_{p,l}^e\}_{l=1}^{\infty}$ of extended current vectors in $\mathcal{K}_a^0$ converging branchwise to $\mathbf{i}_p^e$ and also a sequence $\{c_{k,l}\}_{l=1}^{\infty}$ of current flows in P_k that converge to c_k. With $\sum_{P_k}$ denoting a summation over the branch indices j for P_k, we have

$$\sum_{P_k} r_j (i_{p,j}^e - i_{p,l,j}^e)^2 \;=\; (c_k - c_{k,l})^2 \sum_{P_k} r_j. \tag{7.16}$$

By the permissivity of P_k, $\sum_{P_k} r_j < \infty$. Consequently, the right-hand side and thereby the left-hand side of (7.16) tend to 0 as $l \to \infty$. But that left-hand side arises from the restriction of the squared norm of $\mathcal{K}_a$ to the branch currents of P_k, from which comes our assertion.

Moreover, the norm for $\mathcal{K}_{pa}$ is the restriction of the norm for $\mathcal{K}_a$ to the branches of $\mathbf{M}_{pa}$. We can conclude that $\mathbf{i}_p^e$ is the limit in $\mathcal{K}_a$ of a sequence of vectors in $\mathcal{K}_a^0$. Hence, $\mathbf{i}_p^e \in \mathcal{K}_a$. ♣

As was indicated above, Lemma 7.5-1 establishes (7.15).

To finally establish (7.14), we have to show that

$$\langle \mathbf{v}_p^e, \mathbf{i} \rangle \;=\; 0 \tag{7.17}$$

whenever $\mathbf{i} \in \mathcal{K}_a$. By definition, $\mathbf{v}_p^e$ vanishes throughout $\mathbf{A}_p$. Hence, the left-hand side of (7.17) is equal to $\langle \mathbf{v}_p, \mathbf{i}_s \rangle$, where $\mathbf{v}_p$ is the voltage vector dictated by Theorem 6.2-4 as applied to $\mathbf{M}_{pa}$ and $\mathbf{i}_s$ is the restriction of $\mathbf{i} \in \mathcal{K}_a$ to the branches of $\mathbf{M}_{pa}$.

Lemma 7.5-2. $\mathbf{i}_s \in \mathcal{K}_{pa}$.

Proof. Again let $\mathbf{C}_p$ be the cut at $\mathcal{W}_p$. Now, any loop current in $\mathbf{M}_a$ that passes through some branches of $\mathbf{C}_p$ can be truncated into a loop current in $\mathbf{M}_{pa}$ that flows along the short at $\mathcal{W}_p$. Thus, when $\mathbf{i} \in \mathcal{K}_a^0$, we have $\mathbf{i}_s \in \mathcal{K}_{pa}^0$. We can now use a limiting process as in the preceding proof to conclude that $\mathbf{i}_s \in \mathcal{K}_{pa}$ whenever $\mathbf{i} \in \mathcal{K}_a$. ♣

By Theorem 6.2-4 as applied to $\mathbf{M}_{pa}$ (in particular, by (6.5) with $\mathbf{s} = \mathbf{i}_s$), $\langle \mathbf{v}_p, \mathbf{i}_s \rangle = 0$. This establishes (7.17). Altogether then, we have established (7.14).

The components of $\mathbf{v}$ and $\mathbf{v}_p^e$ for the source branch b_0 are both $-e_0 = -1$. Thus, the left-hand side of (7.14) can be written as a summation just for the branches in $\mathbf{M}$. In fact, upon rearranging that summation into two sums, one over the branch indices for $\mathbf{M} \backslash \mathbf{A}_p$ and the other for $\mathbf{A}_p$, and using the branch conductances g_j, we can rewrite (7.14) as

$$\sum_{\mathbf{M}\backslash\mathbf{A}_p} g_j(v_j - v^e_{p,j})^2 + \sum_{\mathbf{A}_p}(v_j - v^e_{p,j})(i_j - i^e_{p,j}) = 0.$$

Since $\mathbf{v}^e_p$ vanishes on $\mathbf{A}_p$, the second summation is equal to

$$\begin{aligned}\sum_{\mathbf{A}_p} v_j(i_j - i^e_{p,j}) &= \sum_{\mathbf{A}_p} g_j v_j^2 - \sum_{\mathbf{A}_p} v_j i^e_{p,j} \\ &= \sum_{\mathbf{A}_p} g_j(v_j - v^e_{p,j})^2 - \sum_{\mathbf{A}_p} v_j i^e_{p,j}.\end{aligned}$$

Thus, (7.14) becomes

$$\sum_{\mathbf{M}} g_j(v_j - v^e_{p,j})^2 = \sum_{\mathbf{A}_p} v_j i^e_{p,j}.$$

By Schwarz's inequality,

$$\sum_{\mathbf{M}} g_j(v_j - v^e_{p,j})^2 = \sum_{\mathbf{A}_p} \sqrt{g_j} v_j \sqrt{r_j} i^e_{p,j} \leq \left[\sum_{\mathbf{A}_p} g_j v_j^2 \sum_{\mathbf{A}_p} r_j (i^e_{p,j})^2\right]^{1/2}. \tag{7.18}$$

Since the voltage-current regime for $\mathbf{M}$ is of finite power, $\sum_{\mathbf{A}_p} g_j v_j^2 \to 0$ as $p \to \infty$. Our next objective is to show that $\sum_{\mathbf{A}_p} r_j (i^e_{p,j})^2$ remains bounded as $p \to \infty$.

For this purpose, we use the current-maximum principle of Theorem 6.3-10 and the resistance-monotonicity law of Lemma 6.2-6 as follows. Let h_a of Theorem 6.3-10 be the current h_p in the pure voltage source $e_0 = 1$ V appended to $\mathbf{M}_p$ between the short n_1 at $\mathcal{N}_e$ and the short n_2 at $\mathcal{N}_g$. We are free to view that source as a pure current source of value h_p. Next, choose any node n^0_k of $\mathcal{V}_p$ and let it be the node n_3 of Theorem 6.3-10. The node n_4 of that theorem will be the short between the nodes of $\mathcal{V}_p \backslash \{n^0_k\}$. Then, upon shorting n_3 and n_4, we obtain $\mathbf{M}_p$ again; moreover, the current in that short between between n_3 and n_4 is equal to the current $c_{p,k}$ in the path P_k used when we were constructing $\mathbf{i}^e_p$. By Theorem 6.3-10, $|c_{p,k}| \leq |h_p|$. Remember that within $\mathbf{A}_p$, $i^e_{p,j}$ equals $c_{p,k}$ for every branch in P_k, this being so for every chosen contraction path P_k; moreover, $i^e_{p,j}$ equals 0 for every branch of $\mathbf{A}_p$ not on some P_k. We can conclude that the component $i^e_{p,j}$ of $\mathbf{i}^e_p$ for each branch b_j in $\mathbf{A}_p$ satisfies $|i^e_{p,j}| \leq |h_p|$.

Next, let p_0 be a fixed natural number and let $p > p_0$. We get $\mathbf{M}_{p_0}$ from $\mathbf{M}_p$ by shorting the branches in $\mathbf{A}_{p_0} \backslash \mathbf{A}_p$. Refer to Lemma 6.2-6 and the paragraph preceding it, wherein symbols are defined. We can let $\mathbf{M}_p$ take the role of $\mathbf{N}^{\nu}$ and can let $\mathbf{M}_{p_0}$ take the role of $\mathbf{N}'$. Let z (and z') be the input resistance for $\mathbf{M}_p$ (resp. for $\mathbf{M}_{p_0}$) at n_1 and n_2. By Lemma 6.2-6, $z \geq z' > 0$. Since $|h_p| = 1/z$ and $|h_{p_0}| = 1/z'$, we have $|h_p| \leq |h_{p_0}|$ for all $p > p_0$. Consequently, $|i^e_{p,j}| \leq |h_p| \leq |h_{p_0}|$ for every branch b_j in $\mathbf{A}_p$ and for all $p > p_0$. Since all the contraction paths P_k are permissive, we can conclude that the last summation $\sum_{\mathbf{A}_p} r_j (i^e_{p,j})^2$ in (7.18) remains bounded as $p \to \infty$. Consequently, the right-hand side of (7.18) tends to 0 as $p \to \infty$. Therefore, so too does its left-hand side.

Finally, let u^β_s be the voltage at n^β_s under a formal application of Rule 7.3-1 to $\mathbf{M}$, as before. Also, for $\mathbf{M}_p$, let v_p be the voltage at the node n^0_p produced by shorting all of $\mathbf{A}_p$, again as before. Let P be a permissive path in $\mathbf{M}$ between the node n^β_s and the short at $\mathcal{N}_g$. Since $\mathbf{v}^e_p$ vanishes on $\mathbf{A}_p$, we can write the following, where $\sum_P$ is a summation over the branch indices for P:

$$|u^\beta_s - v_p| = \left| \sum_P \pm v_j - \sum_P \pm v^e_{p,j} \right| = \left| \sum_P \pm \sqrt{g_j}(v_j - v^e_{p,j})\sqrt{r_j} \right|$$

$$\leq \left[\sum_P g_j (v_j - v^e_{p,j})^2 \sum_P r_j \right]^{1/2} \leq \left[\sum_M g_j (v_j - v^e_{p,j})^2 \sum_P r_j \right]^{1/2}.$$

Since P is permissive, $\sum_P r_j < \infty$. Thus, $v_p \to u^\beta_s$ as $p \to \infty$. This is what we needed to show in order to justify an application of Rule 7.3-1 for a random walker Ψ starting at a β-node and reaching $\mathcal{N}_e \cup \mathcal{N}_g$, when $\mathcal{N}_e$ and $\mathcal{N}_g$ are specified as in the first paragraph of this section. We summarize all this through

Theorem 7.5-3. *For n^β_s, $\mathcal{N}_e$, $\mathcal{N}_g$, and $\mathbf{A}_p$ defined as in the first paragraph of this section and with n^0_p being the 0-node obtained by shorting all of $\mathbf{A}_p$, the voltages of Rule 7.3-1 extend continuously as $p \to \infty$ from the case where Ψ starts at n^0_p and reaches $\mathcal{N}_e \cup \mathcal{N}_g$ to the case where Ψ starts at n^β_s and reaches $\mathcal{N}_e \cup \mathcal{N}_g$. That is, for a roving Ψ and as $p \to \infty$,*

$$Prob(\mathrm{s}n^0_p, \mathrm{r}\mathcal{N}_e, \mathrm{b}\mathcal{N}_g) \to Prob(\mathrm{s}n^\beta_s, \mathrm{r}\mathcal{N}_e, \mathrm{b}\mathcal{N}_g).$$

In short, we are defining the limiting voltage $\lim_{p\to\infty} v_p = u^\beta_s$ as the value for $Prob(\mathrm{s}n^\beta_s, \mathrm{r}\mathcal{N}_e, \mathrm{b}\mathcal{N}_g)$ for a roving Ψ that reaches $\mathcal{N}_e \cup \mathcal{N}_g$ after starting from n^β_s.

7.6 From a β-Node to a β-Adjacent $(\beta+)$-Node

So far in our recursive development we have examined transitions from within a $(\beta-)$-subsection to a bordering $(\beta+)$-node and from a β-node to within a $(\beta-)$-subsection. As the next step, we discuss transitions from a β-node to its β-adjacent $(\beta+)$-nodes. The argument now needed is the same as that leading up to Theorem 7.4-2.

Let the roving random walker Ψ start at the β-node n_s^β, and let $n_k^{\beta+}$ $(k = 1, \ldots, K)$ be the $(\beta+)$-nodes that are β-adjacent to n_s^β. (See Figure 7.4.) Choose a permissive contraction $\{\mathcal{W}_{k,p_k}\}_{p_k=1}^{\infty}$ for each $n_k^{\beta+}$ within the $(\beta-)$-subsections incident to both n_s^β and $n_k^{\beta+}$. (There may be more than one such $(\beta-)$-subsection, in which case $\mathcal{W}_{k,p_k}$ is the union of the corresponding isolating sets.) Also, let $\mathbf{H}(p_1, \ldots, p_K)$ be the subnetwork induced by all branches that are not separated from n_s^β by $\bigcup_{k=1}^{K} \mathcal{W}_{k,p_k}$. $\mathbf{H}(p_1, \ldots, p_K)$ is of rank β because it contains n_s^β — the one and only $(\beta+)$-node in it. Moreover, $\mathbf{H}(p_1, \ldots, p_K)$ will be β-connected when the $p_1, \ldots, p_K$ are chosen sufficiently large. $\mathbf{H}(p_1, \ldots, p_K)$ plays the same role as the α-network $\mathbf{F}(p_1, \ldots, p_K)$ did before. We define $\mathcal{N}_e$ and $\mathcal{N}_g$ as two finite disjoint sets of nodes within $\mathbf{H}(p_1, \ldots, p_K)$ such that $n_s^\beta \notin (\mathcal{N}_e \cup \mathcal{N}_g)$ and, for each k, $\mathcal{W}_{k,p_k}$ lies entirely within $\mathcal{N}_e$ or alternative within $\mathcal{N}_g$. (It will be understood that as $p_k \to \infty$, $\mathcal{W}_{k,p_k}$ remains in $\mathcal{N}_e$ or in $\mathcal{N}_g$; the latter sets adjust accordingly.) Furthermore, $\mathcal{M}_e$ and $\mathcal{M}_g$ are two finite disjoint sets of nodes that are the same as those of $\mathcal{N}_e$ and $\mathcal{N}_g$, respectively, except that for each k the nodes of $\mathcal{W}_{k,p_k}$ are replaced by the single node $n_k^{\beta+}$.

Under the present meanings of these symbols, we can use the argument of Section 7.4 — virtually word for word — to get Lemma 7.4-1 again and thereby the following extension of Rule 7.3-1.

Theorem 7.6-1. *As the $p_k \to \infty$, the voltages of Rule 7.3-1 extend continuously to a limiting case, whereby the result achieved in Theorem 7.5-3 extends continuously to the following result: For a roving Ψ that reaches $\mathcal{M}_e \cup \mathcal{M}_g$ after starting from n_s^β, $Prob(\mathrm{s}n_s^\beta, \mathrm{r}\mathcal{M}_e, \mathrm{b}\mathcal{M}_g)$ is the voltage at n_s^β when $\mathcal{M}_e$ is held at 1 V and $\mathcal{M}_g$ is held at 0 V.*

Once again, we take this as the definition of probabilities of transitions from a β-node to its β-adjacent $(\beta+)$-nodes.

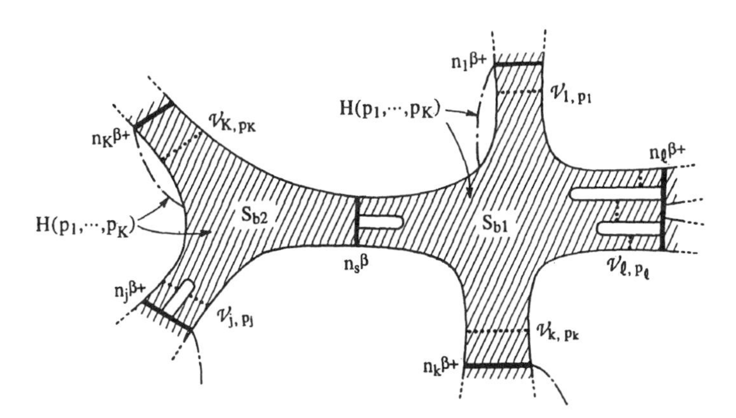

Figure 7.4. A β-node n_s^β, its incident $(\beta-)$-subsections $\mathbf{S}_{b1}$ and $\mathbf{S}_{b2}$, and its β-adjacent $(\beta+)$-nodes $n_1^{\beta+}, \ldots n_K^{\beta+}$. The notation is the same as that used in Figure 7.2, but now $\mathbf{H}(p_1, \ldots p_K)$ replaces $\mathbf{F}(p_1, \ldots, p_K)$.

7.7 Wandering on a ν-Network

In Section 7.4 we started our recursive development of transfinite random walks by assuming that Rule 7.3-1 holds for any truncation of a $(\beta-)$-subsection. Now, we will complete one cycle of that recursion by showing that Rule 7.3-1 holds for any truncation $\mathbf{T}$ of a β-subsection $\mathcal{S}_b^\beta$. This is all we need to do in order to establish that rule for all ranks β with $1 \leq \beta \leq \nu$ because the case where β is replaced by 0 is the established Nash-Williams rule (Theorem 7.1-4). A consequence of this will be that Rule 7.3-1 will become completely established as the governing equation for transfinite random walks on permissively finitely structured ν-networks ($1 \leq \nu \leq \omega, \nu \neq \vec{\omega}$). This generalizes nearest-neighbor random walks on finite 0-networks.

We proceed by setting up a "surrogate" 0-network $\mathbf{T}^{\beta \mapsto 0}$, whose behavior mimics that of $\mathbf{T}$ in a certain way, and then by applying the Nash-Williams rule to $\mathbf{T}^{\beta \mapsto 0}$. $\mathbf{T}^{\beta \mapsto 0}$, in turn, is obtained from a Markov chain encompassing the probabilities for transitions among finitely many maximal nodes of $\mathbf{T}$. Two cases arise. If $\beta = \nu$, we take $\mathbf{T}$ to be $\mathbf{N}^\nu$ itself. Indeed, we are free to view $\mathbf{N}^\nu$ as a truncation of a ν-subsection in a larger $(\nu + 1)$-network. On the other hand, if $\beta < \nu$, we truncate $\mathcal{S}_b^\beta$ as follows. Let $n_l^{(\beta+1)+}$ $(l = 1, \ldots, L)$ be the maximal bordering nodes of $\mathbf{S}_b^\beta$. Choose an isolating set $\mathcal{W}_l$ within $\mathbf{S}_b^\beta$ for each $n_l^{(\beta+1)+}$, and let $\mathbf{T}$ be the subnetwork induced by all branches of $\mathbf{S}_b^\beta$ that are not in the arms corresponding to the $\mathcal{W}_l$. As before, it is understood that the $\mathcal{W}_l$ are chosen sufficiently close to the bordering nodes to ensure that $\mathbf{T}$ is β-connected. $\mathbf{T}$ will have only finitely many β-nodes; this fact is a consequence of Lemma 4.2-4. We now treat $\mathbf{T}$ as a network by itself; that is, we open all branches that are not in $\mathbf{T}$.

As usual, let n_s be any maximal node of $\mathbf{T}$, and let $\mathcal{N}_e$ and $\mathcal{N}_g$ be two disjoint finite sets of maximal nodes of $\mathbf{T}$ such that $n_s \notin (\mathcal{N}_e \cup \mathcal{N}_g)$ and $\mathcal{N}_e \cup \mathcal{N}_g$ contains $\bigcup_{l=1}^L \mathcal{W}_l$ and possibly other nodes as well. Furthermore, let $\mathcal{M}$ be a finite set of maximal nodes in $\mathbf{T}$ consisting of n_s, all the nodes of $\mathcal{N}_e \cup \mathcal{N}_g$, and all the β-nodes of $\mathbf{T}$. $\mathcal{M}$ will be the state space of a certain Markov chain. Two nodes n_a and n_b of $\mathcal{M}$ will be called *$\mathcal{M}$-adjacent* if they are totally disjoint and if there is a path in $\mathbf{T}$ that terminates at n_a and n_b and does not meet any other node of $\mathcal{M}$. Thus, each node n_a of $\mathcal{M}$ will have a unique set $\mathcal{M}_a$ of nodes $\mathcal{M}$-adjacent to n_a.

Lemma 7.7-1. *If two nodes of $\mathcal{M}$ are $\mathcal{M}$-adjacent, then they are also β-adjacent.*

Proof. Let n_a and n_b be $\mathcal{M}$-adjacent nodes of $\mathcal{M}$. If n_a and n_b are not β-adjacent, they are not incident to the same $(\beta-)$-subsection. Consequently, every path in $\mathbf{T}$ connecting n_a and n_b must meet a β-node distinct from n_a and n_b. Since $\mathcal{M}$ contains all the β-nodes of $\mathbf{T}$, n_a and n_b are not $\mathcal{M}$-adjacent. ♣

Let us say that Ψ makes a *one-step $\mathcal{M}$-transition from n_a to n_b* if $n_b \in \mathcal{M}_a$ and if Ψ starts at n_a and reaches n_b before reaching any other node of $\mathcal{M}_a$. To construct our desired Markov chain, we need the probabilities $P_{a,b}$ of the one-step $\mathcal{M}$-transitions. By virtue of Lemma 7.7-1, we can apply to $\mathbf{T}$ Rule 7.3-1 as extended by Theorem 7.6-1 to get those probabilities.

But, first of all, we set $P_{a,a} = 0$. Thus, even though Ψ, after starting from n_a, may return to n_a finitely many times before reaching $\mathcal{M}_a$ (Lemma 7.2-2), we will simply ignore such returns when setting up our Markov chain. Thus, we restrict ourselves to the one-step $\mathcal{M}$-transitions from n_a to its $\mathcal{M}$-adjacent nodes when assigning positive probabilities to the one-steps of the Markov chain. (In this case, the roving of Ψ means that Ψ is restricted to walks that do reach $\mathcal{M}_a$ from n_a.)

Next, let n_a and n_b be any two nodes of $\mathcal{M}$, where now n_a and n_b need not be $\mathcal{M}$-adjacent. We let $P_{a,b}$ be the probability of a one-step $\mathcal{M}$-transition from n_a to n_b. $P_{a,b} = 0$ if n_a and n_b are not $\mathcal{M}$-adjacent because it is then impossible for Ψ to make a one-step $\mathcal{M}$-transition from n_a to n_b. On the other hand, if there is only one node n_b in $\mathcal{M}_a$, we have $P_{a,b} = 1$. However, if there are many nodes in $\mathcal{M}_a$, we use the extension of Rule 7.3-1 given by Theorem 7.6-1 to obtain the probability $P_{a,b}$ of a one-step $\mathcal{M}$-transition from n_a to n_b in $\mathcal{M}_a$. In this last case, we have $0 < P_{a,b} < 1$ by virtue of Corollary 6.3-8 and the fact that there is a path from n_a to each node of $\mathcal{M}_a$ that does not meet any other node of $\mathcal{M}_a$. Finally, to conclude that we have the one-step probabilities of a Markov chain, we have to show that $\sum P_{a,b} = 1$, where the summation is for all $n_b \in \mathcal{M}_a$. This follows immediately from the fact that we are restricting our roving Ψ to walks that do reach $\mathcal{M}_a$ after starting from n_a. It can also be shown electrically as follows: Measure the voltage u_a at n_a when one node n_b of $\mathcal{M}_a$ is held at 1 V and when $\mathcal{M}_a \backslash \{n_b\}$ is held at 0 V. Then sum the various values of u_a obtained as n_b varies through $\mathcal{M}_a$. By the superposition principle, that sum is the voltage at n_a when all of $\mathcal{M}_a$ is held at 1 V and all the other nodes of $\mathbf{T}$ float (i.e., are not incident to any source). Consequently, that sum equals 1. This confirms our assertion.

Thus, for any choice as specified above of the finite set $\mathcal{M}$, we have a Markov chain with $\mathcal{M}$ as its state space. We denote that chain by $\mathrm{M}(\mathcal{M})$. We can examine the wanderings of Ψ among the nodes of $\mathcal{M}$ by analyzing $\mathrm{M}(\mathcal{M})$ — but only up to the point where Ψ arrives at a node of $\bigcup_{l=1}^{L} \mathcal{W}_l$. After that Ψ may leave $\mathbf{T}$ when wandering in $\mathbf{N}^\nu$, in which case $\mathrm{M}(\mathcal{M})$ is no longer relevant. For our purposes, this is of no concern because we are interested only in the wanderings of Ψ from the point where it starts at a node n_s of $\mathbf{T}$ up to the point where it first reaches $\mathcal{N}_e \cup \mathcal{N}_g$. The last set contains $\bigcup_{l=1}^{L} \mathcal{W}_l$. Also, note that when $\mathbf{T} = \mathbf{N}^\nu$, $\bigcup_{l=1}^{L} \mathcal{W}_l$ is void.

Theorem 7.7-2. *The Markov chain* $\mathrm{M}(\mathcal{M})$ *is irreducible and reversible.*

Proof. The case where $\mathcal{M}$ has just two nodes is trivial. So, let $\mathcal{M}$ have more than two nodes.

For any two $\mathcal{M}$-adjacent nodes n_a and n_b of $\mathcal{M}$, $P_{a,b} > 0$ — as was noted above. The irreducibility of $\mathrm{M}(\mathcal{M})$ now follows from the β-connectedness of $\mathbf{T}$ and the fact that $\mathcal{M}$ is a finite set of nodes in $\mathbf{T}$. Indeed, between any two nodes of $\mathcal{M}$, there is a path connecting them and containing only finitely many nodes of $\mathcal{M}$. Therefore, there is a positive probability that Ψ will traverse that path.

As for reversibility, we start by recalling the definition of a *cycle in* $\mathcal{M}$ (see Appendix C5). This is a finite sequence

$$C = (n_1, n_2, \ldots, n_c, n_{c+1} = n_1)$$

of nodes n_k in $\mathcal{M}$ with the following properties: All the nodes of C are distinct except for the first and last; there are at least three nodes in C (i.e., $c > 2$); consecutive nodes in C are $\mathcal{M}$-adjacent. A Markov chain is *reversible* if and only if, for every cycle C, the product $\prod_{k=1}^{c} P_{k,k+1}$ of $\mathcal{M}$-transition probabilities $P_{k,k+1}$ from n_k to n_{k+1} remains the same when every $P_{k,k+1}$ is replaced by $P_{k+1,k}$. Thus, we need only show that

$$P_{1,2}P_{2,3}\cdots P_{c,1} \;=\; P_{1,c}\cdots P_{3,2}P_{2,1}. \tag{7.19}$$

According to Rule 7.3-1 as extended by Theorem 7.6-1, $P_{k,k+1}$ is the voltage u_k at n_k obtained by holding n_{k+1} at 1 V and by holding all the other nodes of $\mathcal{M}$ that are $\mathcal{M}$-adjacent to n_k at 0 V. For this situation, u_k will remain unchanged when the voltages at still other nodes of $\mathcal{M}$ are held at any chosen voltages.

To simplify notation, let us denote n_k by m_0 and n_{k+1} by m_1. Thus, m_0 and m_1 are $\mathcal{M}$-adjacent. Also, let $m_2, \ldots, m_K$ denote all the nodes of $\mathcal{M}$ that are different from n_k and n_{k+1} but are $\mathcal{M}$-adjacent to either n_k or n_{k+1} or both. Since n_k and n_{k+1} are members of a cycle having at least three nodes, we have $K \geq 2$. Now, consider the K-port obtained from $\mathbf{T}$ by choosing m_k, m_0 as the pair of terminals for the kth port ($k = 1, \ldots, K$) with m_0 being the common ground for all ports. To obtain the required node voltages for measuring $P_{k,k+1}$, we externally connect a 1-V voltage source to m_1 from all of the $m_2, \ldots, m_K$, with m_0 left floating (i.e., m_0 has no external connections). The resulting voltage u_0 at m_0 is $P_{k,k+1}$.

With respect to m_0 (taken as the ground node), the voltage at m_1 is $1 - u_0$ and the voltage at m_k ($k = 2, \ldots, K$) is $-u_0$. Moreover, with i_k denoting the current entering m_k ($k = 1, \ldots, K$), the sum $i_1 + \cdots + i_K$ is 0. (Apply Kirchhoff's current law at m_1.) Furthermore, the port currents and voltages are related by $\mathbf{i} = Y\mathbf{u}$, where $\mathbf{i} = (i_1, \ldots, i_K)$, $\mathbf{u} = (1 - u_0, -u_0, \ldots, -u_0)$, and $Y = [Y_{a,b}]$ is a $K \times K$ matrix of real numbers that is symmetric (Lemma 6.2-3). Upon expanding $\mathbf{i} = Y\mathbf{u}$ and adding the i_k, we get

$$0 = i_1 + \cdots + i_K = \sum_{a=1}^{K} Y_{a,1} - u_0 \sum_{a=1}^{K} \sum_{b=1}^{K} Y_{a,b}.$$

Therefore,

$$P_{k,k+1} = u_0 = \frac{\sum_{a=1}^{K} Y_{a,1}}{\sum_{a=1}^{K} \sum_{b=1}^{K} Y_{a,b}}. \tag{7.20}$$

Upon setting $G_k = \sum_{a=1}^{K} \sum_{b=1}^{K} Y_{a,b}$, we can rewrite (7.20) as

$$G_k P_{k,k+1} = \sum_{a=1}^{K} Y_{a,1}. \tag{7.21}$$

Now, $\sum_{a=1}^{K} Y_{a,1}$ is the sum $i_1 + \cdots + i_K$ when $\mathbf{u} = (1, 0, \ldots, 0)$; that is, $\sum_{a=1}^{K} Y_{a,1}$ is the sum of the currents entering $m_1, m_2, \ldots, m_K$ from external connections when 1-V voltage sources are connected to m_1 from all of the $m_0, m_2, \ldots, m_K$.

By reversing the roles of m_0 and m_1, we have by the same analysis that $G_{k+1} P_{k+1,k}$ is the sum $i_0 + i_2 + \cdots + i_K$ of the currents entering $m_0, m_2, \ldots, m_K$ from external connections when 1-V voltage sources are connected to m_0 from all of the $m_1, m_2, \ldots, m_K$. With respect to m_0 acting as the ground node again, we now have $u_1 = \cdots = u_K = -1$, and therefore $i_1 = -\sum_{a=1}^{K} Y_{1,a}$. Moreover, under this latter connection, the sum $-i_1 - i_2 - \cdots - i_K$ of the currents leaving $m_1, m_2, \ldots m_K$ is equal to the current i_0 entering m_0. Hence, $-i_1 = i_0 + i_2 + \cdots + i_K$. Thus,

$$G_{k+1} P_{k+1,k} = -i_1 = \sum_{a=1}^{K} Y_{1,a}. \tag{7.22}$$

Since the matrix Y is symmetric, we have $Y_{1,a} = Y_{a,1}$. So, by (7.21) and (7.22),

$$G_{k+1} P_{k+1,k} = G_k P_{k,k+1}. \tag{7.23}$$

Finally, we may now write

$$P_{1,2} P_{2,3} \cdots P_{c,1} = \frac{G_2}{G_1} P_{2,1} \frac{G_3}{G_2} P_{3,2} \cdots \frac{G_1}{G_c} P_{1,c} = P_{2,1} P_{3,2} \cdots P_{1,c}.$$

This verifies (7.19) and completes the proof. ♣

Because the Markov chain $\mathbf{M}(\mathcal{M})$ is irreducible and reversible, we can synthesize a finite connected 0-network $\mathbf{T}^{\beta\mapsto 0}$ whose 0-nodes correspond bijectively to the nodes of $\mathcal{M}$ and whose random 0-walks are governed by the same one-step transition probability matrix as that for $\mathbf{M}(\mathcal{M})$ (see Appendix C6). We call $\mathbf{T}^{\beta\mapsto 0}$ the *surrogate network for* $\mathbf{T}$ *with respect to* $\mathcal{M}$. A realization for $\mathbf{T}^{\beta\mapsto 0}$ can be obtained by connecting a conductance $g_{k,l} = g_{l,k}$ between the 0-nodes x_k^0 and x_l^0 $(k \neq l)$ in $\mathbf{T}^{\beta\mapsto 0}$, where $g_{k,l}$ is determined as follows: Let $n_k \mapsto x_k^0$ denote the bijection from the nodes of $\mathcal{M}$ to the 0-nodes of $\mathbf{T}^{\beta\mapsto 0}$. If n_k and n_l are not $\mathcal{M}$-adjacent in $\mathbf{T}$, set $g_{k,l} = 0$. If n_k and n_l are $\mathcal{M}$-adjacent in $\mathbf{T}$, relabel n_k as m_0, relabel n_l as m_1, and let $m_2, \ldots, m_K$ be the other nodes of $\mathcal{M}$ that are $\mathcal{M}$-adjacent to either m_0, or m_1, or both. Then, with our prior notation, set $G_k = \sum_{a=1}^{K}\sum_{b=1}^{K} Y_{a,b}$. Also set $G = \sum_k G_k$, where the latter sum is over all indices for all the nodes of $\mathcal{M}$. Finally, set $g_{k,l} = P_{k,l}G_k/G$, where $P_{k,l}$ is the probability of a one-step $\mathcal{M}$-transition from n_k to n_l (before the relabeling). By (7.23), $g_{k,l} = g_{l,k}$. This yields the surrogate network $\mathbf{T}^{\beta\mapsto 0}$.

The one-step transition probabilities for a random 0-walk on $\mathbf{T}^{\beta\mapsto 0}$ that follows the nearest-neighbor rule are the same as the one-step $\mathcal{M}$-transition probabilities $P_{k,l}$ for $\mathbf{M}(\mathcal{M})$. Indeed, the nearest-neighbor rule asserts that the probability of a one-step transition in $\mathbf{T}^{\beta\mapsto 0}$ from a node x_k^0 to an adjacent node x_l^0 is the ratio $g_{k,l}/\sum_{\lambda=1}^{L} g_{k,\lambda}$, where $\sum_{\lambda=1}^{L} g_{k,\lambda}$ is the sum of all conductances incident to x_k^0. Since $g_{k,l} = P_{k,l}G_k/G$, that ratio is equal to $P_{k,l}/\sum_{\lambda=1}^{L} P_{k,\lambda}$. But, as we have noted earlier, $\sum_{\lambda=1}^{L} P_{k,\lambda} = 1$, from which comes our assertion.

(It may be worth noting at this point that $\mathbf{w} = (w_1, \ldots, w_K)$, where $w_k = G_k/G$ and K is the number of nodes in $\mathcal{M}$, is the equilibrium distribution for $\mathbf{M}(\mathcal{M})$. See Appendix C6 again.)

We have now established that the relative multistep transition probabilities for the roving of Ψ among the nodes of $\mathcal{M}$ are the same as those for a random walker on $\mathbf{T}^{\beta\mapsto 0}$. Consequently, the impressed and measured node voltages required by Rule 7.3-1 correspond bijectively to those required by the Nash-Williams rule for $\mathbf{T}^{\beta\mapsto 0}$. This completes our recursive argument for the continuous extension of Rule 7.3-1 to the wandering of Ψ on $\mathbf{N}^{\nu}$ itself.

Note that, by applying Lemma 7.1-1 to $\mathbf{T}^{\beta\mapsto 0}$ or alternatively by invoking Lemma 7.2-2, we can conclude that the roving Ψ, after starting from any node of $\mathcal{M}$, will reach any other node of $\mathcal{M}$ after passing through nodes of $\mathcal{M}$ at most finitely many times.

We summarize and conclude with

Theorem 7.7-3. *Let* $\mathbf{N}^\nu$ $(0 \leq \nu \leq \omega, \nu \neq \vec{\omega})$ *be a sourceless permissively finitely structured* ν*-network. Let a random walker* Ψ *wander through any 0-subsection of* $\mathbf{N}^\nu$ *according to the nearest-neighbor rule. Let* Ψ *rove. Then, Rule 7.3-1 can be continuously extended through subsections of increasing ranks to get a rule governing the wandering of* Ψ *throughout* $\mathbf{N}^\nu$*. Furthermore, let* $\mathcal{M}$ *be any finite nonvoid set of maximal nodes in* $\mathbf{N}^\nu$*. Then, the* $\mathcal{M}$*-transitions of* Ψ *between the nodes of* $\mathcal{M}$ *are governed by an irreducible and reversible Markov chain. With* n_s *being any member of* $\mathcal{M}$ *and with* $\mathcal{N}_e$ *and* $\mathcal{N}_g$ *being any two disjoint nonvoid subsets of* $\mathcal{M}$ *such that* $n_s \notin (\mathcal{N}_e \cup \mathcal{N}_g)$*,* $Prob(\mathrm{s}n_s, \mathrm{r}\mathcal{N}_e, \mathrm{b}\mathcal{N}_g)$ *is the voltage at* n_s *when* $\mathcal{N}_e$ *is held at 1 V and* $\mathcal{N}_g$ *is held at 0 V. In particular, this formula gives the probability for a one-step* $\mathcal{M}$*-transition from* n_s *to any other node* n_0 *of* $\mathcal{M}$ *(i.e., hold* n_0 *at 1 V, hold* $\mathcal{M}\backslash\{n_0\}$ *at 0 V, and measure the voltage at* n_s*).*

APPENDIX A

ORDINAL AND CARDINAL NUMBERS

Presented here is a short description of "ordinal numbers" and "cardinal numbers." These are also called "ordinals" and "cardinals." Just a brief review of these numbers is all that is needed to provide an understanding of how they are used in this book. Moreover, a familiarity with the basic ideas of set theory is all that is needed to follow this appendix. For more information see any one of the introductory textbooks [1], [6], [11], [18], [24].

A1. The natural numbers can be defined in terms of finite sets as follows: 0 is taken to be the empty set $\emptyset$. 1 is defined as the set $\{\emptyset\}$ that contains the empty set. 2 is the set whose members are $\emptyset$ and $\{\emptyset\}$; that is, $2 = \{0, 1\}$. In general the natural number n is the set that contains all the natural numbers from 0 to $n - 1$. Thus,

$$\begin{aligned} 0 &= \emptyset \\ 1 &= \{\emptyset\} = \{0\} \\ 2 &= \{\emptyset, \{\emptyset\}\} = \{0, 1\} \\ 3 &= \{\emptyset, \{\emptyset\}, \{\emptyset, \{\emptyset\}\}\} = \{0, 1, 2\} \\ &\vdots \\ n &= \{0, \ldots, n - 1\} \\ &\vdots \end{aligned}$$

The natural numbers are called the *finite ordinals*. Note that they possess a total ordering induced by the "subset of" relation $\subset$; the same ordering is also induced by the "member of" relation $\in$.

A2. This process of appending more elements to generate still larger sets can be continued onto infinite sets. This yields the *transfinite ordinals*. The first of these is denoted by ω and is defined as the set that contains all the natural numbers. The next one $\omega + 1$ is the set that contains all the natural numbers and ω as well. Thus,

$$\omega = \{0, 1, \ldots\}$$

$$\begin{aligned} \omega+1 &= \{0,1,\ldots,\omega\} \\ \omega+2 &= \{0,1,\ldots,\omega,\omega+1\} \\ &\vdots \end{aligned}$$

Continuing onward, we pass through all the ordinals $\omega + n$, where $n = 0, 1, 2, \ldots$, and then reach $\omega + \omega$, which is also denoted by $\omega \cdot 2$ (i.e., ω "times" 2):

$$\omega+\omega = \omega\cdot 2 = \{0,1,\ldots,\omega,\omega+1,\ldots\}.$$

(We should pause now to point out that $\omega \cdot 2$ and $2 \cdot \omega$ are defined to be two different ordinals; in particular, $\omega \cdot 2$ is the ordered pair of two infinite sequences, whereas $2 \cdot \omega$ is taken to be an infinite sequence of ordered pairs. It is a fact that $2 \cdot \omega = \omega \neq \omega \cdot 2$. This is just one example showing that the arithmetic of the ordinals and the arithmetic of the real numbers differ, but they do agree for the finite ordinals. We do not make any use of ordinal arithmetic in this book.)

Returning to the procedure for generating ordinals, we can suggest what still larger ordinals look like:

$$\begin{aligned} \omega\cdot 2+1 &= \{0,1,\ldots,\omega,\omega+1,\ldots,\omega\cdot 2\} \\ \omega\cdot 2+2 &= \{0,1,\ldots,\omega,\omega+1,\ldots,\omega\cdot 2,\omega\cdot 2+1\} \\ &\vdots \end{aligned}$$

Continuing through $\omega \cdot 2 + n$ for all n, we reach $\omega \cdot 2 + \omega = \omega \cdot 3$. After adding the natural numbers again, we reach $\omega \cdot 4$. Then, $\omega \cdot 5, \ldots, \omega \cdot n, \ldots$ arise similarly. This leads to $\omega \cdot \omega = \omega^2$. The process goes on to $\omega^2 + 1, \ldots, \omega^2 + \omega, \ldots, \omega^2 + \omega^2 = \omega^2 \cdot 2, \ldots, \omega^2 \cdot n, \ldots, \omega^3, \ldots, \omega^4, \ldots, \omega^n, \ldots, \omega^\omega, \ldots, \omega^\omega + \omega^\omega = \omega^\omega \cdot 2, \ldots, \omega^\omega \cdot \omega = \omega^{\omega+1}, \ldots, \omega^{\omega^\omega}$. Still further on is the strange ordinal

$$\epsilon_0 = \omega^{\omega^{\omega^{\omega^{\cdot^{\cdot^{\cdot}}}}}}$$

where the exponentiations by ω continue without end. This is also written as ${}^{\omega}\omega$ and is called "ω tetrated to ω." But having become inured to repetition, we should now expect that there is an ordinal of the form

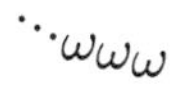

where the tetrations continue without end. There are always more ordinals. What we run out of are names and symbols for them.

We have been looking at the "countable ordinals." Indeed, each of them is a set, which can be counted by recursively applying the principle that a countable set of countable sets is countable. But, there are "uncountable ordinals" as well. For example, the set of all subsets of ω can be shown to be uncountable. There is an ordinal $\mathbf{c}$ corresponding to it. That ordinal is also a "cardinal" (see A8 below).

A3. All the ordinals beyond 0 are of two types: A *successor ordinal* is an ordinal having an immediate predecessor, such as $\omega + p$ and $\omega^{\omega} \cdot m + p$, where p and m are positive natural numbers. A *limit ordinal* is an ordinal having no immediate predecessor; ω and $\omega^{\omega} \cdot m$ are examples of this.

The ordinals serve as ranks for our transfinite graphs. Actually, how transfinite graphs can be constructed has been explicated only up to the ordinal rank ω. Moreover, our construction for the ω-graphs requires the introduction of an arrow rank $\vec{\omega}$, which is not an ordinal rank; it is nonetheless the rank of another kind of transfinite graph. Within the totally ordered set of ranks, $\vec{\omega}$ immediately precedes ω and is larger than every natural-number rank. In fact, there is an arrow rank immediately preceding every limit-ordinal rank.

Furthermore, our constructions of transfinite graphs can be continued beyond ω through ranks of the form $\ldots, \omega + p, \ldots, \omega \cdot \vec{2} = \omega + \vec{\omega}, \omega \cdot 2, \ldots, \omega \cdot 2 + p, \ldots, \omega \cdot \vec{m}, \omega \cdot m, \ldots, \omega \cdot m + p, \ldots, \vec{\omega^2}, \omega^2, \ldots$; see Section 2.5. But it is not clear that our constructions can be continued through all the countable ordinals.

A4. We turn now to the definition of all the ordinals, uncountable as well as countable ones. We shall denote any totally ordered set by the pair $(\mathcal{A}, <)$, where $\mathcal{A}$ is the set and $<$ is the order relation for it.

A *well-ordered set* is a totally ordered set such that every one of its nonvoid subsets has a least member. For instance, the set of ranks for transfinite graphs is well-ordered. Two well-ordered sets $(\mathcal{A}, <)$ and $(\mathcal{D}, \prec)$ are called *order-isomorphic* if there is a bijection $f : \mathcal{A} \rightsquigarrow \mathcal{D}$ such that $f(a_1) \prec f(a_2)$ whenever $a_1 < a_2$.

Let $(\mathcal{A}, <)$ be any well-ordered set, and let $a \in \mathcal{A}$. The *segment* $\mathcal{S}_a$ *of* $\mathcal{A}$ *determined by* a is the set $\mathcal{S}_a = \{x \in \mathcal{A} : x < a\}$. The segments of $\mathcal{A}$ comprise another well-ordered set $(\mathcal{S}, \subset)$ ordered by inclusion. $(\mathcal{A}, <)$ and $(\mathcal{S}, \subset)$ are order-isomorphic; the bijection f establishing this is $f(a) = \mathcal{S}_a$.

An *ordinal* is by definition a well-ordered set $(\mathcal{A}, <)$ such that $\mathcal{S}_a = a$

for every $a \in \mathcal{A}$. Thus, $\mathcal{A}$ consists of all its segments ordered by inclusion. We shall denote ordinals by lower-case Greek letters.

Some consequences of this definition are the following:

- The first element of any ordinal is the empty set $\emptyset$. Indeed, that first element is by definition the segment consisting of all elements preceding the first element; however, there are no such elements — whence our assertion. Consequently, the second element of any ordinal is the set that contains $\emptyset$. And so on — we progress through the countable ordinals listed in A2. But, the present definition allows uncountable ordinals as well. We now have the general definition of an ordinal.

- If α is an ordinal and if $\beta \in \alpha$, then β is an ordinal.

- Two ordinals are either the same ordinal or one is a member of the other.

- Every ordinal is either a successor ordinal or a limit ordinal, defined as before.

A5. Every set can be well-ordered. This is an important fact. It is obtained from the axioms of set theory (including the axiom of choice). A consequence of it is that every well-ordered set is order-isomorphic to a unique ordinal. Thus, for every set $\mathcal{A}$ there is at least one ordinal α and a bijection $f : \alpha \rightsquigarrow \mathcal{A}$. However, there can be many well-orderings for a given set $\mathcal{A}$. In fact, if $\mathcal{A}$ is an infinite set, there are many ordinals corresponding to $\mathcal{A}$ in this way. For example, there is a bijection from the ordinal $\omega = \{0, 1, 2, \ldots\}$ to the ordinal $\omega + 1 = \{0, 1, 2, \ldots, \omega\}$ given by $0 \mapsto \omega$, $1 \mapsto 0$, $2 \mapsto 1$, $3 \mapsto 2$, ... More generally, for any transfinite ordinal β there is a bijection between β and $\beta + n$ whatever be the natural number n.

A6. Let us now consider the cardinal numbers. The ordinal numbers abstract and extend the idea "first, second, third, ..." The cardinal numbers do the same for "one, two, three, ..." Conceptually, the ordinals assign rankings, and the cardinals measure sizes. As with ordinals, there are finite and transfinite cardinals; however, each finite cardinal corresponds to exactly one finite ordinal, but each transfinite cardinal corresponds to infinitely many transfinite ordinals. The important point is that every set $\mathcal{A}$ (in particular, every ordinal) has a unique cardinal number; it is denoted by $\overline{\overline{\mathcal{A}}}$.

More specifically, the sizes of sets can be compared by using injections between them, this being true for infinite sets as well as finite ones. Let $\mathcal{A}$ and $\mathcal{D}$ be sets. If there is a bijection between them, $\mathcal{A}$ and $\mathcal{D}$ are said to have the *same cardinality*. However, if there is an injection from all of $\mathcal{A}$ into a proper subset of $\mathcal{D}$ but no bijection between $\mathcal{A}$ and $\mathcal{D}$, $\mathcal{A}$ is said to have a *smaller cardinality than* $\mathcal{D}$.

For example, the set of all integers and the set of all even integers have the same cardinality because $e \mapsto e/2$ provides a bijection from the latter set onto the former set. It is a fact that the set of all natural numbers and the set of all rational numbers have the same cardinality. It is also true that the set all natural numbers has a smaller cardinality than the set of all real numbers.

The cardinal number $\overline{\overline{\mathcal{F}}}$ of a finite set $\mathcal{F}$ is simply the number of elements in $\mathcal{F}$, and that number is the unique finite ordinal n for which there is a bijection between $\mathcal{F}$ and n.

What we have for an arbitrary infinite set $\mathcal{A}$ is an infinite subset of the transfinite ordinals, each ordinal of which corresponds bijectively to $\mathcal{A}$. (An indication of this was given in A5). Since the ordinals comprise a well-ordered set, there is a least ordinal α bijectively corresponding to $\mathcal{A}$. That ordinal is called the *cardinal number* or simply the *cardinal* for $\mathcal{A}$. Thus, $\overline{\overline{\mathcal{A}}} = \alpha$.

Cardinality is synonymous with cardinal number and cardinal.

A7. Some facts about the cardinals are the following:

- The finite ordinals and the finite cardinals are the same entities.

- ω is the first transfinite cardinal; it is countable. Moreover, all the transfinite ordinals considered in A2 are countable. Thus, the cardinal number of each of them is ω. It is conventional to use $\aleph_0$ in place of ω when viewing ω as a cardinal, but $\aleph_0$ and ω are the same.

- Every transfinite cardinal is a limit ordinal, but not all limit ordinals are cardinals.

- Since any set of ordinals is well-ordered and since every cardinal is an ordinal, every set of cardinals is well-ordered, too. In the order of increasing cardinality, the transfinite cardinals are indexed by the ordinals and denoted as follows:

$$\aleph_0, \aleph_1, \aleph_2, \ldots, \aleph_\omega, \aleph_{\omega+1}, \ldots, \aleph_{\omega\cdot 2}, \ldots, \aleph_{\omega\cdot n}, \ldots, \aleph_{\omega^2}, \ldots, \aleph_{\omega^\omega}, \ldots,$$

$$\aleph_{\omega_1}, \ldots, \aleph_{\omega_2}, \ldots, \aleph_{\omega_\omega}, \ldots, \aleph_{\omega_{\omega_\omega}}, \ldots .$$

This use of the alephs and omegas is conventional, but the notation $\omega_\alpha = \aleph_\alpha$ is also used. Thus, the subscripts in the second row are themselves cardinals and represent respectively $\aleph_1, \aleph_2, \aleph_\omega = \aleph_{\aleph_0}$, and $\aleph_{\aleph_{\aleph_0}}$. (Observe, however, that within the "dots" of the second row there are subscripts that are not cardinals, as in the first row.)

- A set is uncountable if its cardinality is $\aleph_1$ or larger.
- Given any two sets $\mathcal{A}$ and $\mathcal{D}$, exactly one of the following three possibilities holds: $\overline{\overline{\mathcal{A}}} < \overline{\overline{\mathcal{D}}}, \overline{\overline{\mathcal{A}}} = \overline{\overline{\mathcal{D}}}, \overline{\overline{\mathcal{A}}} > \overline{\overline{\mathcal{D}}}$.

A8. The power set $P(\mathcal{A})$ of any set $\mathcal{A}$ is the set of all subsets of $\mathcal{A}$ (including $\emptyset$ and $\mathcal{A}$). It is a fact that $\overline{\overline{P(\mathcal{A})}} > \overline{\overline{\mathcal{A}}}$ if $\mathcal{A}$ is nonvoid. A particular example is the power set $P(\aleph_0)$ of $\aleph_0$, the set of all subsets of the set of natural numbers. Its cardinal number is denoted by $2^{\aleph_0}$, a notation motivated by the fact that a binary choice is made of each natural number when constructing any member of $P(\aleph_0)$. $2^{\aleph_0}$ is the cardinal number of the set of all real numbers; indeed, any real number can be expanded into a binary series, that is, into a series of positive and negative powers of 2 with 0s and 1s as the coefficients — whence our assertion. $2^{\aleph_0}$ is therefore called the *cardinal number of the continuum* and is also denoted by $\mathbf{c} = 2^{\aleph_0}$. Thus, $\mathbf{c} > \aleph_0$.

A conjecture, the so-called *continuum hypothesis*, regarding $\mathbf{c}$ is that $2^{\aleph_0} = \aleph_1$. The *generalized continuum hypothesis* is that $2^{\aleph_\alpha} = \aleph_{\alpha+1}$ for every ordinal α. However, K.Gödel [10] has shown that these hypotheses are consistent with the Zermelo-Fraenkel axioms (including the axiom of choice) in the sense that they cannot be disproven from those axioms. Moreover, P.J.Cohen [5] has shown that they are independent of those axioms in the sense that they cannot be proven from those axioms.

A9. *Sums and products of cardinals:* Let β be any ordinal. A *β-family* is a mapping from β into a set $\mathcal{A}$; thus, it can be written as $\{a_\alpha : \alpha < \beta\}$. (Here, $a_\alpha \in \mathcal{A}$; also, a_α and a_γ need not be distinct when $\alpha \neq \gamma$.)

Let $\{\kappa_\alpha : \alpha < \beta\}$ be a β-family of cardinal numbers κ_α. The *cardinal sum* $\sum_{\alpha<\beta} \kappa_\alpha$ of these κ_α is the cardinality of $\bigcup_{\alpha<\beta} \mathcal{A}_\alpha$, where the $\mathcal{A}_\alpha$ are any pairwise disjoint sets of cardinality κ_α: $\overline{\overline{\mathcal{A}_\alpha}} = \kappa_\alpha$. We write $\kappa_0 + \kappa_1$ for $\sum_{\alpha<2} \kappa_\alpha$, and similarly for sums of finitely many terms. This is *cardinal addition*. It is commutative and associative.

Let $\{\mathcal{A}_\alpha : \alpha < \beta\}$ be a β-family of sets $\mathcal{A}_\alpha$. The *Cartesian product* $\prod_{\alpha<\beta} \mathcal{A}_\alpha$ is the set of all mappings f of β into $\bigcup_{\alpha<\beta} \mathcal{A}_\alpha$ such that, for each $\alpha < \beta$, $f(\alpha) \in \mathcal{A}_\alpha$.

Now consider again a β-family $\{\kappa_\alpha : \alpha < \beta\}$ of cardinal numbers κ_α. The *cardinal product* $\prod^*_{\alpha<\beta} \kappa_\alpha$ of these κ_α is the cardinality of the Cartesian product $\prod_{\alpha<\beta} \mathcal{A}_\alpha$, where the $\mathcal{A}_\alpha$ are now any sets with $\overline{\overline{\mathcal{A}_\alpha}} = \kappa_\alpha$. For two cardinals κ_0 and κ_1, we write $\kappa_0 \cdot \kappa_1 = \prod^*_{\alpha<2} \kappa_\alpha$, and similarly for finitely many terms. This is *cardinal multiplication*. It, too, is commutative and associative. Moreover, it is distributive over addition: $\kappa \cdot (\gamma + \delta) = \kappa \cdot \gamma + \kappa \cdot \delta$.

If γ and δ are nonzero cardinals with $\gamma \leq \delta$ and $\delta \geq \aleph_0$, then $\gamma + \delta = \gamma \cdot \delta = \delta$. Thus, the addition and multiplication of transfinite cardinals is quite trivial.

In this book, we have used cardinal division in the following way. Since $\aleph_0 < \mathbf{c}$, we have $\mathbf{c} \cdot \aleph_0 = \mathbf{c}$. In fact, we can write $\mathbf{c} \div \aleph_0 = \mathbf{c}$ because there is no cardinal, other than $\mathbf{c}$, which when multiplied by $\aleph_0$ yields $\mathbf{c}$ [25, pages 277 and 299].

APPENDIX B

SUMMABLE SERIES

The idea of a "summable series" extends the concept of an absolutely convergent series from a sum of countably many complex numbers to a sum of possibly uncountably many complex numbers. An exposition of this idea is given in [26, pages 11-22]. Listed below are just those definitions and results that provide an understanding of the summable series used in this book. For the sake of generality, complex numbers are considered, even though our use of summable series requires only real numbers.

B1. *Definition of a summable series:* Let K be any infinite set, not necessarily countable, and let $\{u_k\}_{k\in K}$ be a set of complex numbers indexed by the members of K. The set $\{u_k\}_{k\in K}$ is called *summable* and the symbol $\sum_{k\in K} u_k$ is called a *summable series* if there exists a complex number S such that, for each $\epsilon > 0$, there is a finite subset $I \subset K$ such that, for all finite subsets J with $I \subset J \subset K$, we have $|S - \sum_{k\in J} u_k| \leq \epsilon$. In this case, we write $S = \sum_{k\in K} u_k$, and we call S the *sum* of the set $\{u_k\}_{k\in K}$ or of the summable series.

Since K can have any cardinality, there is no natural order for the terms in the summable series, in contrast to conventional series indexed by the natural numbers. In fact, any rearrangement of a given ordering of a summable series does not alter its sum S.

B2. *Cauchy's criterion for summability:* In order for $\sum_{k\in K} u_k$ to be summable, it is necessary and sufficient that the following be satisfied: For each $\epsilon > 0$, there exists a finite subset I of K such that $|\sum_{k\in M} u_k| \leq \epsilon$ for every finite subset M of K that is disjoint from I.

B3. The sum of a summable series is unique. In particular, the terms of the series may be rearranged and added in any order without altering the sum.

B4. If $\sum_{k\in K} u_k$ is summable and if I is any infinite subset of K, then $\sum_{k\in I} u_k$ is summable.

B5. If $S = \sum_{k\in K} u_k$ and $T = \sum_{k\in K} v_k$ are two summable series with the same index set K, then for any two complex numbers a and b, $\sum_{k\in K}(au_k + bv_k)$ is a summable series and has the sum $aS + bT$.

B6. In order for $\sum_{k\in K} u_k$ to be summable, it is necessary and sufficient that $\sum_{k\in K} |u_k|$ be summable. In this case, $|\sum_{k\in K} u_k| \leq \sum_{k\in K} |u_k|$.

B7. In order for $\sum_{k\in K} u_k$ to be summable in the special case where all the u_k are real and nonnegative, it is necessary and sufficient that every partial sum $S_J = \sum_{k\in J} u_k$, where J is a finite subset of K, be bounded by a fixed number that is independent of the choice of J. In this case, $\sum_{k\in K} u_k$ is the supremum of all such S_J.

B8. It follows from B2 that $|u_k| \leq \epsilon$ whenever $k \notin I$, where ϵ and I are as stated in B2. Because of this, we can view the u_k as tending to 0 as I expands through finite subsets of K.

B9. It follows from B8 that only countably many of the terms of a summable series can be nonzero. Indeed, replace ϵ by $\epsilon_n = 1/n$ and let I_n be the finite set that replaces I. Then, $|u_k| > \epsilon_n$ implies that $k \in I_n$; that is, only finitely many of the u_k satisfy $|u_k| > \epsilon_n$. Thus, by letting $n = 1, 2, \ldots$ successively, we can count all the nonzero u_k in $\sum_{k\in K} u_k$.

B10. By virtue of B9, the summable series $\sum_{k\in K} u_k$ is equal to the conventional absolutely convergent series $\sum_{k\in J} u_k$, where $J = \cup_{n=1}^{\infty} I_n$.

B11. Since B10 is true, one may wonder why we bother with summable series. Why not simply replace the summable series $\sum_{k\in K} u_k$ by the conventional series whose terms are all the nonzero terms in $\sum_{k\in K} u_k$?

The answer is that we may have an uncountable set of entities such as the currents i_k in an uncountable cut (i.e., in an uncountable set of branches) within an uncountable network. With all branches oriented in the same way in the cut and with K being the index set for those currents, Kirchhoff's current law would assert that $\sum_{k\in K} i_k = 0$. Just which branches carry nonzero currents will depend in general upon the choice of sources in the uncountable network. The use of summable series allows us to write that law once and for all such choices.

As another example, consider an inseparable Hilbert space $\mathcal{H}$ and an uncountable orthonormal basis $\{b_k\}_{k\in K}$ for $\mathcal{H}$. Any element a of $\mathcal{H}$ has the Fourier expansion $a = \sum_{k\in K}(a, b_k)b_k$, which is a summable series. If we change the choice of a, we will change in general the countable subset of K consisting of those k for which $(a, b_k) \neq 0$. Nonetheless, the same summable Fourier expansion can be written whatever be the choice of a.

B12. The set $\{u_k\}_{k\in K}$ of complex numbers u_k is called *square-summable* (some say *quadratically summable*) if $\sum_{k\in K} |u_k|^2$ is summable. Given any two square-summable sets $\{u_k\}_{k\in K}$ and $\{v_k\}_{k\in K}$, we have the

Schwarz inequality:

$$\sum_{k \in K} |u_k v_k| \leq \left[\sum_{k \in K} |u_k^2| \sum_{k \in K} |v_k|^2 \right]^{1/2}$$

and also the *Minkowski inequality*:

$$\left[\sum_{k \in K} |u_k + v_k|^2 \right]^{1/2} \leq \left[\sum_{k \in K} |u_k|^2 \right]^{1/2} + \left[\sum_{k \in K} |v_k|^2 \right]^{1/2}.$$

APPENDIX C

IRREDUCIBLE AND REVERSIBLE MARKOV CHAINS

The Markov chains of interest to us are those that represent a random walk on a sourceless, connected, finite 0-network with no self-loops. Not all Markov chains can serve this purpose. This appendix presents a brief survey of the characteristics of those Markov chains that do. A good source for more information about such Markov chains is [13].

C1. We start with a general definition of a Markov chain having a finite state space. Consider a chance process having a finite number of possible outcomes m_k, called *states*. Together they comprise the *state space* $\mathcal{M} = \{m_1, \ldots, m_K\}$. Here, K is a natural number no less than 2. The chance process occurs at an instant of time and repeats at discrete time points t, which we take to be the integers $t = \ldots, -1, 0, 1, \ldots$.

A *probability vector* is a row vector $\mathbf{p} = (p_1, \ldots, p_K)$ of K nonnegative real numbers p_k that sum to 1: $\sum_{k=1}^{K} p_k = 1$. Now, let $\mathbf{w}_t = (w_{t,1}, \ldots, w_{t,K})$ be a probability vector, each of whose components $w_{t,k}$ represent a probability that the outcome at t of the chance process is the state m_k. $w_{t,k}$ will depend in general on prior conditions. We call $\mathbf{w}_t$ a *probability distribution at time t*.

Let $P = [P_{k,l}]$ be a $K \times K$ matrix, each row of which is a probability vector. P is called a *transition probability matrix*, and its entries are called *transition probabilities* or, more specifically, *one-step transition probabilities*. We require henceforth that $P_{k,k} = 0$ for every k.

Given a probability distribution $\mathbf{w}_t$ at t, let us assume that the probability distribution $\mathbf{w}_{t+1}$ at time $t+1$ is given by $\mathbf{w}_{t+1} = \mathbf{w}_t P$. It is easily checked that $\mathbf{w}_{t+1}$ is also a probability vector. Thus, if it is a certainty that the state is m_k at t, then $P_{k,l}$ is the probability that the state is m_l at $t+1$.

A *Markov chain* M is a sequence $\{\ldots, \mathbf{w}_t, \mathbf{w}_{t+1}, \ldots\}$ of probability distributions governed by a given transition probability matrix P according to $\mathbf{w}_{t+1} = \mathbf{w}_t P$. Thus, with P given, the probability distribution $\mathbf{w}_{t+1}$ is uniquely determined by the preceding $\mathbf{w}_t$. The prior probability distributions $\mathbf{w}_s$ $(s < t)$ affect $\mathbf{w}_{t+1}$ only through their influence upon $\mathbf{w}_t$. This is the *Markov property*. Moreover, for any positive natural number τ, we have that $\mathbf{w}_{t+\tau} = \mathbf{w}_t P^\tau$, and each row of P^τ is also a probability vector.

The entries of P^τ are the *τ-step transition probabilities.*

C2. We have tacitly assumed in the preceding definitions that P does not depend upon t. When this is so, the Markov chain is called *stationary.* It will always be assumed henceforth (and tacitly so) that the Markov chain at hand is stationary.

C3. A Markov chain is called *irreducible* if, for some sufficiently large natural number τ, P^τ has only positive entries. This means that any state can be reached from any other state. More precisely, whatever be the choice of the probability distribution $\mathbf{w}_0$, $\mathbf{w}_0 P^\tau$ will have only positive components if τ is chosen large enough.

C4. Every irreducible Markov chain $\mathbf{M}$ has a unique fixed point; that is, there exists a unique probability distribution $\mathbf{w}$ such that $\mathbf{w}P = \mathbf{w}$. $\mathbf{w}$ is called the *equilibrium distribution* of $\mathbf{M}$. It is a fact that every component of $\mathbf{w}$ is positive. Moreover, if $\mathbf{w}_0$ is any probabilty distribution, then $\mathbf{w}_0 P^\tau$ converges componentwise to $\mathbf{w}$ as $\tau \to 0$.

C5. A Markov chain is called *reversible* if $w_k P_{k,l} = w_l P_{l,k}$ for every k and l, where w_k is the kth component of the fixed point $\mathbf{w}$ of $P = [P_{k,l}]$. A characterization of a reversible Markov chain can be given in terms of the "cycles" in the state space $\mathcal{M}$. A *cycle* in $\mathcal{M}$ is a finite sequence

$$C = \{m_1, m_2, \ldots, m_c, m_{c+1} = m_1\}$$

of members of $\mathcal{M}$ having the following properties. (Here we have renumbered the states in the cycle C in order to avoid a double-subscript notation.) All the members of C are distinct except for the first and last, which are the same; C has at least three distinct members (i.e., $c > 2$); for each pair of consecutive members m_k and m_{k+1} in C, the one-step transition probability $P_{k,k+1}$ is positive. It is a fact that a (stationary) Markov chain is reversible if and only if for every cycle C the product $\prod_{k=1}^{c} P_{k,k+1}$ remains unchanged when every $P_{k,k+1}$ is replaced by $P_{k+1,k}$, that is, if and only if

$$P_{1,2}P_{2,3}\ldots P_{c,1} = P_{1,c}\ldots P_{3,2}P_{2,1}$$

for every choice of cycle [13, page 23].

C6. Given an irreducible and reversible Markov chain $\mathbf{M}$ with a finite state space $\mathcal{M}$ and with $P_{k,k} = 0$ for every k, there is a sourceless, connected, finite 0-network $\mathbf{N}^0$ with no self-loops such that there is a bijection between $\mathcal{M}$ and the node set $\mathcal{N}$ of $\mathbf{N}^0$ under which the one-step transition probabilities for $\mathbf{M}$ are equal to the one-step transition probabilities for the random walk on $\mathbf{N}^0$ following the nearest-neighbor rule. (See Section 7.1 for the definition of such a walk.) This network $\mathbf{N}^0$ can be constructed as follows.

Let $\mathbf{w}$ be the equilibrium distribution for $\mathbf{M}$, and let $P = [P_{k,l}]$ be the transition probability matrix for $\mathbf{M}$. For each k and l with $k \neq l$, set $g_{k,l} = w_k P_{k,l}$. By reversibility, $g_{k,l} = g_{l,k}$. Also, $g_{k,k} = 0$ for every k because $P_{k,k} = 0$. Since $w_k \neq 0$ for every k, $g_{k,l} \neq 0$ if and only if $P_{k,l} \neq 0$. Let the node set $\mathcal{N}$ correspond bijectively with $\mathcal{M}$. If $P_{k,l} \neq 0$, we take it that there is a branch incident to the nodes of indices k and l having the conductance $g_{k,l}$. On the other hand, if $P_{k,l} = 0$, we take it that there is no branch incident to those two nodes. Since $\sum_{m=1}^{K} P_{k,m} = 1$ for every k, it follows immediately that $P_{k,l}$ is given by the nearest-neighbor rule: $P_{k,l} = g_{k,l} / \sum_{m=1}^{K} g_{k,m}$. Finally, since $g_{k,k} = 0$ for every k, there are no self-loops.

Actually, the network we have obtained is unique only up to a multiplicative constant; that is, we can multiply all the $g_{k,l}$ by the same positive number to obtain another electrical network yielding the same transition probabilities under the nearest-neighbor rule.

Bibliography

[1] A. Abian, *The Theory of Sets and Transfinite Arithmetic*, W.B. Saunders Co., Philadelphia, 1965.

[2] M. Behzad and G. Chartrand, *Introduction to the Theory of Graphs*, Allyn and Bacon, Boston, 1971.

[3] B. Bollobas, *Graph Theory*, Springer-Verlag, New York, 1979.

[4] G. Cantor, *Grundlagen einer allgemeinen Mannigfaltigkeitslehre. Ein mathematische-philosophischer Versuch in der Lehre des Unendlichen*, B.G. Teubner, Leipzig, 1883. An English translation exists: Foundations of a general theory of manifolds, *The Campaigner*, The Theoretical Journal of the National Caucus of Labor Committees, **9** (January and February, 1976), 69-96.

[5] P.J. Cohen, The independence of the continuum hypothesis, *Proc. Nat. Acad. Sci.*, **50** (1963), 1143-1148 and **51** (1964), 105-110.

[6] K. Devlin, *The Joy of Sets*, Second Edition, Springer Verlag, New York, 1979.

[7] P.G. Doyle and J.L. Snell, *Random Walks and Electric Networks*, The Carus Mathematical Monographs, The Mathematical Association of America, 1984.

[8] L. Euler, Solutio problematis ad geometriam situs pertinentis, *Commentarii Academiae Scientiarum Imperialis Petropolitanae*, **8** (1736), 128-140.

[9] H. Flanders, Infinite networks: I—Resistive networks, *IEEE Trans. Circuit Theory*, **CT-18** (1971), 326-331.

[10] K. Gödel, *The Consistency of the Axiom of Choice and the Generalized Continuum Hypothesis with the Axioms of Set Theory*, Annals of Mathematics Studies, Princeton, 1940; revised edition 1951.

[11] P.R. Halmos, *Naive Set Theory*, Springer-Verlag, New York, 1974.

[12] H.J. Keisler, *Foundations of Infinitesimal Calculus*, Prindle, Weber, and Schmidt, Boston, 1976.

[13] F.P. Kelly, *Reversibility and Stochastic Networks*, John Wiley, New York, 1979.

[14] G. Kirchhoff, Ueber die Auflösung der Gleichungen, auf welche man bei der Untersuchung dere linearen Vertheilung galvanischer Ströme geführt wird, *Annalen der Physik und Chemie*, **72** (1847), 497-508.

[15] D. König, *Theorie der endlichen und unendlichen Graphen*, Akademische Verlagsgesellschaft M. B. H., Leipzig, 1936.

[16] T. Lindstrom, An invitation to nonstandard analysis, in *Nonstandard Analysis and Its Applications*, (N.Cutland, Editor), Cambridge University Press, Cambridge, England, 1988, 1-105.

[17] B. Mohar and W. Woess, A survey on spectra of infinite graphs, *Bull. London Math. Soc.*, **21** (1989), 209-234.

[18] J.D. Monk, *Introduction to Set Theory*, McGraw-Hill Book Co., New York, 1969.

[19] C. St. J.A. Nash-Williams, Random walk and electric currents in networks, *Proc. Cambridge Phil. Soc.*, **55** (1959), 181-194.

[20] C. St. J.A. Nash-Williams, Infinite graphs — A survey, *Journal of Combinatorial Theory*, **3** (1967), 286-301.

[21] B. Peikari, *Fundamentals of Network Analysis and Synthesis*, Prentice-Hall, Englewood Cliffs, New Jersey, 1974.

[22] P. Penfield, Jr., R. Spence, and S. Duinker, *Tellegen's Theorem and Electrical Networks*, M.I.T. Press, Cambridge, Massachusetts, 1970.

[23] A. Robinson, *Nonstandard Analysis*, North Holland Publishing Co., Amsterdam, 1966.

[24] B. Rotman and G.T. Kneebone, *The Theory of Sets and Transfinite Numbers*, Oldbourne Book Co., London, 1966.

[25] J.E. Rubin, *Set Theory*, Holden Day, San Francisco, 1967.

[26] L. Schwartz, *Méthodes Mathématiques pour les Sciences Physiques*, Hermann, Paris, 1961.

[27] P.M. Soardi, *Potential Theory on Infinite Networks*, Lecture Notes in Mathematics 1590, Springer Verlag, New York, 1994.

[28] B.D.H. Tellegen, A general network theorem, with applications, *Philips Res. Rept.*, **7** (1952), 259-269.

[29] C. Thomassen, Infinite graphs, *Graph Theory*, **2** (1983), 129-160.

[30] H.S. Wall, *Analytic Theory of Continued Fractions*, Van Nostrand, New York, 1948.

[31] R.J. Wilson, *Introduction to Graph Theory*, Oliver and Boyd, Edinburgh, 1972.

[32] W. Woess, Random walks on infinite graphs and groups — a survey on selected topics, *Bull. London Math. Soc.*, **26** (1994), 1-60.

[33] A.H. Zemanian, The connections at infinity of a countable resistive network, *Circuit Theory and Applications*, **3** (1975), 333-337.

[34] A.H. Zemanian, Infinite electrical networks with finite sources at infinity, *IEEE Trans. Circuits and Systems*, **CAS-34** (1987), 1518-1534.

[35] A.H. Zemanian, *Infinite Electrical Networks*, Cambridge University Press, New York, 1991.

[36] A.H. Zemanian, *Transfinite Random Walks Based on Electrical Networks: II*, CEAS Technical Report 604, University at Stony Brook, Stony Brook, N.Y., 1991.

[37] A.H. Zemanian, Transfinite graphs and electrical networks, *Trans. Amer. Math. Soc.*, **334** (November 1992), 1-36.

[38] A.H. Zemanian, Random walks on ω-networks, in *Harmonic Analysis and Discrete Potential Theory*, (M.Picardello, Editor), Plenum Publishing Co., London, 1992, 249-257.

[39] A.H. Zemanian, Uncountable Transfinite Graphs and Electrical Networks, *Internat. Jour. Circuit Theory and Appl.*, **22** (1994), 387-398.

[40] A.H .Zemanian, *The Existence and Uniqueness of Node Voltages in a Nonlinear Resistive Transfinite Electrical Network*, CEAS Technical Report 626, University at Stony Brook, Stony Brook, N.Y., 1992.

[41] A.H. Zemanian, Random walks on finitely structured transfinite networks, *Potential Analysis*, in press.

INDEX OF SYMBOLS

This is a list of symbols — along with the pages where they are defined. A few symbols, such as $\mathcal{V}$, have more than one meaning, but the contexts in which they are used should clarify which meaning is appropriate.

INDEX